APPLIED RASCH MEASUREMENT: A BOOK OF EXEMPLARS

EDUCATION IN THE ASIA-PACIFIC REGION: ISSUES, CONCERNS AND PROSPECTS

Volume 4

Applied Rasch Measurement: A Book of Exemplars

Papers in Honour of John P. Keeves

Edited by

SIVAKUMAR ALAGUMALAI

DAVID D. CURTIS

and

NJORA HUNGI

Flinders University, Adelaide, Australia
School of Oriental and Studies,
University of London

Springer

A C.I.P. Catalogue record for this book is available from the Library of Congress.

ISBN 1-4020-3072-X (HB)
ISBN 1-4020-3076-2 (e-book)

Published by Springer,
P.O. Box 17, 3300 AA Dordrecht, The Netherlands.

Sold and distributed in North, Central and South America
by Springer,
101 Philip Drive, Norwell, MA 02061, U.S.A.

In all other countries, sold and distributed
by Springer,
P.O. Box 322, 3300 AH Dordrecht, The Netherlands.

Printed on acid-free paper

Printed in the Netherlands.

SERIES SCOPE

The purpose of this Series is to meet the needs of those interested in an in-depth analysis of current developments in education and schooling in the vast and diverse Asia-Pacific Region. The Series will be invaluable for educational researchers, policy makers and practitioners, who want to better understand the major issues, concerns and prospects regarding educational developments in the Asia-Pacific region.

The Series complements the *Handbook of Educational Research in the Asia-Pacific Region*, with the elaboration of specific topics, themes and case studies in greater breadth and depth than is possible in the Handbook.

Topics to be covered in the Series include: secondary education reform; reorientation of primary education to achieve education for all; re-engineering education for change; the arts in education; evaluation and assessment; the moral curriculum and values education; technical and vocational education for the world of work; teachers and teaching in society; organisation and management of education; education in rural and remote areas; and, education of the disadvantaged.

Although specifically focusing on major educational innovations for development in the Asia-Pacific region, the Series is directed at an international audience.

The Series *Education in the Asia-Pacific Region: Issues, Concerns and Prospects*, and the *Handbook of Educational Research in the Asia-Pacific Region*, are both publications of the Asia-Pacific Educational Research Association.

Those interested in obtaining more information about the Monograph Series, or who wish to explore the possibility of contributing a manuscript, should (in the first instance) contact the publishers.

Books published to date in the series:

1. Young People and the Environment:
   An Asia-Pacific Perspective
   Editors: John Fien, David Yenken and Helen Sykes

2. Asian Migrants and Education:
   The Tensions of Education in Immigrant Societies and among Migrant Groups
   Editors: Michael W. Charney, Brenda S.A. Yeoh and Tong Chee Kiong

3. Reform of Teacher Education in the Asia-Pacific in the New Millennium:
   Trends and Challenges
   Editors: Yin.C. Cheng, King W. Chow and Magdalena M. Mok

Contents

Preface

While the primary purpose of the book is a celebration of John's contributions to the field of measurement, a second and related purpose is to provide a useful resource. We believe that the combination of the developmental history and theory of the method, the examples of its use in practice, some possible future directions, and software and data files will make this book a valuable resource for teachers and scholars of the Rasch method.

This book is a tribute to Professor John P Keeves for the advocacy of the Rasch model in Australia. Happy 80th birthday John!

There are good introductory texts on Item Response Theory, Objective Measurement and the Rasch model. However, for a beginning researcher keen on utilising the potentials of the Rasch model, theoretical discussions of test theory and associated indices do not meet their pragmatic needs. Furthermore, many researchers in measurement still have little or no knowledge of the features of the Rasch model and its use in a variety of situations and disciplines. This book attempts to describe the underlying axioms of test theory, and, in particular, the concepts of objective measurement and the Rasch model, and then link theory to practice. We have been introduced to the various models of test theory during our graduate days. It was time for us to share with those keen in the field of measurement in education, psychology and the social sciences the theoretical and practical aspects of objective measurement. Models, conceptions and applications are refined continually, and this book seeks to illustrate the dynamic evolution of test theory and also highlight the robustness of the Rasch model.

Part 1

The volume has an introductory section that explores the development of measurement theory. The first chapter on classical test theory traces the developments in test construction and raises issues associated with both terminologies and indices to ascertain the stability of tests and items. This chapter leads to a rationale for the use Objective Measurement and deals specifically with the Rasch Simple Logistic Model. Chapters by Geoff Masters and David Andrich highlight the fundamental principles of the Rasch model and also raise issues where misinterpretations may occur.

Part 2

This section of the book includes a series of chapters that present applications of the Rasch measurement model to a wide range of data sets. The intention in including these chapters is to present a diverse series of case studies that illustrate the breadth of application of the method.

Of particular interest will be contact details of the authors of articles in Parts 2A and 2B. Sample data sets and input files may be requested from these contributors so that students of the Rasch method can have access to both the raw materials for analyses and the results of those analyses as they appear in published form in their chapter.

Part 3

The final section of the volume includes reviews of recent extensions of the Rasch method which anticipate future developments of it. Contributions by Luo Guanzhong (unfolding model) and Mark Wilson (Multitrait Model) raise issues about the dynamic developments in the application and extensions of the Rasch model. Trevor Bond's conclusion in the final chapter raises possibilities for users of the principles of objective measurement, and its use in social sciences and education.

Appendix

This section introduces the software packages that are available for Rasch analysis. Useful resource locations and key contact details are made available for prospective users to undertake self-study and explorations of the Rasch model.

August 2004

Sivakumar Alagumalai
David D. Curtis
Njora Hungi

The Contributors

Contributors are listed in alphabetical order, together with their affiliations and email addresses. Titles of chapters that they have authored are in alphabetical order, together with the respective page numbers. An asterisk preceding the chapter title indicates joint-authored chapters.

Afrassa, T.M.
South Australian Department of Education and Children's Services
[email: Afrassa.Tilahun@saugov.sa.gov.au]
Chapter 4: Monitoring Mathematics Achievement over Time

Alagumalai, S.
School of Education, Flinders University, Adelaide, South Australia
[email: sivakumar.alagumalai@flinders.edu.au]
* Chapter 1: Classical Test Theory
* Epilogue: Our Experiences and Conclusion
Appendix: IRT Software

Andrich, D.
Murdoch University, Murdoch, Western Australia
[email: D.Andrich@murdoch.edu.au]
Chapter 3: The Rasch Model explained
* Chapter 17: Information Functions for the General Dichotomous Unfolding Model

Barrett, S.
University of Adelaide, Adelaide, South Australia
[email: steven.barrett@adelaide.edu.au]
Chapter 9: Raters and Examinations

Blackman, I.
School of Nursing, Flinders University, Adelaide, South Australia
[email: Ian.Blackman@flinders.edu.au]
Chapter 14: Estimating the Complexity of Workplace Rehabilitation Task using Rasch

Bond, T.
School of Education, James Cook University, Queensland, Australia
[email: trevor.bond@jcu.edu.au]

Chapter 18: Past, present and future: An idiosyncratic view of Rasch measurement

Curtis, D.D.
School of Education, Flinders University, Adelaide, South Australia
[email: david.curtis@flinders.edu.au]
* Chapter 1: Classical Test Theory
Chapter 10: Comparing Classical and Contemporary Analyses and Rasch Measurement
* Epilogue: Our Experiences and Conclusion

Hoskens, M.
University of California, Berkeley, California, United States
[email: hoskens@socrates.berkeley.edu]
* Chapter 16: Multidimensional Item Responses: Multimethod-multitrait perspectives

Hungi, N.
School of Education, Flinders University, Adelaide, South Australia
[email: njora.hungi@flinders.edu.au]
Chapter 8: Applying the Rasch Model to Detect Biased Items
* Epilogue: Our Experiences and Conclusion

I Gusti Ngurah, D.
School of Education, Flinders University, Adelaide, South Australia; Pendidikan Nasional University, Bali, Indonesia
[email: ngurah.darmawan@flinders.edu.au]
Chapter 15: Creating a Scale as a General Measure of Satisfaction for Information and Communications Technology use

Kotte, D.
Casual Impact, Germany
[email: dieter.kotte@causalimpact.com]
* Chapter 5: Manual and Automatic Estimates of Growth and Gain Across Year Levels: How Close is Close?

Lietz, P.
International University Bremen, Germany
[email: p.lietz@iu-bremen.de]
* Chapter 5: Manual and Automatic Estimates of Growth and Gain Across Year Levels: How Close is Close?

Luo, Guanzhong
Murdoch University, Murdoch, Western Australia
[email: G.Luo@murdoch.edu.au]
* Chapter 17: Information Functions for the General Dichotomous Unfolding Model

Masters, G.N.
Australian Council for Educational Research, Melbourne, Victoria
[email: Masters@acer.edu.au]
Chapter 2: Objective Measurement

Taguchi, K.
Flinders University, South Australia; University of Adelaide, South Australia
[email: kazuyo.taguchi@adelaide.edu.au]
Chapter 6: Japanese Language Learning and the Rasch Model

Tedman, D.K.
St John's Grammar School, Adelaide, South Australia
[email: raymond.tedman@adelaide.edu.au]
Chapter 13: Science Teachers' Views on Science, Technology and Society Issues

Thompson, M.
University of Adelaide Senior College, Adelaide, South Australia
[email: dtmt@senet.com.au]
Chapter 11: Combining Rasch Scaling and Multi-level Analysis

Wilson, M.
University of California, Berkeley, California, United States
[email: mrwilson@socrates.Berkeley.EDU]
* Chapter 16: Multidimensional Item Responses: Multimethod-multitrait perspectives

Yates, S.M.
School of Education, Flinders University, Adelaide, South Australia
[email: Shirley.Yates@flinders.edu.au]
Chapter 12: Rasch and Attitude Scales: Explanatory Style

Yuan, Ruilan
Oxley College, Victoria, Australia
[email: yuan-ru@oxley.vic.edu.au]
Chapter 7: Chinese Language Learning and the Rasch Model

Chapter 1

CLASSICAL TEST THEORY

Sivakumar Alagumalai and David D. Curtis
Flinders University

Abstract: Measurement involves the processes of description and quantification. Questionnaires and test instruments are designed and developed to measure conceived variables and constructs accurately. Validity and reliability are two important characteristics of measurement instruments. Validity consists of a complex set of criteria used to judge the extent to which inferences, based on scores derived from the application of an instrument, are warranted. Reliability captures the consistency of scores obtained from applications of the instrument. Traditional or classical procedures for measurement were based on a variety of scaling methods. Most commonly, a total score is obtained by adding the scores for individual items, although more complex procedures in which items are differentially weighted are used occasionally. In classical analyses, criteria for the final selection of items are based on internal consistency checks. At the core of these classical approaches is an idea derived from measurement in the physical sciences: that an observed score is the sum of a true score and a measurement error term. This idea and a set of procedures that implement it are the essence of Classical Test Theory (CTT). This chapter examines underlying principles of CTT and how test developers use it to achieve measurement, as they have defined this term. In this chapter, we outline briefly the foundations of CTT and then discuss some of its limitations in order to lay a foundation for the examples of objective measurement that constitute much of the book.

Key words: classical test theory; true score theory; measurement

1. AN EVOLUTION OF IDEAS

The purpose of this chapter is to locate Item Response Theory (IRT) in relation to CTT. In doing this, it is necessary to outline the key elements of CTT and then to explore some of its limitations. Other important concepts,

S. Alagumalai et al. (eds.), Applied Rasch Measurement: A Book of Exemplars, 1–14.

specifically measurement and the construction of scales, are also implicated in the emergence of IRT and so these issues will be explored, albeit briefly. Our central thesis is that the families of IRT models that are being applied in education and the social sciences generally represent a stage in the evolution of attempts to describe and quantify human traits and to develop laws that summarise and predict observations. As in all evolving systems, there is at any time a status quo; there are forces that direct development; there are new ideas; and there is a changing environmental context in which existing and new ideas may develop and compete.

1.1 Measurement

Our task is to trace the emergence of IRT families in a context that was substantially defined by the affordances of CTT. Before we begin that task, we need to explain our uses of the terms IRT and measurement.

1.1.1 Item Response Theory

IRT is a complex body of methods used in the analysis of test and attitude data. Typically, IRT is taken to include one-, two- and three-parameter item response models. It is possible to extend this classification by the addition of even further parameters. However, the three-parameter model is often considered the most general, and the others as special cases of it. When the pseudo-guessing parameter is removed, the two-parameter model is left, and when the discrimination parameter is removed from that, the one-parameter model remains. If mathematical formulations for each of these models are presented, a sequence from a general case to special cases becomes apparent. However, the one-parameter model has a particular and unique property: it embodies measurement, when that term is used in a strict axiomatic sense. The Rasch measurement model is therefore one member of a family of models that may be used to model data: that is, to reflect the structure of observations. However, if the intention is to measure (strictly) a trait, then one of the models from the Rasch family will be required. The Rasch family includes Rasch's original dichotomous formulation, the rating scale (Andrich) and partial credit (Masters) extensions of it, and subsequently, many other developments including facets models (Linacre), the Saltus model (Wilson), and unfolding models (Andrich, 1989).

1.1.2 Models of measurement

In the development of ideas of measurement in the social sciences, the environmental context has been defined in terms of the suitability of available methods for the purposes observers have, the practicability of the range of methods that are available, the currency of important mathematical and statistical ideas and procedures, and the computational capacities available to execute the mathematical processes that underlie the methods of social inquiry. Some of the ideas that underpin modern conceptions of measurement have been abroad for many years, but either the need for them was not perceived when they were first proposed, or they were not seen to be practicable or even necessary for the problems that were of interest, or the computational environment was not adequate to sustain them at that time. Since about 1980, there has been explosive growth in the availability of computing power, and this has enabled the application of computationally complex processes, and, as a consequence, there has been an explosion in the range of models available to social science researchers.

Although measurement has been employed in educational and psychological research, theories of measurement have only been developed relatively recently Keats (1994b). Two approaches to measurement can be distinguished. Axiomatic measurement evaluates proposed measurement procedures against a theory of measurement, while pragmatic measurement describes procedures that are employed because they appear to work and produce outcomes that researchers expect. Keats (1994b) presented two central axioms of measurement, namely transitivity and additivity. Measurement theory is not discussed in this chapter, but readers are encouraged to see Keats and especially Michell (Keats, 1994b; Michell, 1997, 2002).

The term 'measurement' has been a contentious one in the social sciences. The history of measurement in the social sciences appears to be one punctuated by new developments and consequent advances followed by evolutionary regression. Thorndike (1999) pointed out that E.L. Thorndike and Louis Thurstone had recognised the principles that underlie IRT-based measurement in the 1920s. However, Thurstone's methods for measuring attitude by applying the law of comparative judgment proved to be more cumbersome than investigators were comfortable with, and when, in 1934, Likert, Roslow and Murphy (Stevens, 1951) showed that an alternative and much simpler method was as reliable, most researchers adopted that approach. This is an example of retrograde evolution because Likert scales produce ordinal data at the item level. Such data do not comply with the measurement requirement of additivity, although in Likert's procedures, these ordinal data were summed across items and persons to produce scores.

Stevens (1946) is often cited as the villain responsible for promulgating a flawed conception of measurement in psychology and is often quoted out of context. He said:

> But measurement is a relative matter. It varies in kind and degree, in type and precision. In its broadest sense measurement is the assignment of numerals to objects or events according to rules. And the fact that numerals can be assigned under different rules leads to different kinds of scales and different kinds of measurement. The rules themselves relate in part to the concrete empirical operations of our experimental procedures which, by their sundry degrees of precision, help to determine how snug is the fit between the mathematical model and what it stands for. (Stevens, 1951, p. 1)

Later, referring to the initial definition, Stevens (p. 22) reiterated part of this statement ('measurement is the assignment of numerals to objects or events according to rules'), and it is this part that is often recited. The fuller definition does not absolve Stevens of responsibility for a flawed definition. Clearly, even his more fulsome definition of measurement admits that some practices result in the assignment of numerals to observations that are not quantitative, namely nominal observations. His definition also permitted the assignment of numerals to ordinal observations. In Stevens' defence, he went to some effort to limit the mathematical operations that would be permissible for the different kinds of measurement. Others later dispensed with these limits and used the assigned numerals in whatever way seemed convenient.

Michell (2002) has provided a brief but substantial account of the development and subsequent use of Stevens' construction of measurement and of the different types of data and the types of scales that may be built upon them. Michell has shown that, even with advanced mathematical and statistical procedures and computational power, modern psychologists continue to build their work on flawed constructions of measurement. Michell's work is a challenge to psychologists and psychometricians, including those who advocate application of the Rasch family of measurement models.

The conception of measurement that has been dominant in psychology and education since the 1930s is the version formally described by Stevens in 1946. CTT is compatible with that conception of measurement, so we turn to an exploration of CTT.

2. TRUE SCORE THEORY

2.1.1 Basic assumptions

CTT is a psychometric theory that allows the prediction of outcomes of testing, such as the ability of the test-takers and the difficulty of items. Charles Spearman laid the foundations of CTT in 1904. He introduced the concept of an observed score, and argued that this score is composed of a true score and an error. It is important to note that the only element of this relation that is manifest is the observed score: the true score and the error are latent or not directly observable. Information from observed score can be used to improve the reliability of tests. CTT is a relatively simple model for testing which is widely used for the construction and evaluation of fixed-length tests.

Keeves and Masters (1999) noted that CTT pivots on true scores as distinct from raw scores, and that the true scores can be estimated by using group properties of a test, test reliability and standard errors of estimates. In order to understand better the conceptualisation of error and reliability, it is useful to explore assumptions of the CTT model.

The most basic equation of CTT is:

$$S_i = \tau_i + e_i \qquad (1)$$

Where S = raw score in test
τ = true score (not necessarily a perfectly valid score)
e = error term (test score deviance from true score).

It is important to note that the errors are assumed to be random in CTT and not correlated with τ or S. The errors in total cancel one another out and the error axiom can be represented as: Expectancy value of error = 0.

These assumptions about errors are discussed in Keats (1997). They are some of a series of assumptions that underpin CTT, but that, as Keats noted, do not hold in practice. These 'old' assumptions of CTT are contrasted with those of Item Response Theory (IRT) by Embretson and Hershberger (1999, pp. 11–14).

The above assumption leads to the decomposition of variances. Observed score variance comprises true score variance and error variance and this relation can be represented as:

$$\sigma^2_s = \sigma^2_\tau + \sigma^2_e \qquad (2)$$

Recall that τ and e are both latent variables, but the purpose of testing is to draw inferences about τ, individuals' true scores. Given that the observed score is known, something must be assumed about the error term in order to estimate τ.

Test reliability (ρ) can be defined formally as the ratio of true score variance to raw score variance: that is:

$$\rho = \sigma^2_{\tau}/\sigma^2_{s} \tag{3}$$

But, since τ cannot be observed, its variance cannot be known directly. However, if two equal length tests that tap the same construct using similar items are constructed, the correlation of persons' scores on them can be shown to be equal to the test reliability. This relationship depends on the assumption that errors are randomly distributed with a mean of 0 and that they are not correlated with τ or S. Knowing test reliability provides information about the variance of the true score, so knowing a raw score permits the analyst to say something about the plausible range of true scores associated with the observed score.

2.1.2 Estimating test reliability in practice

The formal definition of reliability depends on two ideal parallel tests. In practice, it is not possible to construct such tests, and a range of alternative methods to estimate test reliability has emerged. Three approaches to establishing test reliability coefficients have been recognised (American Educational Research Association, American Psychological Association, & National Council on Measurement in Education, 1999). Closest to the formal definition of reliability are those coefficients derived from the administration of parallel forms of an instrument in independent testing sessions and these are called alternative forms coefficients. Correlation coefficients obtained by administration of the same instrument on separate occasions are called test-retest or stability coefficients. Other coefficients are based on relationships among scores derived from individual items or subsets of items within a single administration of a test, and these are called internal consistency coefficients.

Two formulae in common use to estimate internal test reliability are the Kuder-Richardson formula 20 (KR20) and Cronbach's alpha.

A consequence of test reliability as conceived within CTT is that longer tests are more reliable than shorter ones. This observation is captured in the Spearman-Brown prophecy formula, for which the special case of predicting the full test reliability from the split half correlation is:

$$R = 2r / (1+r) \tag{4}$$

Where R is the reliability of the full test, and
r is the reliability of the equivalent test halves.

2.1.3 Implications for test design and construction

Two essential processes of test construction are standardisation and calibration. Standardisation and calibration present the same argument that the result of a test does not give an absolute measurement of one's ability nor latent traits. However, the results do allow performances to be compared. Stage (2003, p. 2) indicated that CTT has been a productive model that led to the formulation of a number of useful relationships:

- the relation between test length and test reliability;
- estimates of the precision of difference scores and change scores;
- the estimation of properties of composites of two or more measures; and
- the estimation of the degree to which indices of relationship between different measurements are attenuated by the error of measurement in each.

It is necessary to ensure that the test material, the test administration situation, test sessions and methods of scoring are comparable to allow optimal standardisation. With standardisation in place, calibration of the test instrument enables one person to be placed relative to others.

2.1.4 Item level analyses

Although the major focus of CTT is on test-level information, item statistics, specifically item difficulty and item discrimination, are also important. Basic strategies in item analysis and item selection include identifying items that conform to a central concept, examining item-total and inter-item correlations, checking the homogeneity of a scale (the scale's internal consistency) and finally appraising the relationship between homogeneity and validity. Indications of scale dimensionality arise from the latter process.

There are no complex theoretical models that relate an examinee's ability to success on a particular item. The p-value, which is the proportion of a well-defined group of examinees that answers an item correctly, is used as the index for the item difficulty. A higher value indicates easier items. (Note the counter-intuitive use of the term 'difficulty'). The item discrimination index is the correlation coefficient between the scores on the item and the scores on the total test and indicates the extent to which an item discriminates between high ability examinees and low ability examinees. This can be represented as:

$$\text{Item discrimination index} = \text{p (Upper)} - \text{p (Lower)} \qquad (5)$$

Similarly, the point-biserial correlation coefficient is the Pearson r between the dichotomous item variable and the (almost) continuous total score variables (also called the item-total Pearson r). Arbitrary interpretations have been made on ranges of values of the point-biserial correlation. They range from Very Good (>0.40), Good (<0.39, >0.30), Fair (<0.29, >0.20), Non-discriminating (<0.19, >0.00), Need attention (<0.00).

Item discrimination is optimal when the item facility is 0.50. Removing non-discriminating items will improve test reliability.

The discussions above highlight ways to improve the reliability of test scores in CTT. They include:

- increasing the number of items to adequately sample an identified construct or behaviour;
- deletion of items that do not discriminate or that are imprecise;
- identical test conditions for all examines and explicitly stated;
- objective-type questions; and
- heterogeneous group.

Analysis using the CTT model aims to eliminate items whose functions are incompatible with the psychometric characteristics described above. There may be several reasons for rejection:

- item has a high success or failure rate (very low or very high *p*);
- item has low discrimination;
- item key is incorrect or correct answer is not selected; and
- distracters do not work.

This systematic 'cleaning process' seeks to ensure that the test measures one and only one trait by using measures of internal consistency to estimate reliability, usually by seeking to maximise the Cronbach alpha statistic, and by the application of other techniques such as factor analysis.

2.1.5 Classical concept of reliability

A number of factors affect the reliability of a test: namely, the measurement precision, group heterogeneity, test length and time limit. The reliability is higher where each item has measurement precision and produces more stable responses. Furthermore, when the group of examinees is heterogeneous, the reliability is relatively higher. A large number of items and ample time given to complete a test raise its reliability.

The reliability coefficient (ρSS), which is the proportion of true variance in obtained test scores, can be represented as:

$$\rho_{SS} = \sigma^2_\tau / \sigma^2_s \tag{5}$$

To illustrate the effect of measurement precision, consider two tests administered to the same group of examinees:

	σ^2_τ	σ^2_e
Test A:	50	20
Test B:	50	10

Using Equation (5),
Reliability Coefficient of Test A = (50) / (50 + 20) = 0.71
Reliability Coefficient of Test B = (50) / (50 + 10) = 0.83

Hence, a relatively small measurement error, which is indicative of high precision, leads to better reliability coefficient. Guilford (1954, p. 351) argued that most scores are fallible, and a useful transformation of Equation (4) would lead to obtaining information about the error variance from experimental data.

The error term, e_i, too is useful in test interpretation. The standard error of measurement (σ_e) is the standard deviation of the errors of measurements in a test. The standard error of measurement is a measure of the discrepancies between obtained scores and true scores, and indicates how much a test score might differ from true scores. It also depends on the confidence (expectancy) interval for the true score, and flags the range of test scores a given true score might produce. The standard error of measurement can be represented as:

$$\sigma^2_e = \sigma^2_\tau (1 - \rho_{SS}) \tag{6}$$

This is a direct indicator of the probable extent of error in any score in a test to which it applies. Hopkins (1998) highlights that the value of the standard error of measurement is completely determined by the test's reliability index, and vice versa. Cronbach's Alpha, Kuder-Richardson's formulae (20 and 21) and Spearman-Brown's formula are useful indices of reliability.

2.1.6 Validity under the CTT model

Validity is probably the most important criterion in judging the effectiveness of an instrument. In the past, various types of validity, such as face validity, content validity, criterion validity and others, were identified, and an assessment would have been labelled valid if it met all of these to a substantial extent (Burns, 1997; Zeller, 1997). More recently, rather than a characteristic of an instrument, validity has come to be viewed as an ongoing process in which an argument, along with supportive evidence, is advanced that the intended interpretations of test scores are warranted (American Educational Research Association et al., 1999). The forms of evidence that need to be set out correspond closely with the types of validity that have been described in former representations of this construct.

The aspects of validity amenable to investigation through CTT are those claims that depend on comparisons with similar instruments or intended outcomes: namely, concurrent and criterion validity.

2.2 A critique of CTT

CTT has limited effectiveness in educational measurement. When different tests that seek to measure the same content are administered to different cohorts of students, comparisons of test items and examinees are not sound. Various equating processes which make assumptions about ability distributions have been implemented, but there is little theoretical justification for them. Raw scores add further ambiguity to measurement as student abilities, which are based on the total score obtained on a test, cannot be compared. Although z-score is used as standardisation criteria to overcome the problem, it is assumed that the examinees are from the same population.

Under CTT, item difficulty and item discrimination indices are group dependent: the values of these indices depend on the group of examinees in which they have been obtained. Another shortcoming is that observed and true test scores are test dependent. Observed and true scores rise and fall with changes in test difficulty. Another shortcoming is the assumption of equal errors of measurement for all examinees. In practice, ability estimates are less precise both for low and high ability students than for students whose ability is matched to the test average.

Wright (2001, p. 786) argued that the traditional conception of reliability is incorrect in that it assumes that 'sample and test somehow match'. The skewness of the empirical data set has been overlooked in computing reliability coefficients. Schumaker (2003, p. 6) indicated that CTT reliability coefficients 'do not always behave as expected because of sample dependency, non-linear raw scores, a restriction in the range of scores, offsetting negative and positive inter-item correlations, and the scoring rubric'.

3. CONCLUSION

We have argued in this chapter that theories of measurement and theories of testing have both shown an evolutionary pattern. The central ideals of measurement were recognised early in the 20th century, but tools for applying these principles were not readily available. Processes to achieve true measurement were developed, but were found to be impracticable, given the limited computational resources available. Since the 1950s, models applying axiomatic measurement principles have emerged, and in the last two decades especially, computing power and its widespread availability have made these methods practicable.

In education and psychology, many of the most interesting variables are latent. Factors such as student ability, social class, self-efficacy and many others cannot be observed. These variables are operationalised on the basis of theory and observable indicators of them proposed. Applying the principle of falsifiability, that, unless an observation is capable of refuting a proposition, it is of no use in supporting it, and therefore it requires strong theory to ensure that observations that are inconsistent with theory, or theory inconsistent with observed data, can be identified. CTT is built upon true score theory, and Keats described it as weak true score theory (Keats, 1994a). CTT is compatible with a particular conception of measurement, attributed to Stevens, that we have labelled 'pragmatic measurement'. This might be described as a weak measurement theory. An alternative, axiomatic measurement theory, and a development of this, namely simultaneous conjoint measurement, have been articulated, and are now practicable in educational and psychological measurement. Parallel with this advance, several groups of item response theories (IRTs) have developed. One in particular, the one-parameter or Rasch measurement model, is uniquely compatible with axiomatic measurement. The remaining entries in this book discuss the development and characteristics of this model and provide many examples of its application in education and psychology.

A final note of caution is warranted. Measurement in the social sciences has taken some retrograde evolutionary steps in the past because pragmatic considerations have over-ridden theoretical ones, and relatively weak theories and methods have been applied. Now we are at a stage where we believe that our theories and methods are strong. We should be able to make more important contributions to policy and practice in education. But we need also to be aware of new developments, new opportunities and new challenges. Michell (1997) has posed challenges to measurement practitioners. Do we meet the standards that he has set?

A note on further sources

A number of online communities of practice and reference groups archive their ongoing discussions on measurement and issues raised above. These digital repositories serve to continuously refine our understanding. The URLs below are good starting points to connect to these resources.

National Center for Educational Statistics
http://nces.ed.gov/nationsreportcard/

National Association of Test Directors
http://www.natd.org/publications.htm

Tests and Measures in the Social Sciences: March 2004 ed.
http://libraries.uta.edu/helen/Test&meas/testframed.htm

What is Measurement?
http://www.rasch.org/rmt/rmt151i.htm
http://www.rasch.org/rmt/

4. REFERENCES

American Educational Research Association, American Psychological Association, & National Council on Measurement in Education. (1999). *Standards for educational and psychological testing*. Washington, DC: American Educational Research Association.

Burns, R. B. (1997). *Introduction to research methods* (3rd ed.). South Melbourne, Australia: Longman.

Embretson, S. E. (1999). Issues in the measurement of cognitive abilities. In S. E. Embretson & S. L. Hershberger (Eds.), *The new rules of measurement. What every psychologist and educator should know* (pp. 1-15). Mahwah, NJ: Lawrence Erlbaum and Associates.

Guilford, J.P. (1954). Psychometric Methods. (2nd Ed). Tokyo: Kogakusha Company.

Holland, P.W., & Hoskens, M (2002). Classical Test Theory as a First-Order Item Response Theory: Application to True-Score Prediction From a Possibly Nonparallel Test. ETS Research Report. Educational Testing Service. Princeton, NJ.

Hopkins, K.D. (1998). Educational and Psychological Measurement and Evaluation. (8th Ed.). Boston: Allyn and BaconKeats, J. A. (1994a). Classical test theory. In T. Husen & T. N. Postlethwaite (Eds.), *The international encyclopedia of education* (2 ed., Vol. 2nd, pp. 785-792). Amsterdam: Elsevier.

Keats, J. A. (1994b). Measurement in educational research. In T. Husen & T. N. Postlethwaite (Eds.), *The international encyclopedia of education* (2 ed., Vol. 7, pp. 3698-3707). Amsterdam: Elsevier.

Keats, J. A. (1997). Classical test theory. In J. P. Keeves (Ed.), *Educational research, methodology, and measurement: an international handbook* (pp. 713-719). Oxford: Pergamon.

Keeves, J.P. & Masters, G.N. (1999). Introduction. In Masters, G.N. and Keeves, J.P. Advances in Measurement in Educational Research and Assessment. Amsterdam: Pergamon.

Lord, F. M. & Novick, M. R. (1968). Statistical Theories of Mental Test Scores. Reading, MA: Addison-Wesley.

Novick, M. R. (1966). The axioms and principal results of classical test theory. Journal of Mathematical Psychology, 3, 1-18.

Oppenheim, A.N. (1992). Questionnaire design, interviewing and attitude measurement. London: Continuum

Michell, J. (1997). Quantitative science and the definition of measurement in psychology. *British Journal of Psychology, 88*, 355-383.

Michell, J. (2002). Stevens's theory of scales of measurement and its place in modern psychology. *Australian Journal of Psychology, 54*(2), 99-104.

Schumacker, R.E. (2003). Reliability in Rasch Measurement: Avoiding the Rubber Ruler. Paper presented at the Annual Meeting of the American Educational Research Association. Chicago, Illinois. 25 Apr.

Stage, C. (2003). Classical Test Theory or Item Response Theory: The Swedish Experience. Online: Available at www.cepchile.cl

Stevens, S. S. (1951). Mathematics, measurement, and psychophysics. In S. S. Stevens (Ed.), *Handbook of experimental psychology* (pp. 1-49). New York: John Wiley.

Thorndike, R. M. (1999). IRT and intelligence testing: past, present, and future. In S. E. Embretson & S. L. Hershberger (Eds.), *The new rules of measurement. What every psychologist and educator should know* (pp. 17-35). Mahwah, NJ: Lawrence Erlbaum and Associates.

Wright, B. (2001). Reliability! Rasch Measurement Transactions, 14(4).

Zeller, R. A. (1997). Validity. In J. P. Keeves (Ed.), *Educational research, methodology, and measurement: an international handbook* (pp. 822-829). Oxford: Pergamon.

Chapter 2

OBJECTIVE MEASUREMENT

Geoff N. Masters
Australian Council for Educational Research

Abstract: The Rasch model is described as fundamental to objective measurement and this chapter examines key issues of conceptualising variable, especially in education. The notion of inventing units for objective measurement is discussed, and its importance in developmental assessment is highlighted.

Key words: measurement, human variability, units, objectivity, consistency of measure, variables, educational achievement, intervals, comparison

1. CONCEPTUALISING VARIABLES

In life, the most powerful ideas are the simplest. Many areas of human endeavour, including science and religion, involve a search for simple unifying ideas that offer the most parsimonious explanations for the widest variety of human experience.

Early in human history, we found ourselves surrounded by objects of impossible complexity. To make sense of the world we found it useful, and probably necessary, to ignore this complexity and to invent simple ways of thinking about and describing the objects around us. One useful strategy was to focus on particular ways in which objects differed.

The concepts of 'big' and 'small' provided an especially useful distinction. Bigness was an idea that allowed us to ignore the myriad other ways in which objects differed—including colour, shape and texture—and to focus on just *one* feature of an object: its bigness. The abstract notion of 'bigness' was a powerful idea because it could be used in describing objects as different as rivers, animals, rocks and trees.

S. Alagumalai et al. (eds.), Applied Rasch Measurement: A Book of Exemplars, 15–25.

For much of our history, the concept of 'bigness' no doubt served us well. But as we made more detailed observations of objects, and as we reflected on those observations, we realised that the idea of bigness often was less useful than the separate ideas of size and weight, even though size and weight usually were closely related. Later, we developed the ideas of length, area and volume as useful ways of capturing and conveying the notion of size. And, as we grappled with our experience that larger objects were not always heavier, we introduced the more sophisticated concepts of density and specific gravity.

Each of these ideas provided a way of focusing on just *one* way in which objects differed at a time, and so provided a tool for dealing with the otherwise unmanageable complexity of the world around us. Bigness, weight, length, volume and density were just some of our ideas for describing the ways in which objects varied; other 'variables' included hardness, temperature, inertia, speed, acceleration, malleability, and momentum. As our understandings improved and our observations became still more sophisticated, we found it useful to invent new variables subtly distinguished from existing variables: for example, to distinguish mass from weight, velocity from speed, and temperature from heat.

The advantage of a variable was that it allowed us to set aside—at least temporarily—the very complex ways in which objects differed, and to see objects through just *one* lens at a time. For example, objects could be placed in a single order of increasing weight, regardless of their varying shapes, colours, surface areas, volumes, and temperatures. The intention of the weight 'lens' was to allow us to see objects on just *one* of an essentially infinite number of possible dimensions.

We sometimes wondered whether we had invented these variables or simply discovered them. Was the concept of momentum a human invention, or was momentum 'discovered'? Certainly, it was a human decision to focus attention on specific aspects of variability in the world around us and to work to clarify and operationalise variables. The painstaking and relatively recent work of Anders Celsius (1701-44) and Gabriel Fahrenheit (1686-1736) to develop a useful working definition of temperature was testament to that. On the other hand, the variables we developed were intended to represent 'real' differences among objects. Ultimately, the question of whether variables were discovered or invented was of limited philosophical interest: the important question about a variable was whether it was useful in practice.

Human Variability

But it was not only inanimate objects that were impossibly complex; people were too. Again, a strategy for dealing with this complexity was to focus on particular ways in which people varied. Some humans were faster runners than others, some had greater strength, some were better hunters, more graceful dancers, superior warriors, more skilled craftsmen, wiser teachers, more compassionate counsellors, more comical entertainers, greater orators. The list of dimensions on which humans could be compared was unending, and the language we developed to describe this variability was vast and impressive.

In dealing with human complexity, our decision to focus on *one* aspect of variability at a time was at least as important as it was in dealing with the complexity of inanimate objects. To select the best person to lead the hunting party it was desirable to focus on individuals' prowess as hunters, and to recognise that the best hunter was not necessarily the most entertaining dancer around the campfire or the best storyteller in the group. There were times when our very existence depended on clarity about the relative strengths and weaknesses of fellow human beings.

The decision to pay attention to *one* aspect of variability at a time was also important when it came to monitoring the development of skills, understandings, attitudes and values in the young. As adults, we sought to develop different kinds of abilities in children, including skills in hunting, dancing, reading, writing, storytelling, making and using weapons and tools, constructing dwellings, and preparing food. We also sought to develop children's knowledge of local geography, flora and fauna, and their understandings of tribal customs and rituals, religious ceremonies, and oral history. To monitor children's progress towards mature, wise, well-rounded adults, we often found it convenient to focus on just one aspect of their development at a time.

We sometimes wondered whether the variables we used to deal with the complexity of human behaviour were 'real' in the sense that temperature and weight were 'real'. Did children *really* differ in reading ability? Were differences in children's reading abilities 'real' in the sense that differences in objects' potential energy or momentum were 'real'?

Once again, the important question was whether a variable such as reading ability was a useful idea in practice. Common experience suggested that children *did* differ in their reading abilities and that individuals' reading abilities developed over time. But was the idea of a variable of increasing reading competence supported by closer observations of reading behaviour? Did this idea help in understanding and promoting reading development? As with all variables, the most important question about dimensions of human variability was whether they were helpful in dealing with the complexities of human experience.

2. INVENTING UNITS

The second step towards measurement was remarkable because it was taken in relation to the most intangible of variables: time.

Time, unlike other variables such as length and weight, could not be manipulated and was much more difficult to conceptualise. But, amazingly, man found himself living inside a giant clock. By carefully inspecting the rhythmical ticking of the clock's mechanism, man learnt how to measure time, a lesson he then applied to the measurement of other variables.

The regular rotation of the Earth on its axis marked out equal amounts of time and provided humans with our first unit of measurement, the day. By counting days, we were able to replace qualitative descriptions of time ('a long time ago') with quantitative descriptions ('five days ago'). This was the second requirement for measurement: a unit of measurement. A unit was a fixed amount of a variable that could be repeated without modification and counted. The invention of units allowed the question *how much?* to be answered by counting *how many* units.

The regular revolution of the moon around the Earth provided a larger unit of time, the moon or lunar month. And the regular revolution of the Earth around the sun led to the seasons and a still larger unit, the year. The motion of these heavenly bodies provided us with an instrument for marking off equal amounts of time and taught us that units could be combined to form larger units, or subdivided to form still smaller units (hours, minutes, seconds).

Ancient civilisations created ways of tabulating their measurements of time in calendars chiselled in stone, and used moving shadows to invent units smaller than the day. By observing the rhythmical motion of the giant clock in which we lived, humans developed a sophistication in the measurement of time long before we developed a similar sophistication in the measurement of more tangible variables such as length, weight and temperature.

It seems likely that the earliest unit of distance was based on the primary unit of time. In man's early history, 'a long way' became '2-days walk', again allowing the question *how much?* to be answered by counting *how many* units. For shorter distances, we counted paces. One thousand paces we called a mile (*mil*). Other units of length we defined in terms of parts of the body—the foot, cubit (length of forearm), hand—or in terms of objects that could be carried and placed end to end—the chain, link (1/100 of a chain), rod, perch (a pole), and yard (a stick).

Our recent and continuing use of many of these units is a reminder of how recently we mastered the measurement of length. The same is true of the units used to measure some other variables (eg, 'stones' to measure weight). And still other units were invented so recently that we know the names of their inventors (eg, Celsius and Fahrenheit).

3. PURSUING OBJECTIVITY

The invention of units such as paces, feet, spans, cubits, chains, stones, rods and poles which could be repeated without modification provided humans with instruments for measuring. An important question in making measurements was whether different instruments provided numerically equivalent measures of the same object.

If two instruments did not provide numerically equivalent measures, then one possibility was that they were not calibrated in the same unit. It was one thing to agree on the use of a foot to measure length, but whose foot? What if my stone was heavier than yours? What if your chain was longer than mine? A fundamental requirement for useful measurement was that the resulting measures were independent of the measuring instrument and of the person doing the measuring: in other words, that they were *objective.*

To achieve this kind of objectivity, it was necessary to establish and share common, or standard, units of measurement. For example, in 1790 it was agreed to measure length in terms of a 'metre', defined as one ten-millionth of the distance from the North Pole to the Equator. After the 1875 Treaty of the Metre, a metre was re-defined as the length of a platinum-iridium bar kept at the International Bureau of Weights and Measures near Paris, and from 1983, a metre was defined as the distance travelled by light in a vacuum in 1/ 299,792,458 of a second. All measuring sticks marked out in metres and centimetres were calibrated against this standard unit.

Bureaus of weights and measures were established to ensure that standards were maintained, and that instruments were calibrated accurately against standard units. In this way, measures could be compared directly from instrument to instrument—an essential requirement for accurate communication and for the successful conduct of commerce, science and industry.

If two instruments did not provide numerically equivalent measures, then a second, more serious, possibility was that they were not providing measures of the same variable. The simplest indication of this problem was when two instruments produced significantly different orderings of a set of objects.

For example, two measuring sticks, one calibrated in centimetres, the other calibrated in inches, provided different numerical measures of an object. But when a number of objects were measured in both inches and centimetres and the measures in inches were plotted against the measures in centimetres, the resulting points approximated a straight line (and with no measurement error, would have formed a perfect straight line). In other words, the two measuring sticks provided *consistent* measures of length.

However, if on one instrument, Object A was measured to be significantly greater than Object B, but on a second instrument, Object B was measured to be significantly greater than Object A, then that would be evidence of a basic inconsistency. What should I conclude about the relative standings of Objects A and B on my variable?

A fundamental requirement for measurement was that it should not matter which instrument was used, or who was doing the measuring (ie, the requirement of objectivity/impartiality). Only if different instruments provided *consistent* measurements was it possible to achieve this kind of objectivity in our measures.

4. EDUCATIONAL VARIABLES

In educational settings it is common to separate out and pay attention to one aspect of a student's development at a time.

When a teacher seeks to establish the stage a student has reached in his or her learning, to monitor that student's progress over time, or to make decisions about the most appropriate kinds of learning experiences for individuals, these questions usually are addressed in relation to *one* area of learning at a time. For example, it is usual to assess a child's attainment in numerical reasoning separately from the many other dimensions along which that child might be progressing (such as reading, writing, and spoken language), even though those aspects of development may be related.

Most educational variables can be conceptualised as aspects of learning in which students make progress over a number of years. Reading is an example. Reading begins in early childhood, but continues to develop through the primary years as children develop skills in extracting increasingly subtle meanings from increasingly complex texts. And, for most children, reading development does not stop there: it continues into the secondary years.

Teachers and educational administrators use measures of educational attainment for a variety of purposes.

Measures on educational variables are sought whenever there is a desire to ensure that limited places in educational programs are offered to those who are most deserving and best able to benefit. For example, places in medical schools are limited because of the costs of providing medical programs and because of the limited need for medical practitioners in the community. Medical schools seek to ensure that places are offered to applicants on the basis of their likely success in medical school and, where possible, on the extent to which applicants appear suited to subsequent medical practice. To allocate places fairly, medical schools go to some trouble to identify and measure relevant attributes of applicants. Universities and schools offering scholarships on the basis of merit similarly go to some trouble to identify and measure candidates on appropriate dimensions of achievement.

Measures of educational achievement and competence also are sought at the completion of education and training programs. Has the student achieved a sufficient level of understanding and knowledge by the end of a course of instruction to be considered to have satisfied the objectives of that course? Has the student achieved a sufficient level of competence to be allowed to practice (eg, as an accountant? a lawyer? a paediatrician? an airline pilot?). Decisions of this kind usually are made by first identifying the areas of knowledge, skill and understanding in which some minimum level of competence must be demonstrated, and by then measuring candidates' levels of competence or achievement in each of these areas.

Measures of educational achievement also are required to investigate ways of improving student learning: for example, to evaluate the impact of particular educational initiatives, to compare the effectiveness of different ways of structuring and managing educational delivery, and to identify the most effective teaching strategies and most cost-effective ways of lifting the achievements of under-achieving sections of the student population. Most educational research, including the evaluation of educational programs, depends on reliable measures of aspects of student learning. The most informative studies often track student progress on one or more variables over a number of years (ie, longitudinal studies).

The intention to separate out and measure variables in education is made explicit in the construction and use of educational tests. The intention to obtain only *one* test score for each student so that all students can be placed in a single score order reflects the intention to measure students on just *one* variable, and is called the intention of unidimensionality. On such a test, higher scores are intended to represent more of the variable that the test is designed to measure, and lower scores are intended to represent less. The use of an educational test to provide just *one* order of students along an educational variable is identical in principle to the intention to order objects along a single variable of increasing heaviness.

Occasionally, tests are constructed with the intention not of providing one score, but of providing several scores. For example, a test of reasoning might be constructed with the intention of obtaining both a verbal reasoning score and a quantitative reasoning score for each student. Or a mathematics achievement test might be constructed to provide separate scores in Number,

Measurement and Space. Tests of this kind are really composite tests. The set of verbal reasoning items constitutes one measuring instrument; the set of quantitative reasoning items constitutes another. The fact that both sets of items are administered in the same test sitting is simply an administrative convenience.

Not every set of questions is constructed with the intention that the questions will form a measuring instrument. For example, some questionnaires are constructed with the intention of reporting responses to each question separately, but with no intention of combining responses across questions (eg, How many hours on average do you spend watching television each day? What type of book or magazine do you most like to read?). Questions of this kind are asked not because they are intended to provide evidence about the same underlying variable, but because there is an interest in how some population of students responds to each question separately. The best check on whether a set of questions is intended to form a measuring instrument is to establish whether the writer intends to combine responses to obtain a total score for each student.

The development of every measuring instrument begins with the concept of a variable. The intention underlying every measuring instrument is to assemble a set of items capable or providing evidence about the variable of interest, and then to combine responses to these items to obtain measures of that variable. This intention raises the question of whether the set of items assembled to measure each variable work together to form a useful measuring instrument.

5. EQUAL INTERVALS?

When a student takes a test, the outcome is a test score, intended as a measure of the variable that the test was designed to measure. Test scores provide a single order of test takers—from the lowest scorer (the person who answers fewest items correctly or who agrees with fewest statements on a questionnaire) to the highest scorer. Because scores order students along a variable, they are described as having 'ordinal' properties.

In practice, it is common to assume that test scores also have 'interval' properties: that is, that equal differences in scores represent equal differences in the variable being measured (eg, that the difference between scores of 25 and 30 on a reading comprehension test represents the same difference in reading ability as the difference between scores of 10 and 15). The attempt to attribute interval properties to scores is an attempt to treat them as though they were measures similar to measures of length in centimetres or measures of weight in kilograms. But scores are not counts of equal units of measurement, and so do not share the interval properties of measures.

Scores are counts of items answered correctly and so depend on the particulars of the items counted. A score of 16 out of 20 easy items does not have the same meaning as a score of 16 out of 20 hard items. In this sense, a score is like a count of objects. A count of 16 potatoes is not a 'measure' because it is not a count of *equal units*. Sixteen small potatoes do not represent the same amount of potato as 16 large potatoes. When we buy and sell potatoes, we use and count a unit (kilogram or pound) which maintains its meaning across potatoes of different sizes.

A second reason why ordinary test scores do not have the properties of measures is that they are bounded by upper and lower limits. It is not possible to score below zero or above the maximum possible score on a test. The effect of these so-called 'floor' and 'ceiling' effects is that equal differences in test score do not represent equal differences in the variable being measured. For example, on a 30-item mathematics test, a difference of one score point at the extremes of the score range (eg, the difference between scores of 1 and 2, or between scores of 28 and 29) represents a larger difference in mathematics achievement than a difference of one score point near the middle of the score range (eg, the difference between scores of 14 and 15).

Nevertheless, users of test results regularly assume that test scores have the interval properties of measures. Often users are unaware that they are making this assumption. But interval properties are assumed whenever test scores are used in simple statistical procedures such as the calculation of means and standard deviations, or in more sophisticated statistical procedures such as regression analyses or analyses of variance. In these common procedures, users of test scores treat them as though they have the interval properties of inches, kilograms and hours.

6. OBJECTIVITY

Every test constructor knows that, in themselves, individual test items are unimportant. No item is indispensable: items are constructed merely as opportunities to collect evidence about some variable of interest, and every test item could be replaced by another, similar item. More important than individual test items is the variable about which those items are intended to provide evidence.

A particular item developed as part of a calculus test, for example, is not in itself significant. Indeed, students may never again encounter and have to solve that particular item. The important question about a test item is not whether it is significant in its own right, but whether it is a useful vehicle for collecting evidence about the variable to be measured (in this case, calculus ability).

Another way of saying this is that it should not matter to our conclusion about a student's ability in calculus which particular items the student is given to solve. When we construct a test it is our intention that the results will have a generality beyond the specifics of the test items. This intention is identical to our intention that measures of height should not depend on the details of the measuring instrument (eg, whether we use a steel rule, a wooden rule, a builder's tape measure, a tailor's tape, etc). It is a fundamental intention of all measures that their meaning should relate to some general variable such as height, temperature, manual dexterity or empathy, and should not be bound to the specifics of the instrument used to obtain them.

The intention that measures of educational variables should have a general meaning independent of the instrument used to obtain them is especially important when there is a need to compare results on different tests. A teacher or school wishing to administer a test prior to a course of instruction (a pre-test) and then after a course of instruction (a post-test) to gauge the impact of the course, often will not wish to use the same test on both occasions. A medical school using an admissions test to select applicants for entry often will wish to compare results obtained on different forms of the admissions test at different test sittings. Or a school system wishing to monitor standards over time or growth across the years of school will wish to compare results on tests used in different years or on tests of different difficulty designed for different grade levels.

There are many situations in education in which we seek measures that are freed of the specifics of the instrument used to obtain them and so are comparable from one instrument to another.

It is also the intention when measuring educational variables that the resulting measures should not depend on the persons doing the measuring. This consideration is especially important when measures are based on judgements of student work or performance. To ensure the objectivity of measures based on judgements it is usual to provide judges with clear guidelines and training, to provide examples to illustrate rating points (eg, samples of student writing or videotapes of dance performances), to use multiple judges, procedures for identifying and dealing with discrepancies, and statistical adjustments for systematic differences in judge harshness/leniency.

Although it is clearly the intention that educational measures should have a meaning freed of the specifics of particular tests, ordinary test scores (eg, number of items answered correctly) are completely test bound. A score of

29 on a particular test does not have a meaning similar to a measure of 29 centimetres or 29 kilograms. To make an sense of a score of 29 it is necessary to know the total number of test items: 29 out of 30 items? 29 out of 40? 29 out of 100? Even knowing that a student scored 29 out of 40 is not very helpful. Success on 29 easy items does not represent the same ability as success on 29 difficult items. To understand completely the meaning of a score of 29 out of 40 it would be necessary to consider each of the 40 items attempted.

The longstanding dilemma in educational testing has been that, while particular test items are never of interest in themselves, but are intended only as indicators of the variable of interest, the meaning of number-right scores is always bound to some particular set of items. Just as we intend the measure of a student's writing ability to be independent of the judges who happen to assess that student's writing, so we seek measures of variables such as numerical reasoning which are neutral with respect to, and transcend, the particular items that happen to be included in a test. This dilemma was addressed and resolved in the work of Danish mathematician Georg Rasch in the 1950s.

The statistical model developed by Rasch (1960) provides practitioners with a basis for:

- establishing the extent to which a set of test items work together to provide measures of just *one* variable;
- defining a *unit* of measurement for the construction of interval-level measures of educational variables; and
- constructing numerical measures that have a meaning *independent* of the particular set of items used.

Rasch's measurement model, which is described and applied in the chapters of this book, made objective measurement a possibility in the social and behavioural sciences.

7. REFERENCES

Rasch, G (1960). *Probabilistic Models for Some Intelligence and Attainment Tests.* Copehanhagen: Danish Institute for Educational Research.

Chapter 3

THE RASCH MODEL EXPLAINED

David Andrich
Murdoch University

Abstract: This Chapter explains the Rasch model for ordered response categories by demonstrating the latent response structure and process compatible with the model. This is necessary because there is some confusion in the interpretation of the parameters and the possible response process characterised by the model. The confusion arises from two main sources. First, the model has the initially counterintuitive properties that (i) the values of the estimates of the thresholds defining the boundaries between the categories on the latent continuum can be reversed relative to their natural order, and (ii) that adjacent categories cannot be combined in the sense that their probabilities can be summed to form a new category. Second, two *identical* models at the level of a single person responding to a single item, the so called *rating* and *partial credit models*, have been portrayed as being different in the response structure and response process compatible with the model. This Chapter studies the structure and process compatible with the Rasch model, in which subtle and unusual distinctions need to be made between the *values* and *structure* of response probabilities and between *compatible* and *determined* relationships. The Chapter demonstrates that the response process compatible with the model is one of classification in which a response in any category implies a latent response at every threshold. The Chapter concludes with an example of a response process that is compatible with the model and one that is incompatible.

Key words: rating credit models, partial credit models, Guttman structure, combing categories

1. INTRODUCTION

This Chapter explains the Rasch model for ordered response categories in standard formats by demonstrating the latent response structure and process compatible with the model. Standard formats involve one response in one of

S. Alagumalai et al. (eds.), Applied Rasch Measurement: A Book of Exemplars, 27–59.

the categories deemed a-priori to reflect *increasing* levels of the latent trait common in quantifying attitude, performance, and status in the social sciences. Table 3-1 shows such formats for four ordered categories. Later in the paper, a response format not compatible with the model is also shown.

Table 3-1. Standard response formats for the Rasch model

Fail	<	Pass	<	Credit	<	Distinction
Never	<	Sometimes	<	Often	<	Always
Strongly Disagree	<	Disagree	<	Agree	<	Strongly Agree

The response structure and process characterized by the Rasch model concerns the response of *one* person to *one* item in *one* of the categories. These categories are deemed *a-priori* to reflect increasing levels of the latent trait. This is evident in each of the formats in Table 3-1. The first example in Table 3-1, the assessment of performance in the ordered categories of *Fail < Pass < Credit < Distinction,* is used for further illustrations in the Chapter.

1.1 The original form of the model

The model of concern was derived originally (Andrich,1978a) by expressing the general model studied by Rasch (1961) and Andersen (1977),

$$\Pr\{X_{ni} = x\} = \frac{1}{\gamma_{ni}} \exp(\kappa_x + \varphi_x(\beta - \delta)) \tag{1}$$

in terms of thresholds resolved from the general scoring functions φ_x and category coefficients κ_x as[1]

$$\Pr\{X = x, x > 0\} = \frac{1}{\gamma} \exp(-\sum_{k=1}^{x} \tau_k + x(\beta - \delta)); \Pr\{X = 0\} = \frac{1}{\gamma} \tag{2}$$

where (i) $X = x$ is an integer random variable characterizing $m+1$ successive categories which imply increasing values on the latent trait, (ii) β and δ are respectively locations on the same latent continuum of a person and an item, (iii) τ_k, $k = 1,2,3,...m$ are m thresholds which divide the continuum into to $m+1$ ordered categories and which, without loss of

[1] $\varphi_x = x$; $\kappa_x = -\sum_{k=1}^{x} \tau_k ; \kappa_0 \equiv 0$

generality, have the constraint $\sum_{k=1}^{m}\tau_k = 0$, and (iv) $\gamma = 1 + \sum_{x=1}^{m}\exp(-\sum_{k=1}^{x}\tau_k + x(\beta - \delta))$ is a normalizing factor that ensures that the probabilities in (2) sum to 1. The thresholds are points at which the probabilities of responses in one of the two adjacent categories are equal.

Figure 3-1 shows the probabilities of responses in each category, known as category characteristic curves (CCCs) for an item with three thresholds and four categories, together with the location of the thresholds on the latent trait.

In addition to ensuring the probabilities sum to 1, it is important to note this normalising factor contains all thresholds. This implies that the response in any category is a function of the location of all thresholds, not just of the thresholds adjacent to the category. Thus even though the numerator contains only the thresholds $\tau_k, k = 1,2,...x$, that is, up to the successful response x, the denominator contains all of the thresholds m. Therefore a change of value for any threshold, implies a change of the probabilities of a response in every category. In particular, a change in the value of the last threshold m changes the probability of the response in the first category. This feature constrains the kind of response process that is compatible with the model and is considered further in the Chapter.

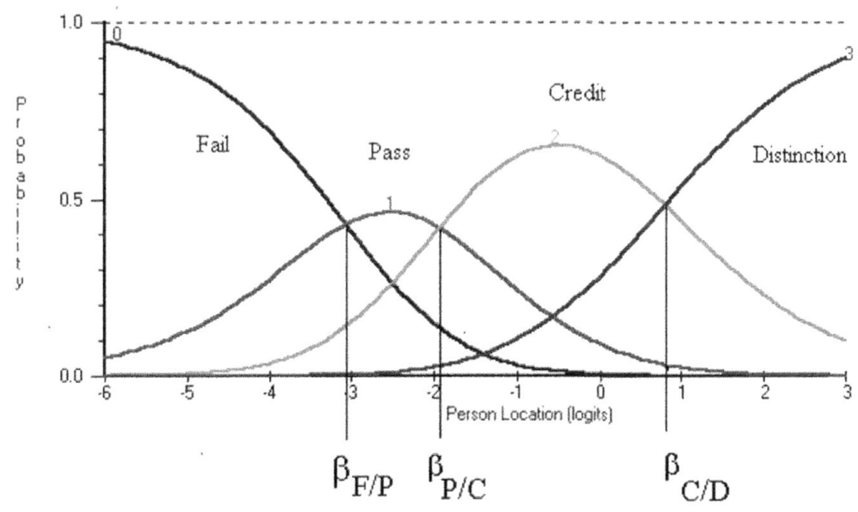

Figure 3-1. Probabilities of responses in each of four ordered categories showing the thresholds between the successive categories for an item in performance assessment

Note that in Eqs. (1) and (2), the person and item were not subscripted. This is because the response is concerned only with one person responding to one item and subscripting was unnecessary. The first application of the model (Andrich, 1978b), was to a case in which all items had the same response format and in which, therefore, the model applied specified that all items had the same thresholds. With explicit subscripts, this model takes the form

$$\Pr\{X_{ni} = x, x > 0\} = \frac{1}{\gamma_{ni}} \exp(-\sum_{k=1}^{x} \tau_k + x(\beta_n - \delta_i)); \Pr\{X_{ni} = 0\} = \frac{1}{\gamma_{ni}} \quad (3)$$

where n is a person and i is an item. Because the thresholds τ_k, $k = 1,2,3,...m$ were taken to be identical across items, they were not subscripted by i.

For subsequent convenience, let $\tau_0 \equiv 0$. Then Eq. (3) can be written as the single expression

$$\Pr\{X_{ni} = x,\} = \frac{1}{\gamma_{ni}} \exp(-\sum_{k=0}^{x} \tau_k + x(\beta_n - \delta_i)) \quad (4)$$

In a subsequent application (Wright and Masters, 1982), the model was written in the form equivalent to

$$\Pr\{X_{ni} = x\} = \frac{1}{\gamma_{ni}} \exp(-\sum_{k=1}^{x} \tau_{ki} + x(\beta_n - \delta_i)) \quad (5)$$

in which the thresholds $\tau_{ki}, k = 1,2,3...m_i$, $\sum_{k=1}^{m_i} \tau_{ki} = 0$, were taken to be different among items, and were therefore subscripted by i as well as k. $\tau_{0i} \equiv 0$ remains.

These models have become known as the *rating scale* model (Eq. 4) and the *partial credit* model (Eq. 5) respectively. This is unfortunate because it gives the impression that models (3) and (5) are different in their response structure and process for a single person responding to a single item, rather than in merely the parameterisation in the usual situation where the number of items is greater than 1. Therefore this is the first point of clarification and emphasis – that the so called *rating scale* and *partial credit* models, at the level of one person responding to one item, are identical in their structure and in the response process they can characterise. The only difference is that in the Eq. (3) the thresholds among items are identical and in Eq. (5) they are different. Some item formats are more likely to have an identical response structure across items than others. In this Chapter, the model will be referred to simply as the Rasch model (RM) with the model for dichotomous responses being just a further special case.

1.2 The alternate form of the model

Wright and Masters (1982) expressed the model of Eq. (5) effectively in the form

$$\Pr\{X_{ni}=x, x>0\}=\frac{\exp\sum_{k=1}^{x}(\beta_n-\delta_{ki})}{1+\sum_{x=1}^{m}\exp\sum_{k=1}^{x}(\beta_n-\delta_{ki})};\Pr\{X_{ni}=0\}=\frac{1}{1+\sum_{x=1}^{m}\exp\sum_{k=1}^{x}(\beta_n-\delta_{ki})} \quad (6)$$

As shown below, Eq (6) can be derived directly from Eq. (5). However, this difference in form has also contributed to confusing the identity of the models. To derive Eq. (6) from Eq. (5), first recall that $\sum_{k=1}^{m}\tau_{ki}=0$ in Eq. (5), implying that the thresholds $\tau_{ki}, k=1,2,3,...m_i$, are deviations from δ_i.

Let

$$\delta_{ki}=\delta_i+\tau_{ki}. \quad (7)$$

Then without loss of generality, $\delta_i=\overline{\delta}_{ki}$ and clearly $\tau_{ki}=\delta_{ki}-\delta_i=\delta_{ki}-\overline{\delta}_{ki}$. Second, consider the expression $(-\sum_{k=1}^{x}\tau_{ki}+x(\beta_n-\delta_i))$ in Eq. (5). It can be reexpressed as

$$\begin{aligned}&(-\sum_{k=1}^{x}\tau_{ki}+x(\beta_n-\delta_i))=x(\beta_n-\delta_i))-\sum_{k=1}^{x}\tau_{ki}\\&=x\beta_n-x\delta_i-\sum_{k=1}^{x}\tau_{ki}=\sum_{k=1}^{x}\beta_n-\sum_{k=1}^{x}\delta_i-\sum_{k=1}^{x}\tau_{ki}=\sum_{k=1}^{x}\beta_n-\sum_{k=1}^{x}(\delta_i+\tau_{ki})=\sum_{k=1}^{x}\beta_n-\sum_{k=1}^{x}\delta_{ki}\\&=\sum_{k-1}^{x}(\beta_n-\delta_{ki}).\end{aligned} \quad (8)$$

Substituting Eq. (7) into Eq. (5) gives

$$\Pr\{X_{ni} = x\} = \frac{1}{\gamma_{ni}} \exp(\sum_{k=1}^{x} (\beta_n - \delta_{ki}); \Pr\{X_{ni} = 0\} = \frac{1}{\gamma_{ni}}. \quad (9)$$

where $\gamma_{ni} = 1 + \sum_{x=1}^{m} \exp \sum_{k=1}^{x} (\beta_n - \delta_{ki})$ is the normalizing factor made explicit giving Eq. (6).

By analogy to $\tau_{0i} \equiv 0$, *let* $\delta_{0i} \equiv 0$. *Then Eq. (9) can be written as the single expression*

$$\Pr\{X_{ni} = x\} = \frac{1}{\gamma_{ni}} \exp(\sum_{k=1}^{x} (\beta_n - \delta_{ki}). \quad (10)$$

2. DERIVATIONS OF THE MODEL

The original derivation of the models by Andrich and Wright and Masters was also different, the one by Wright and Masters taking a reverse process to that taken by Andrich. However, these two derivations give an identical structure and response process when derived in full. Both derivations are now explained.

Before proceeding, and because the RM for dichotomous responses is central to the derivations, it is noted that it takes the form

$$\Pr\{X_{ni} = x\} = \frac{\exp x(\beta_n - \delta_i)}{1 + \exp(\beta_n - \delta_i)}; x \in \{0,1\} \quad (11)$$

where in this case there is only the one threshold, the location of the item, δ_i.

2.1 The original derivation

In the original derivation of the model (Andrich, 1978a), an instantaneous latent dichotomous response process $\{Y_{nki} = y\}, y \in \{0,1\}$ was postulated at each threshold k with the probability of the response given by

$$\Pr\{Y_{nki} = y\} = \exp y(\alpha_k(\beta_n - (\delta_i + \tau_{ki}))) / \{1 + \exp \alpha_k(\beta_n - (\delta_i + \tau_{ki}))\} \quad (12)$$

where (i) α_k characterises the discrimination at threshold k and τ_{ki} qualifies the single location δ_i to characterise the m thresholds. In fact, the items and persons, as indicated earlier, were not subscripted, but in order not to have to stress the issue of the identity of Eqs. (5) and (7) at the level of a single person and a single item, these subscripts are included in Eq. (12). Notice that the response process in Eq. (12) is not the dichotomous RM of Eq. (11) which including τ_{ki} would take the form

$$\Pr\{Y_{nki} = y\} = \exp y(\beta_n - (\delta_i + \tau_{ki})) / \{1 + \exp(\beta_n - (\delta_i + \tau_{ki}))\} \tag{13}$$

and in which the discrimination parameter across thresholds is assumed to be identical and effectively have a value of 1. The dichotomous RM was specialised at a subsequent point in the derivation in order to interpret Eq. (1) and two of its properties identified by Andersen (1977).

This specialisation is critical but is generally ignored; ignoring this specialisation is a another source of confusion as it fails to acknowledge that categories cannot be combined in the usual way by pooling frequencies of responses in adjacent categories in the data or equivalently by adding the probabilities of adjacent categories in the model following estimation of the parameters, except in one special case that is commented upon later in the Chapter.

2.1.1 Latent unobservable responses at the thresholds

Although instantaneously assumed to be independent, it is not possible for the latent dichotomous response processes at the thresholds to be either observable or independent – there is only *one* response in *one* of $m+1$ categories. Therefore the responses must be *latent*. Furthermore, the categories are deemed to be ordered – thus if a response is in category x, then this response is deemed to be in a category *lower* than categories $x+1$ or greater, and at the same time, in a category greater than categories $x-1$ or lower. Therefore the responses must be *dependent* and a *constraint* must be placed on any process in which the latent responses at the thresholds are instantaneously considered independent. This constraint ensures taking account of the substantial dependence. The Guttman structure provides this constraint.

2.1.2 The Guttman structure

For I independent dichotomous items there are 2^I possible response patterns. These are shown in Table 3-2. The top part of Table 3-2 shows the subset of *Acceptable* responses according to the Guttman structure. The number of these patterns is $I+1$.

Table 3-2. The Guttman Structure with dichotomous items in difficulty order

Items	1	2	3	.	.	.	I–2	I–1	I
I+1 Acceptable response patterns in the Guttman structure									
	0	0	0	.	.	.	0	0	0
	1	0	0	.	.	.	0	0	0
	1	1	0	.	.	.	0	0	0
	1	1	1	.	.	.	0	0	0
	.	.	.	.	.	.	.	.	.
	.	.	.	.	.	.	.	.	.
	.	.	.	.	.	.	.	.	.
	1	1	1	.	.	.	1	0	0
	1	1	1	.	.	.	1	1	0
	1	1	1	.	.	.	1	1	1
2^I– I–1 Unacceptable response patterns for the Guttman structure									
	0	1	0	.	.	.	0	0	0
	0	1	1				1	1	1
	0	0	1				1	1	1
	.	.	.	.	.	.	.	.	.
	.	.	.	.	.	.	.	.	.
	.	.	.	.	.	.	.	.	.
	0	0	0	.	.	.	1	1	1
	0	0	0	.	.	.	0	1	1
	0	0	0	.	.	.	0	0	1
2^I total number of patterns under independence									

The rationale for the Guttman patterns in Table 3-2 (Guttman,1950) is that for unidimensional responses across items, if a person succeeds on an item, then the person should succeed on all items that are easier than that item and that if a person failed on an item, then the person should fail on all items more difficult than that item. Put another way, if person A is more able than person B, then person A should answer all items correctly that person B does, and in addition at least the next most difficult item. Of course, with experimentally independent items, that is, where each item is responded to independently of every other item, it is possible that the Guttman structure will not be observed in data.

2.1.3 The rationale for the latent Guttman structure at the thresholds of polytomous items

Returning to the latent dichotomous responses at the thresholds of an item with ordered categories, if the responses at m thresholds were independent, then as in the whole of Table 3-2, there would be 2^m possible patterns. However, there is in fact only one of $m+1$ possible responses, a response in one of the categories. Therefore the total number 2^m of response patterns under independence must be *reduced* to just $m+1$ responses under some kind of dependence. The Guttman structure provides the reduction of this space in exactly the way required.

The rationale for the Guttman structure, as with the ordering of items in terms of their difficulty, is that the thresholds of an item are required to be ordered, that is

$$\tau_1 < \tau_2 < \tau_3 .... < \tau_{m-1} < \tau_m \qquad (14)$$

This requirement of ordered thresholds is independent of the RM – it is required by the Guttman structure. However, as shown later in the Chapter, this requirement is completely *compatible* with the structure of the responses in the RM.

The requirement of threshold order implies that a person at the threshold of two higher categories is required to be at a higher location than a person at the boundary of two lower categories. For example, it requires that a person who is at the threshold between a Credit and Distinction in the first row in Table 3-1, and reflected in Figure 3-1, should have a higher ability than a person who is a the threshold between a Fail and a Pass. To make this requirement more concrete, the person who has equal chance of receiving Credit or Distinction should have a higher ability than a person who has an equal chance of receiving a Fail or a Pass. In short, if the categories are to be ordered, then the thresholds that define them should also be ordered.

More formally, this implies that the latent successes at the latent dichotomous responses at the thresholds, should become more difficult for successive pairs of adjacent categories, that is, if Eq. (14) holds, then $\Pr\{Y_{nki} = y+1\} < \Pr\{Y_{nki} = y\}$ for all β. The condition in Eq.(14) operationalises the meaning of ordered categories in terms of the thresholds between the categories. Following Guttman's rationale, a person who succeeds at a particular threshold, must succeed on all thresholds of lesser difficulty; and a person who fails at a particular threshold must fail on all thresholds of greater difficulty. Thus for the Guttman rationale to be consistent, it must be increasingly difficult to succeed on the successive thresholds.

2.1.4 Constraining the response probabilities to the Guttman Structure

To obtain the probability of a response pattern according to the Guttman structure, two simplifications are made to Eq. (12). First, let $\delta_{ki} = \delta_i + \tau_{ki}$ as in Eq. (7), and second, let

$$\eta_{nki} = 1/\{1 + \exp \alpha_k (\beta_n - \delta_{ki})\} \qquad (15)$$

giving

$$\Pr\{Y_{nki} = 1\} = \exp(1\alpha_k(\beta_n - \delta_{ki}))\} / \eta_{nki} \tag{16}$$

and

$$\Pr\{Y_{nki} = 0\} = \exp(0\alpha_k(\beta_n - \delta_{ki}))\} / \eta_{nki} \tag{17}$$

Note that Eq. (17) shows the response of $Y_{nki} = 0$ on the right side in the same form as Eq. (16) which shows the response of $Y_{nki} = 1$. For further exposition, and to simplify the expressions, suppose that the number of thresholds is $m = 3$, as in Table 3-1.

On the initial assumption of independence of responses at the thresholds (which we subsequently constrain not to be independent), the probabilities of the latent dichotomous responses at the thresholds are given by

$$\Pr\{y_{n1i}, y_{n2i}, y_{n3i}\} = \Pr\{y_{n1i}\} \Pr\{y_{n2i}\} \Pr\{y_{n3i}\}. \tag{18}$$

The patterns and probabilities of Eq. (18) (under independence) are shown in Table 3-3 with the Guttman response patterns at the top of the table as in Table 3-2.

Table 3-3. Probabilities of all response patterns for an item with three thresholds under the assumption of independence: Guttman patterns in the top of the Table

	$\Pr\{y_{n1i}, y_{n2i}, y_{n3i}\}$	$\Pr\{y_{n1i}\}$	$\Pr\{y_{n2i}\}$	$\Pr\{y_{n3i}\}$
Guttman Patterns	$\Pr\{0,0,0\}$ =	$(\frac{e^{0(\alpha_1(\beta_n-\tau_{1i}))}}{\eta_{n1i}})$	$(\frac{e^{0(\alpha_2(\beta_n-\tau_{2i}))}}{\eta_{n2i}})$	$(\frac{e^{0(\alpha_2(\beta_n-\tau_{3i}))}}{\eta_{n3i}})$
	$\Pr\{1,0,0\}$ =	$(\frac{e^{1(\alpha_1(\beta_n-\tau_{1i}))}}{\eta_{n1i}})$	$(\frac{e^{0(\alpha_2(\beta_n-\tau_{2i}))}}{\eta_{n2i}})$	$(\frac{e^{0(\alpha_2(\beta_n-\tau_{3i}))}}{\eta_{n3i}})$
	$\Pr\{1,1,0\}$ =	$(\frac{e^{1(\alpha_1(\beta_n-\tau_{1i}))}}{\eta_{n1i}})$	$(\frac{e^{1(\alpha_2(\beta_n-\tau_{2i}))}}{\eta_{n2i}})$	$(\frac{e^{0(\alpha_2(\beta_n-\tau_{3i}))}}{\eta_{n3i}})$
	$\Pr\{1,1,1\}$ =	$(\frac{e^{1(\alpha_1(\beta_n-\tau_{1i}))}}{\eta_{n1i}})$	$(\frac{e^{1(\alpha_2(\beta_n-\tau_{2i}))}}{\eta_{n2i}})$	$(\frac{e^{1(\alpha_2(\beta_n-\tau_{3i}))}}{\eta_{n3i}})$
	$\sum_{\text{Guttman patterns}} \Pr\{y_{n1i}, y_{n2i}, y_{n3i}\} < 1$			
Non Guttman Patterns	$\Pr\{0,1,0\}$ ..=	$(\frac{e^{0(\alpha_1(\beta_n-\tau_{1i}))}}{\eta_{n1i}})$	$(\frac{e^{1(\alpha_2(\beta_n-\tau_{2i}))}}{\eta_{n2i}})$	$(\frac{e^{0(\alpha_2(\beta_n-\tau_{3i}))}}{\eta_{n3i}})$
	$\Pr\{0,0,1\}$ ..=	$(\frac{e^{0(\alpha_1(\beta_n-\tau_{1i}))}}{\eta_{n1i}})$	$(\frac{e^{0(\alpha_2(\beta_n-\tau_{2i}))}}{\eta_{n2i}})$	$(\frac{e^{1(\alpha_2(\beta_n-\tau_{3i}))}}{\eta_{n3i}})$
	$\Pr\{1,0,1\}$ ..=	$(\frac{e^{1(\alpha_1(\beta_n-\tau_{1i}))}}{\eta_{n1i}})$	$(\frac{e^{0(\alpha_2(\beta_n-\tau_{2i}))}}{\eta_{n2i}})$	$(\frac{e^{1(\alpha_2(\beta_n-\tau_{3i}))}}{\eta_{n3i}})$
	$\Pr\{0,1,1\}$ =	$(\frac{e^{0(\alpha_1(\beta_n-\tau_{1i}))}}{\eta_{n1i}})$	$(\frac{e^{1(\alpha_2(\beta_n-\tau_{2i}))}}{\eta_{n2i}})$	$(\frac{e^{1(\alpha_2(\beta_n-\tau_{3i}))}}{\eta_{n3i}})$
	$\sum_{\text{all 8 patterns}} \Pr\{y_{n1i}, y_{n2i}, y_{n3i}\} = 1$			

Because the total number of patterns in Table 3-3 is $2^m = 2^3 = 8$, which sum to 1 under the assumption of independence, the probabilities of the Guttman $m+1 = 3+1 = 4$ patterns cannot sum to 1: that is, in Table 3-3:

$$\Pr\{0,0,0\} + \Pr\{1,0,0\} + \Pr\{1,1,0\} + \Pr\{1,1,1\} = \Gamma_{ni} < 1. \qquad (19)$$

Ensuring that the probabilities of the Guttman patterns do sum to 1 is readily accomplished by normalisation, that is, simply dividing the probability of each Guttman pattern by the sum Γ_{ni}. Taking the Guttman patterns and normalising their probabilities are the critical moves that account for the dependence of responses at the thresholds.

The probabilities of the Guttman patterns after this normalisation are shown in Table 3-4.

Table 3-4. Probabilities of Guttman response patterns for an item with three thresholds taking account of dependence of responses at the thresholds

$\Pr\{y_{n1i}, y_{n2i}, y_{n3i}\} =$	$[(\Pr\{y_{n1i}\})$	$(\Pr\{y_{n2i}\})$	$(\Pr\{y_{n3i}\})]/\Gamma_{ni}$
$\Pr\{0,0,0\} =$	$[(\dfrac{e^{0(\alpha_1(\beta_n-\tau_{1i}))}}{\eta_{n1i}})$	$(\dfrac{e^{0(\alpha_2(\beta_n-\tau_{2i}))}}{\eta_{n2i}})$	$(\dfrac{e^{0(\alpha_2(\beta_n-\tau_{3i}))}}{\eta_{n3i}})]/\Gamma_{ni}$
$\Pr\{1,0,0\} =$	$[(\dfrac{e^{1(\alpha_1(\beta_n-\tau_{1i}))}}{\eta_{n1i}}))$	$(\dfrac{e^{0(\alpha_2(\beta_n-\tau_{2i}))}}{\eta_{n2i}})$	$(\dfrac{e^{0(\alpha_2(\beta_n-\tau_{3i}))}}{\eta_{n3i}})]/\Gamma_{ni}$
$\Pr\{1,1,0\} =$	$[(\dfrac{e^{1(\alpha_1(\beta_n-\tau_{1i}))}}{\eta_{n1i}})$	$(\dfrac{e^{1(\alpha_2(\beta_n-\tau_{2i}))}}{\eta_{n2i}})$	$(\dfrac{0(\alpha_2(\beta_n-\tau_{3i}))}{\eta_{n3i}})]/\Gamma_{ni}$
$\Pr\{1,1,1\} =$	$[(\dfrac{e^{1(\alpha_1(\beta_n-\tau_{1i}))}}{\eta_{n1i}})$	$(\dfrac{e^{1(\alpha_2(\beta_n-\tau_{2i}))}}{\eta_{n2i}})$	$(\dfrac{e^{1(\alpha_2(\beta_n-\tau_{3i}))}}{\eta_{n3i}})]/\Gamma_{ni}$
$\sum_{\text{4 Guttman patterns}} \Pr\{y_{n1i}, y_{n2i}, y_{n3i}\} = 1$			

Although the normalisation shown in Table 3-4 is straightforward, recognising that it accounts for the dependence of the dichotomous responses at the latent thresholds is critical. The division by Γ_{kni} of the response probabilities under independence, and the reduction of the possible outcomes to those conforming to the Guttman structure, means *that the responses* $\{y_{n1i}, y_{n2i}, y_{n3i}\}$ *must always be taken as a set.* Thus the response $\{y_{n1i}, y_{n2i}, y_{n3i}\} = \{1,0,0\}$ implies a latent success at the first threshold, *and* a latent failure at all subsequent thresholds. This important point is reconsidered later to clarify a further confusion where it is mistakenly considered that the model characterises some kind of sequential step process.

The probabilities of the Guttman patterns in Table 3-4 can be written as the single equation

$$\Pr\{y_{n1i}, y_{n2i}, y_{n3i}\} = \frac{[(e^{y_{n1i}(\alpha_1(\beta_n-\delta_{1i}))} e^{y_{n2i}(\alpha_2(\beta_n-\delta_{1i}))} e^{y_{n3i}(\alpha_3(\beta_n-\delta_{1i}))}) / \eta_{n1i}\eta_{n2i}\eta_{n3i}\Gamma_{ni}}{\sum_G [(e^{y_{n1i}(\alpha_1(\beta_n-\delta_{1i}))} e^{y_{n2i}(\alpha_2(\beta_n-\delta_{1i}))} e^{y_{n3i}(\alpha_3(\beta_n-\delta_{1i}))}) / \eta_{n1i}\eta_{n2i}\eta_{n3i}\Gamma_{ni}} \quad (20)$$

where $\sum_G$ indicates the sum over all Guttman patterns. The denominator of Eq. (20) cancels, reducing it to

$$\Pr\{y_{n1i}, y_{n2i}, y_{n3i}\} = \frac{(e^{y_{n1i}(\alpha_1(\beta_n-\delta_{1i}))} e^{y_{n2i}(\alpha_2(\beta_n-\delta_{2i}))} e^{y_{n3i}(\alpha_3(\beta_n-\delta_{3i}))})}{\sum_G [(e^{y_{n1i}(\alpha_1(\beta_n-\delta_{1i}))} e^{y_{n2i}(\alpha_2(\beta_n-\delta_{2i}))} e^{y_{n3i}(\alpha_3(\beta_n-\delta_{3i}))})}. \quad (21)$$

Let $\gamma_{ni} = \sum_G [(e^{y_{n1i}(\alpha_1(\beta_n-\delta_{1i}))} e^{y_{n2i}(\alpha_2(\beta_n-\delta_{1i}))} e^{y_{n3i}(\alpha_3(\beta_n-\delta_{1i}))})$.

Then

$$\Pr\{y_{n1i}, y_{n2i}, y_{n3i}\} = \frac{1}{\gamma_{ni}} (e^{y_{n1i}(\alpha_1(\beta_n-\delta_{2i}))} e^{y_{n2i}(\alpha_2(\beta_n-\delta_{2i}))} e^{y_{n3i}(\alpha_3(\beta_n-\delta_{3i}))}) \quad (22)$$

Taking advantage of the role of the total score in the Guttman pattern permits a simplification of its representation as shown in Table 3-5. This total score is defined by the integer random variable $X_{ni} = x, x \in \{0,1,2,3\}$. Thus a score of 0 means that all thresholds have been failed, a score of 1 means that the first has been passed and all others failed, and so on.

Table 3-5. Simplifying the probabilities of Guttman response in Table 3-4 taking advantage of the role of total score in the Guttman pattern

$\Pr\{y_{n1i}, y_{n2i}, y_{n3i}\} =$	$[(\Pr\{y_{n1i}\})$	$(\Pr\{y_{n2i}\})$	$(\Pr\{y_{n3i}\})]/\Gamma_{ni}$
$\Pr\{0,0,0\} = \Pr\{X_{ni} = 0\}$	$[(\frac{e^{0(\alpha_1(\beta_n-\tau_{1i}))}}{\eta_{n1i}})$	$(\frac{e^{0(\alpha_2(\beta_n-\tau_{2i}))}}{\eta_{n2i}})$	$(\frac{e^{0(\alpha_2(\beta_n-\tau_{3i}))}}{\eta_{n3i}})]/\Gamma$
$\Pr\{1,0,0\} = \Pr\{X_{ni} = 1\}$	$[(\frac{e^{1(\alpha_1(\beta_n-\tau_{1i}))}}{\eta_{n1i}}))$	$(\frac{e^{0(\alpha_2(\beta_n-\tau_{2i}))}}{\eta_{n2i}})$	$(\frac{e^{0(\alpha_2(\beta_n-\tau_{3i}))}}{\eta_{n3i}})]/\Gamma$
$\Pr\{1,1,0\} = \Pr\{X_{ni} = 2\}$	$[(\frac{e^{1(\alpha_1(\beta_n-\tau_{1i}))}}{\eta_{n1i}})$	$(\frac{e^{1(\alpha_2(\beta_n-\tau_{2i}))}}{\eta_{n2i}})$	$(\frac{0(\alpha_2(\beta_n-\tau_{3i}))}{\eta_{n3i}})]/\Gamma$
$\Pr\{1,1,1\} = \Pr\{X_{ni} - 3\}$	$[(\frac{e^{1(\alpha_1(\beta_n-\tau_{1i}))}}{\eta_{n1i}})$	$(\frac{e^{1(\alpha_2(\beta_n-\tau_{2i}))}}{\eta_{n2i}})$	$(\frac{e^{1(\alpha_2(\beta_n-\tau_{3i}))}}{\eta_{n3i}})]/\Gamma$
$\sum_{4\ Guttman\ patterns} \Pr\{y_{n1i}, y_{n2i}, y_{n3i}\} = 1$			

2.1.5 Specialising the discrimination at the thresholds to 1

As indicated above, Eq (7) for the response at the thresholds is not the RM – it is the two parameter model which includes a discrimination α_k at each threshold *k* (Birnbaum, 1968). The discrimination must be specialised to $\alpha_k = 1$ for all thresholds to produce the RM for ordered categories; specialising the discrimination to $\alpha_k = 0$, considered later, provides another important insight into the model.

Thus let $\alpha_k = 1 \forall k$. Then

$$\Pr\{y_{n1i}, y_{n2i}, y_{n3i}\} = \frac{1}{\gamma_{ni}}(e^{y_{n1i}(\beta_n - \delta_{1i})} e^{y_{n2i}(\beta_n - \delta_{2i})} e^{y_{n3i}(\beta_n - \delta_{3i})}). \quad (23)$$

Table 3-6 shows the specific probabilities of all Guttman patterns with the discriminations $\alpha_k = 1$. The equality of discrimination at the thresholds in the numerator of the right side of Eq. (23) permits it to be simplified considerably as shown in the last row of Table 3-6. In particular, the coefficients of the parameter β_n reduce to successive integers $X_{ni} = x,\, x \in \{0,1,2,3\}$ to give

$$\Pr\{X_{ni} = x\} = e^{(x\beta_n - \sum_{k}^{x} \delta_{ki})} / \gamma_{ni}. \quad (24)$$

Substituting

$$\delta_{ki} = \delta_i + \tau_{ki} \quad (25)$$

again in Eq. (24) gives

$$\Pr\{X = x, x > 0\} = \frac{1}{\gamma}\exp(-\sum_{k=1}^{x} \tau_{ki} + x(\beta_n - \delta_i)); \Pr\{X = 0\} = \frac{1}{\gamma} \quad (26)$$

which is Eq.(5) and defining $\tau_{0i} \equiv 0$ again gives Eq. (5).

It cannot be stressed enough that (i) the simplification on the right side of Eq. (26), which gives integer scoring x as the coefficient of $(\beta_n - \delta_i)$, follows from the equality of discriminations at the thresholds, and (ii) that the integer score reflects in each case a Guttman response pattern of the latent dichotomous responses at the thresholds. A further consequence of the discrimination of the thresholds is considered in Section 3-2. Thus the total score which can be used to completely characterise the Guttman pattern also appears on the right side of the Eq. (26) because of equal discrimination at the thresholds.

Table 3-6. The Guttman patterns represented by the total scores

$\Pr\{y_{n1i}, y_{n2i}, y_{n3i}\} = \Pr\{X_{ni} = x\} =$	$[e^{y_{n1i}(\beta_n - \tau_{1i})}$	$e^{y_{n2i}(\beta_n - \tau_{2i})}$	$e^{y_{n3i}(\beta_n - \tau_{3i})}]/\gamma_{ni}$
$\Pr\{0,0,0\} = \Pr\{X_{ni} = 0\} =$	$[e^{0(\beta_n - \tau_{1i})}$	$e^{0(\beta_n - \tau_{2i})}$	$e^{0(\beta_n - \tau_{3i})}]/\gamma_{ni}$
$\Pr\{1,0,0\} = \Pr\{X_{ni} = 1\} =$	$[e^{1(\beta_n - \tau_{1i})}$	$e^{0(\beta_n - \tau_{2i})}$	$e^{0(\beta_n - \tau_{3i})}]/\gamma_{ni}$
$\Pr\{1,1,0\} = \Pr\{X_{ni} = 2\} =$	$[e^{1(\beta_n - \tau_{1i})}$	$e^{1(\beta_n - \tau_{2i})}$	$e^{0(\beta_n - \tau_{3i})}]/\gamma_{ni}$
$\Pr\{1,1,1\} = \Pr\{X_{ni} = 3\} =$	$[\dfrac{e^{1(\beta_n - \tau_{1i})}}{\eta_{n1i}}$	$\dfrac{e^{1(\beta_n - \tau_{2i})}}{\eta_{n2i}}$	$(\dfrac{e^{1(\beta_n - \tau_{3i}))}}{\eta_{n3i}})]/\gamma_{ni}$
$\Pr\{X_{ni} = x\} =$	$e^{\sum_k y_{nki}(\beta_n - \tau_{ki})}/\gamma_{ni}$	$= e^{(x\beta_n - \sum_k^x \tau_{ki})}/\gamma_{ni}$	

2.2 The reverse derivation

Wright and Masters (1982) derived the same model by proceeding in reverse to the above derivation. Forming the probability of a response in the higher of two adjacent categories in Eq. (5), conditional on the response being in one of those two categories, gives on simplification,

$$\frac{\Pr\{X_{ni} = x\}}{\Pr\{X_{ni} = x-1\} + \Pr\{X_{ni} = x\}} = \frac{\exp(\beta_n - (\delta_i + \tau_x))}{1 + \exp(\beta_n - (\delta_i + \tau_x))}. \tag{27}$$

This conditional latent response is dichotomous. It is latent again, derived as an implication of the model, as in the original derivation, because there is not a sequence of observed conditional responses, but just one response in one of the m categories. Consistent with the implied ordering of the categories, the conditional response in the higher of the two categories, category $X_{ni} = x$ is deemed the *relative success;* the response $X_{ni} = x - 1$ a *relative failure*. The dichotomous responses at the threshold for item i are latent. Because there is only one response in one category they are never observed.

There are two distinct elements in equation (27): first, the structure of the relationship between scores in adjacent categories to give an implied dichotomous response; second the specification of the probability of this response by the dichotomous RM. These two features need separation (before being brought together), analogous to taking the more general two parameter logistic in the original derivation and then specializing it to the dichotomous RM.

First, generalize the probability in Eq. (27) to

$$\frac{\Pr\{X_{ni}=x\}}{\Pr\{X_{ni}=x-1\}+\Pr\{X_{ni}=x\}}=P_x, \tag{28}$$

and its complement

$$\frac{\Pr\{X_{ni}=x-1\}}{\Pr\{X_{ni}=x-1\}+\Pr\{X_{ni}=x\}}=Q_x=1-P_x. \tag{29}$$

To simplify the derivation of the model beginning with Eqs (28) and (29), we ignore the item and person subscripts and let the (unconditional) probability of a response in any category x be given by

$$\Pr\{X_{ni}=x\}=\pi_x. \tag{30}$$

Therefore

$$P_x=\frac{\pi_x}{\pi_{x-1}+\pi_x}, \tag{31}$$

and the task is to show that, from Eq. (31), it follows that

$$\pi_x=\Pr\{X_{ni}=x\}=\frac{1}{\gamma_{ni}}\exp(-\sum_{k=1}^{x}\tau_k+x(\beta_n-\delta_i))$$

of Eq. (5).

Immediately consider that the number of thresholds is m.

From Eq. (31)

$P_x(\pi_{x-1}+\pi_x)=\pi_x$, $\pi_x(1-P_x)=\pi_{x-1}P_x$, that is $\pi_x Q_x=\pi_{x-1}P_x$, and

$$\pi_x = \pi_{x-1}\frac{P_x}{Q_x} \ . \tag{32}$$

Beginning with $\pi_x, x = 1$, the recursive relationship

$$\pi_x = \pi_0 \frac{P_1}{Q_1}\frac{P_2}{Q_2}\frac{P_3}{Q_3}....\frac{P_x}{Q_x} = \pi_0 \prod_{k=1}^{x}\frac{P_k}{Q_k} \tag{33}$$

follows.

However $\sum_{x=0}^{m_i} \pi_x = 1$; therefore

$\pi_0 + \pi_0\frac{P_1}{Q_1} + \pi_0\frac{P_1}{Q_1}\frac{P_2}{Q_2} + \pi_0\frac{P_1}{Q_1}\frac{P_2}{Q_2}...\pi_0\frac{P_1}{Q_1}\frac{P_2}{Q_2}...\frac{P_m}{Q_m} = 1$, and

$\pi_0 = \dfrac{1}{1 + \sum_{x=1}^{m_i}\prod_{k=1}^{x}\frac{P_k}{Q_k}}$. Substituting for π_0 in Eq. (33)

gives

$$\pi_x = \frac{\prod_{k=1}^{x}\frac{P_k}{Q_k}}{1 + \sum_{x=1}^{m}\prod_{k=1}^{x}\frac{P_k}{Q_k}} \ . \tag{34}$$

That is, in full,

$$\pi_x = \frac{\frac{P_1}{Q_1}\frac{P_2}{Q_2}\frac{P_3}{Q_3}...\frac{P_x}{Q_x}}{1 + \frac{P_1}{Q_1} + \frac{P_1}{Q_1}\frac{P_2}{Q_2} + \frac{P_1}{Q_1}\frac{P_2}{Q_2}\frac{P_3}{Q_3} + ...\frac{P_1}{Q_1}\frac{P_2}{Q_2}\frac{P_3}{Q_3}....\frac{P_{m_i}}{Q_{m_i}}} \ , \tag{35}$$

which on simplification by dividing the numerator and denominator by $Q_1Q_2Q_3...Q_m$,

gives $\pi_x = P_1P_2P_3....P_xQ_{x+1}Q_{x+2}...Q_m / D$,

where

$$D = Q_1Q_2Q_3...Q_m + P_1Q_2Q_3...Q_m + P_1P_2Q_3....Q_m + ...P_1P_2P_3...P_m,$$

that is

$$\Pr\{X_{ni} = x\} = P_1P_2P_3....P_xQ_{x+1}Q_{x+2}...Q_m / D \quad (36)$$

It is important to note that the probability of $\Pr\{X_{ni} = x\}$, arises from a probability of a *relative success or failure at all thresholds*. These successes and failures have the very special structure that the probabilities of successes at the first x successive thresholds are followed by the probabilities of failures at the remaining thresholds. The pattern of successes and failures are compatible once again with the Guttman structure. Thus the derivation in both directions results in a Guttman structure of responses at the thresholds as the implied response for a response in any category.

2.2.1 The Guttman structure and the ordering of thresholds

In the original derivation, the thresholds were made to conform to the natural order as a mechanism for imposing the Guttman structure. The Guttman structure it will be recalled was used to reduce the set of possible responses at the m thresholds, 2^m, under the assumption of independence of responses at the thresholds, to a set with dependence and which was compatible with the number of actual independent possible responses, $m+1$.

In the reverse derivation above, which begins with a conditional response at the thresholds relative to the pair of adjacent categories, the Guttman structure for the responses follows. This Guttman structure in turn implies an ordering of the thresholds.

The ordering of the thresholds is compatible with the concept of a ruler marked with lines (thresholds). A response in any category x implies a response that is above all categories scored $0,1,...x-2, x-1$ and below all categories $x+1, x+2,...m$. That is indeed the meaning of order – the responses in the categories cannot be independent and the response in any one category determines the implied response for all of the other categories.

2.2.2 The Rasch model at the thresholds

The above derivation did not require that the RM was imposed on the conditional response at the thresholds. Inserting

$$P_x = \frac{\exp(\beta_n - (\delta_i + \tau_{xi}))}{1 + \exp(\beta_n - (\delta_i + \tau_{xi}))} \tag{37}$$

into Eq. (36) gives

$$\Pr\{X_{ni} = x\} = \prod_{k=1}^{x} \frac{\exp 1(\beta_n - (\delta_i + \tau_{xi}))}{1 + \exp(\beta_n - (\delta_i + \tau_{xi}))} \prod_{k=x+1}^{m} \frac{\exp 0(\beta_n - (\delta_i + \tau_{xi}))}{1 + \exp(\beta_n - (\delta_i + \tau_{xi}))} / D \tag{38}$$

that is,

$$\Pr\{X_{ni} = x\} = \frac{\exp[\sum_{k=1}^{x} 1(\beta_n - (\delta_i + \tau_{xi}))] \exp[\sum_{k=x+1}^{m} 0(\beta_n - (\delta_i + \tau_{xi}))]}{\prod_{k=1}^{m} (1 + \exp(\beta_n - (\delta_i + \tau_{xi})))} / D \tag{39}$$

which on simplification of the product terms gives

$$\Pr\{X_{ni} = x\} = \frac{\exp \sum_{k=1}^{x} (\beta_n - (\delta_i + \tau_{xi}))}{\prod_{k=1}^{m} (1 + \exp(\beta_n - (\delta_i + \tau_{xi})))} / D \tag{40}$$

and on further simplification of the normalising factor gives Eq. (5).

It is once again important to note that in Eq. (37), a response is implied at every threshold.

The next Section considers the slips in the derivation of the model which lead to potential misinterpretations.

2.3 Misinterpretations

In the above derivation, which took the reverse part from the original, care was taken to ensure that the full response structure was evident by separating the response structure at the thresholds from specifying the dichotomous RM into the conditional response at a threshold in Eqs. (28) and (29). If that is done, then it shows that, as in the original derivation, the reverse derivation requires a Guttman response structure at the thresholds.

If this is not done, and in addition the normalising factor is not kept a track of closely, then there is potential for misinterpreting the response process implied by the model. This misinterpretation is exaggerated if the model is expressed in log odds form. Both of these are now briefly considered.

2.3.1 Specifying the Rasch model immediately into the conditional dichotomous response at the threshold

Suppose the RM is specified immediately into Eq. (31) to give

$$\pi_x = \frac{e^{\beta_n - (\delta_i + \tau_{ki})}}{\eta_{nki}} \tag{41}$$

where $\eta_{nki} = 1 + e^{\beta_n - (\delta_i + \tau_{ki})}$ is the normalising factor and

$$1 - \pi_x = \frac{1}{\eta_{nki}} . \tag{42}$$

The probability of the failed response at threshold x, expressed fully is

$$1 - \pi_x = \frac{e^{0(\beta_n - (\delta_i + \tau_{ki}))}}{\eta_{nki}} , \tag{43}$$

but because the exponent in the numerator in (43) is 0, the numerator reduces to 1, and the role of failing at the threshold is obscured.

Pursuing the same derivation as that from Eqs. (31) to (36) using Eqs (41), (42) and (43) immediately gives Eq. (40),

$$\Pr\{X_{ni} = x\} = \frac{\exp \sum_{k=1}^{x} (\beta_n - (\delta_i + \tau_{ki}))}{\prod_{k=1}^{m} (1 + \exp(\beta_n - (\delta_i + \tau_{ki})))} / D , \tag{44}$$

where

$$D = \sum_{x=0}^{m} \frac{\exp \sum_{k=1}^{x} (\beta_n - (\delta_i + \tau_{ki}))}{\prod_{k=1}^{m} (1 + \exp(\beta_n - (\delta_i + \tau_{ki})))} \qquad (45)$$

giving

$$\Pr\{X_{ni} = x\} = \frac{\exp \sum_{k=1}^{x} (\beta_n - (\delta_i + \tau_{xi}))}{\gamma_{ni}} \qquad (46)$$

where $\gamma_{ni} = \sum_{x=0}^{m} \exp \sum_{k=1}^{x} (\beta_n - (\delta_i + \tau_{ki}))$ is the simplified normalizing factor and Eq. (46) is identical to Eq. (5).

2.3.2 Ignoring the probabilities of failing thresholds

If the attention is on the numerator, $\exp \sum_{k=1}^{x} (\beta_n - (\delta_i + \tau_{ki}))$, in Eqs.(44) - (46) without the full derivation, it is easy to consider that the probability of a response in any category $X_{ni} = x$ is only a function of the thresholds $k = 1,2,...x$ up to category x. To stress the point, this occurs because the factor $\exp[\sum_{k=x+1}^{m} 0(\beta_n - (\delta_i + \tau_{xi}))] = 1$, explicit in the numerator of Eq. (39) in the full derivation, simplifies to 1 immediately in Eqs.(44) - (46) and is therefore left only implicit in those equations. Being implicit means that it is readily ignored.

2.3.3 Ignoring the denominator and misinterpretation

The clue that this cannot be the case comes from the normalizing constant, the denominator, $\gamma_{ni} = \sum_{x=0}^{m} \exp \sum_{k=1}^{x} (\beta_n - (\delta_i + \tau_{ki}))$, which as noted earlier, contains all thresholds. However, if that is treated as a normalizing constant, without paying attention to its threshold parameters, further plays into the misinterpretation that a response in category $X_{ni} = x$ depends only on thresholds $k = 1,2,...x$ up to category x. As has also been indicated already, the probability in any category depends on all thresholds, and this is transparent from the normalizing constant, which unlike the numerator, explicitly contains all thresholds.

2.3.4 The log odds form and misinterpretation

The probability of success at a threshold relative to its adjacent categories, which was used earlier, gives Eq. (27):

$$\frac{\Pr\{X_{ni}=x\}}{\Pr\{X_{ni}=x-1\}+\Pr\{X_{ni}=x\}}=\frac{\exp(\beta_n-(\delta_i+\tau_x))}{1+\exp(\beta_n-(\delta_i+\tau_x))} \quad (47)$$

Taking the ratio of the response in two adjacent categories gives the odds of success at the threshold:

$$\Pr\{X_{ni}=x\}/\Pr\{X_{ni}=x-1\}=\exp(\beta_n-(\delta_i+\tau_x)). \quad (48)$$

Taking the logarithm gives

$$\ln(\Pr\{X_{ni}=x\}/\Pr\{X_{ni}=x-1\})=\beta_n-\delta_i-\tau_x. \quad (49)$$

This log odds form of the model, while simple, eschews its richness and invites making up a response process, such as a sequential step response process at the thresholds, which has nothing to do with the model. It does this because it can give the impression that there is an independent response at each threshold, an interpretation which incorrectly ignores that there is only one response among the categories and that the dichotomous responses at the thresholds are latent, only implied, and never observed. Attempting to explain the process and structure of the model from the log odds form of Eq. (49) is fraught with difficulties and misunderstandings.

3. RELATIONSHIPS AMONG THE PROBABILITIES

In the original derivation, the Guttman structure was imposed, and it was justified on two grounds: first that it reduced the sample space from independent responses to the required sample space compatible with just one response in one of the categories; second by postulating that the thresholds are in their natural order. In the reverse derivation carried out by Wright and Masters (1982) and all of their subsequent expositions of the model in this form, no comment is made on the implied Guttman structure of the responses at the thresholds and it is implied consistently that the responses at the thresholds is indepenent . This can be explained by their inserting the RM into Eq. (27) immediately, and incorrectly interpreting the compatible response process as one in which the response $X_{ni} = x$ involves only thresholds $k = 1,2,...x$ and ignoring thresholds $k = x+1, x+2,...m$. In particular, it leads to them to interpret a sequential response process involving “steps”, which is considered futher in the last Section. The complete reverse derivation, as shown above, implies a Guttman structure at the thresholds in their natural order, which in turn once again implies an ordering of the thresholds.

Thus the probability structure of the RM for ordered categories implies a Guttman structure no matter which of the common two ways it is derived. This is a property of the model and not a matter of interpretation. In addition, this Guttman structure implies and follows from an ordering of the thresholds. This will be elaborated further in Section 4. However, for the present, it is noted and stressed, that this does not imply that the *values* of the thresholds in any data set will be ordered.

The ordering of the thresholds is a property of the data, and in this section, the reason for this is consolidated using the above derivations. First, an example of category characteristic curves with reversed thresholds is shown.

Figure 3-2 shows the CCCs of an item in which the last two thresholds are reversed. It is evident that the threshold between Pass and Credit has a greater location than the threshold between Credit and Distinction. It means that if this is accepted, then the person who has 50% chance of being given a Credit or a Distinction has less ability than a person who has 50% chance of getting a Pass or a Credit. This clearly violates any a-priori principle of ordering of the categories. It means that there is a problem with the data.

Other symptoms of this problem is that there is no region which the grade of Credit is most likely and that the region in which Credit should be assigned is undefined in an ordered structure.

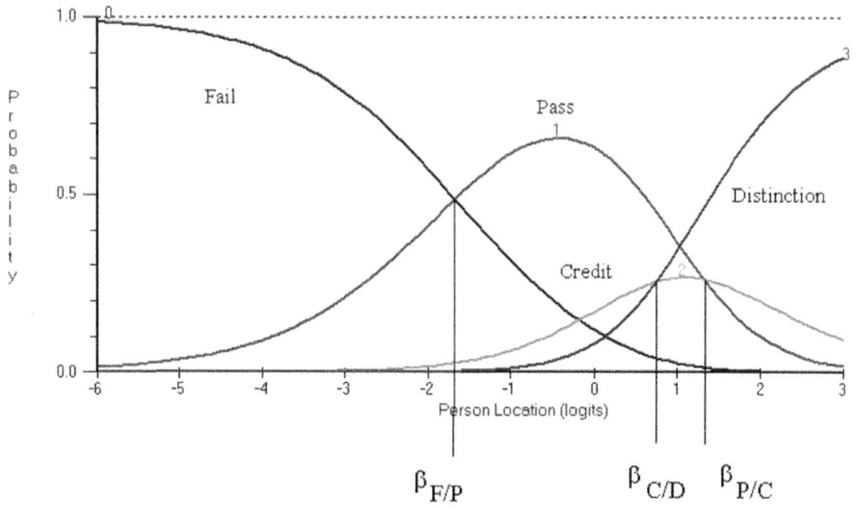

Figure 3-2. Category characteristic curves showing the probabilities of responses in each of four ordered categories when the thresholds are disordered

3.1 The relationship amongst probabilities when thresholds are ordered

In each of the analyses of the model, the focus is on the response of a single person to a single item, and on the probabilities of such a response. Now consider the relationship between the probabilities of pairs of successive categories.

From Eq. (34),

$$\Pr\{X_{ni} = x\} / \Pr\{X_{ni} = x-1\} = \exp(\beta_n - (\delta_i + \tau_x)) \qquad (50)$$

and

$$\Pr\{X_{ni} = x+1\} / \Pr\{X_{ni} = x\} = \exp(\beta_n - (\delta_i + \tau_{x+1}))$$

Therefore

$$\frac{\Pr\{X_{ni} = x\}}{\Pr\{X_{ni} = x-1\}} \frac{\Pr\{X_{ni} = x\}}{\Pr\{X_{ni} = x+1\}} = \exp(\tau_{x+1} - \tau_x) \qquad (52)$$

If $\tau_{x+1} > \tau_x$, then $\tau_{x+1} - \tau_x > 0$, $\exp(\tau_{x+1} - \tau_x) > 1$ and from Eq. (52),

$$[\Pr\{X_{ni} = x\}]^2 > \Pr\{X_{ni} = x - 1\} \Pr\{X_{ni} = x + 1\} \quad (53)$$

This effectively implies that the response distribution among the categories, again for any person to the item, has a single mode.

However, given that there is only one response in one category, there is no constraint on responses in the categories of any person to any item that would ensure this latent relationship holds. Of course, any person responds only once to any one item. However, in analysing the data with the model, in which the threshold values are estimated, gives the parameters of the probabilities of all categories of Eq. (5) for an item for any person location. And, because the item parameters in the Rasch model can be estimated while conditioning out the person parameters, it can do this without making assumptions about the distribution of the persons – it really is, remarkably, an equation about the internal structure of an item as revealed by features of the data. With reversed thresholds, it can be inferred that the implied response structure of any person to the item does not satisfy that relationship in Eq. (53).

Whatever the frequencies in the data, the model tries to estimate parameters compatible with the Guttman structure. If the frequencies in the data are not compatible with Eq. (53) when thresholds are ordered, then the model effectively rearranges the order of the thresholds. Thus the implied response pattern for a score of 2 in a four–category item when the thresholds are ordered, as the items shown in Figure 3-1 is (1,1,0). When thresholds 2 and 3 are reversed, then the implied response pattern remains (1,1,0), but now with respect to reversed thresholds. Then relative to the *intended* ordering, the implied response pattern is (1,0,1). Thus the implied Guttman structure plays a role in the estimation by forcing the threshold values to an order that conforms to it, even if that means reversing the threshold values to accommodate the Guttman structure. The response of 2 to such an item implies that a person has simultaneously passed the first and third thresholds of the a–priori order and failed the second.

3.1.1 The model and relative frequencies

The probability statement of the model as in Eq. (5) corresponds to frequencies. However, it is an implied frequency of responses by persons of the same ability to the item with the given parameters. The probability statement is conditional on ability, and does not refer to frequencies in the sample. Although lack of data in a sample in any category can result in estimates of parameters with large standard errors, the key factor in the estimates is the relationship amongst the categories of the implied probabilities of Eq. (53). These cannot be inferred directly from raw frequencies in categories in a sample. Thus in the case of Fig. 2, any *single* person whose ability estimate is between the thresholds identified by $\beta_{C/D}$ and $\beta_{P/C}$ will, simultaneously, have a higher probability of a getting a Distinction and a Pass than getting a Credit. This is not only incompatible with ordering of the categories, but it is *not a matter of the distribution of the persons in the sample of data analysed.*

To consolidate this point, Table 3-7 shows the frequencies of responses of 1000 persons for two items each with 11 categories. These are simulated data which fit the model. It shows that in the middle categories, the frequencies are very small compared to the extremes, and in particular, the score of 5 has a 0 frequency for item 1. Nevertheless, the threshold estimates shown in Table 3-8, have the natural order. The method of estimation, which exploits the structure of responses among categories to span and adjust for the category with 0 frequency and conditions out the person parameters, is described in Luo and Andrich (2004). The reason that the frequencies in the middle categories are low or even 0 is that they arise from a bimodal distribution of person locations. It is analogous to having heights of a sample of adult males and females. This too would be bimodal and therefore heights somewhere in between the two modes would have, by definition, a low frequency. However, it would be untenable if the low frequencies in the middle heights would reverse the lines (thresholds), which define the units on the ruler. Figure 3-1 shows the frequency distribution of the estimated person parameters and confirms that it is bimodal. Clearly, given the distribution, there would be few cases in the middle categories with scores of 5 and 6.

Table 3-7. Frequencies of responses in two items with 11 categories

Item	0	1	2	3	4	5	6	7	8	9	10	11
I0001	81	175	123	53	15	0	8	11	51	120	165	86
I0002	96	155	119	57	17	5	2	26	48	115	161	87

Table 3-8. Estimates of thresholds for two items with low frequencies in the middle categories

		THRESHOLDS										
Item	Locn	1	2	3	4	5	6	7	8	9	10	11
1	0.002	-3.96	-2.89	-2.01	-1.27	-0.62	-0.02	0.59	1.25	2	2.91	4.01
2	-0.002	-3.78	-2.92	-2.15	-1.42	-0.73	-0.05	0.64	1.36	2.14	2.99	3.94

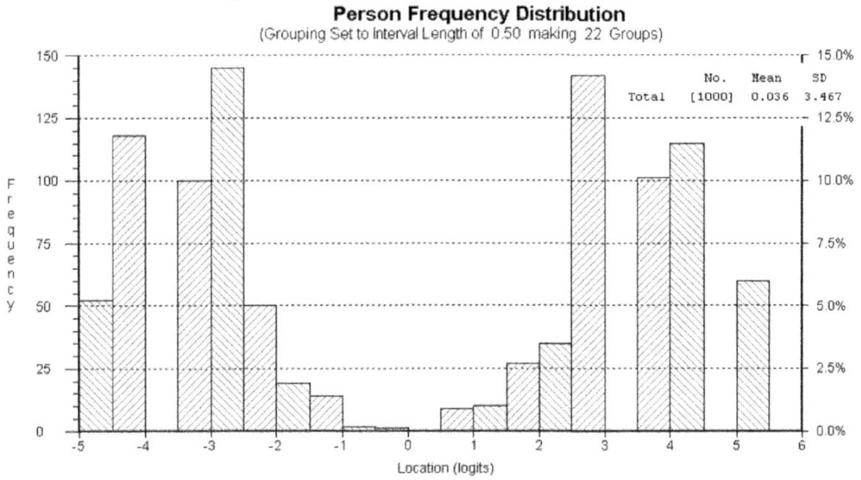

Figure 3-3. A bimodal distribution of person location estimates

3.1.2 Fit and reversed thresholds

The emphasis in the above explanations has been on the structure of the RM. This structure shows that the ordering of thresholds is compatible with the Guttman structure of the implied, latent, dichotomous responses at the thresholds. It has also been explained why the values of the thresholds in any particular data set do not have to conform to the order compatible with the model. One of the consequences of this relationship between the structure and values of the thresholds is that the usual statistical tests of fit of the data to the model are not necessarily violated because the thresholds are reversed. Indeed, data can be simulated according to Eq. (5), and thresholds which are reversed used in the simulation. The data will fit the RM perfectly.

In addition, tests of fit generally involve the estimates of the parameters. By using threshold estimates that are reversed, which arise from the property of the data, any test of fit that recovers the data from those estimates will not reveal any misfit because of the reversals of the thresholds – the test of fit is totally circular on this feature. The key feature, independent of fit, and independent of the distribution of the persons, is the ordering of the thresholds estimates themselves. The ordering of the thresholds is a necessary condition for evidence that the categories are operating as intended and it is a necessary condition for the responses to be compatible with the RM.

Thus although the invariance property of the RM is critical in choosing its application, statistical tests of fit are not the only relevant criteria for its application: in items in which the categories are intended to be ordered, the thresholds defining the categories must also be ordered. The thresholds must be ordered independently of the RM as a whole, but the power of the RM resides in the property that its structure is compatible with this ordering even though the values of the thresholds do not have to be ordered. This is the very reason that the RM is able to detect an empirical problem in the operation of the ordering of the categories.

3.2 The collapsing of adjacent categories

A distinctive feature of the RM is that adding the probabilities of two adjacent categories, produces a model which is no longer a RM. That is, taking Eq. (5), and forming

$$\Pr\{X_{ni} = x\} + \Pr\{X_{ni} = x+1\}$$
$$= \frac{1}{\gamma_{ni}}\exp(-\sum_{k=1}^{x}\tau_{ki} + x(\beta_n - \delta_i)) + \frac{1}{\gamma_{ni}}\exp(-\sum_{k=1}^{x+1}\tau_{ki} + x(\beta_n - \delta_i)) \quad (54)$$

gives Eq. (54) which cannot be reduced to the form of Eq. (5). Thus Eq.(54) is not a RM. It is possible to form such an equation, and this has been done for example in Masters and Wright (1997) in forming an alternate model with different thresholds, specifically the Thurstone model. However, the very action of forming Eq. (54), and then forming a model with new parameters, destroys the RM and its properties, irrespective of how well the data fit the RM. This has been discussed at length in Andrich (1995), Jansen and Roskam (1986) and was noted by Rasch (1966).

Specifically, summing the probabilities of adjacent categories to dichotomise a set of responses is not permissible within the framework of the RM. Thus let

$$P^*_{xni} = \sum_{x}^{m}\Pr\{X_{ni} = x\} \quad . \quad (55)$$

Then P^*_{xni} characterises a dichotomous response in which $1 - P^*_{xni} = 1 - \sum_{x}^{m}\Pr\{X_{ni} = x\}$. A parameterisation of the form

$$P^*_{xni} = \frac{1}{\lambda_{nxi}}\exp(\beta^*_n - \delta^*_{xi}) \quad (56)$$

where $\lambda_{nxi} = 1 + \exp(\beta_n^* - \delta_{xi}^*)$, is a model incompatible with the RM in which the parameters β_n^* and δ_{xi}^* are non linear functions of β_n and δ_i and τ_{ki} of Eq. (5). In Eq. (56), and because by definition and irrespective of the data, $P_{xni}^* > P_{(x+1)ni}^*$, the parameters δ_{xi}^* are ordered irrespective of the relationship among the probabilities of responses in the categories, and have been used to avoid using the parameters of the RM which can show disordering (Masters and Wright, 1997). However, because it is out of the framework of the RM, no property of the RM can be explained or circumvented simply by forming Eq. (56). Thus the results from the parameters of Eq. (56) cannot be parameters of the RM, even if they are formed after the RM is used to estimate the parameters in Eq. (5).

There is one special case when adding the probabilities of two adjacent categories is permissible, but not in the form of the Rasch model. It is permissible from the more general Eq. (22). If a discrimination $\alpha_{x+1i} = 0$, then the probabilities

$$\Pr\{X_{ni} = x\} + \Pr\{X_{ni} = x+1\} = \Pr\{X_{ni} = x'\} \tag{57}$$

can be reduced to the form of Eq. (5) where x' replaces the categories x and $x+1$ and every category greater than $x+1$ is reduced by 1 giving a new random variable $X_{ni}^{'} = x^{'}$ where $x^{'} \in \{0,1,2,...m-1\}$. In this case, where the discrimination at the thresholds is 0, the response in the two adjacent categories is random irrespective of the location of the person. In this case, the two adjacent categories are effectively one category, and to be compatible with the RM, the categories should be combined.

4. THE PROCESS COMPATIBLE WITH THE RASCH MODEL

From the previous algebraic explanations in the derivation of the RM, and some of the above elaborations, it is possible to summarise the response *process* that is compatible with the model. The response process is one of *ordered classification.* That is, the process is one of considering the property of an object, which might be a property of oneself or of some performance, relative to an item with more than two ordered categories, and deciding the category of the response. This process is considered more closely as it is an important part of the RM.

Central in everyday language, in formal activities involving applied problem solving, and in scientific research, is classification. By *classification* is meant that classes are conceptualised so that different entities can be assigned to one of the classes. In everyday language this tends to be carried out implicitly, while in scientific work it is carried out explicitly. Applications of the RM involve a kind of classification system in which the classes form an ordinal structure among themselves.

Ordered classification systems are common in the social sciences, and in particular, where measurement of the kind found in the physical sciences would be desirable, but no measuring instrument exists. Table 3-1 shows three examples of such formats. Some further points of elaboration are now considered. First, the minimum number of meaningful classes is two - a class that contains certain entities and the complementary class that excludes them. In the case where these are ordered, the dichotomous RM is relevant.

4.1 An example of a response format compatible with the model

Table 3-9 shows an example of a set of four ordered classes with operational definitions more detailed than those of Table 3-1. These are termed *inadequate setting, discrete setting, integrated setting,* and *integrated and manipulated setting,* according to which writing samples from students were assessed. While this case is specific, it is the prototype of the kind of classification system for which the RM is relevant and shown in Table 3-1. It shows that each successive category implies the previous category in the order *and in addition*, reflects more of the assessed trait. This is compatible with the Guttman structure.

Table 3-9. Operational definitions of ordered classes for judging essays

0	*Inadequate setting*: Insufficient or irrelevant information given for the story. Or, sufficient elements may be given, but they ar simply listed fromt the task statement, and not linked or logically organised.
1	*Discrete setting* Discrete setting as an introduction, with some details which also show some linkage and organisation. May have an additional element to those listed which is relevant to the story.
2	*Integrated setting:* There is a setting which, rather than simply being at the beginning, is introduced throughout the story.
3	*Integrated and manipulated setting:* In addition to the setting being introduced throughout the story, pertinent information is woven or integrated so that this integration contributes to the story.

Reprinted with permission from Harris, 1991, p.49.

4.2 Reinforcing the simultaneous response across categories and thresholds

That the RM characterises a classification process into ordered categories, with a probabilistic element for each classification, is consolidated by the operation of the implied Guttman structure across thresholds. A response in any category implies a response which is below all other categories above it in the order, and above all categories below the selected category in the order. A category is defined by two adjacent thresholds, and a response in the category implies a success on the lower threshold and failure at the succeeding threshold. Thus a response in a category implies that the latent response was a success at the lower of the two thresholds, and a failure at the greater of the two thresholds. *And this response determines the implied latent responses at all of the other thresholds.* Specifically, the implied responses at all thresholds below the lesser of the pair of thresholds with one success and one failure are also successes, and the implied responses at the threshold above the greater of this pair of thresholds are also failures. This is compatible with the format of Tables 3-1 and 3-8.

Because of the possible structure of the responses, and the implied process whereby a response is between a pair of ordered thresholds at which one response is a success and the other is a failure, and all others are determined by it, confirms that the response process is a *classification* process among ordered categories. As already noted, the probability of a single response depends on the location of all of the thresholds, and not just the pair of thresholds on either side of the response. Therefore, this response process compatible with the RM is not just a classification process, but a *simultaneous classification* process, across the thresholds. The further implication is that when the manifest responses are used to estimate the thresholds, the threshold locations are themselves empirically defined *simultaneously* - that is, the estimates arise from data in which all the thresholds of an item were involved simultaneously in every response. This gives the RM the distinctive feature that it can be used to assess whether or not the categories are working in the intended ordering, or whether on this feature, the empirical ordering breaks down.

To consolidate the understanding of process compatible with the RM, this Chapter concludes with a response structure and process which is *not compatible* with it, even though it has been used as prototypic of the response for which the model is relevant. This has been another point of confusion in the literature.

4.3 A response structure and process incompatible with the Rasch model

Table 3-10 shows the example (Adams, Wilson and Wang, 1997) which is considered by them to be prototypic for the RM. It shows the example of a person taking specified and successive steps towards completing a mathematical problem, and not proceeding when a step has been failed. It is not debated here whether or not students carrying such problems do indeed follow such a sequence of steps. Instead, it should be evident from the simultaneous classification process compatible with the RM described above, that if a person did solve the problem in the way specified in Table 3-10, then the response process could not follow the RM for more than two ordered categories.

Table 3-10. A "partial credit item" and a response structure which is incompatible with the Rasch model

$\sqrt{9.0/0.3-5}$ =?	
0	No steps taken
1	$9.0/0.3=30$
2	30 - 5
3	$\sqrt{25}$

From Adams, Wilson and Wang, 1997, p.13.

The RM of Eq. (5) cannot characterise a sequential process which stops when there is a failure at any threshold as if the response at every threshold is independent of the response at every other threshold. This is the kind of interpretation that can easily be made when the derivation of the model is incomplete as indicated in Section 3.3.

Finally, it is stressed that the RM is a static measurement model used to estimate the location of the entity of measurement and that there is only one fixed person parameter in the model which does not change to characterise changes in location. The model does not and cannot characterise a process of how the entity arrived at its location or category – it can only estimate the location given that the entity has been classified in the category.

5. REFERENCES

Adams, R.J., Wilson, M., and Wang, W. (1997) The multidimensional random coefficients multinomial logit model. *Applied Psychological Measurement,* 21, 1-23.

Andersen, E.B. (1977). Sufficient statistics and latent trait models. *Psychometrika, 42,* 69-81.

Andrich, D. (1978a). A rating formulation for ordered response categories. *Psychometrika, 43*, 357-374.

Andrich, D. (1978b). Application of a psychometric rating model to ordered categories which are scored with successive integers. *Applied Psychological Measurement, 2,* 581-94.

Andrich, D. (1995). Models for measurement, precision and the non-dichotomization of graded responses. *Psychometrika, 60,* 7-26.

Birnbaum, A. (1968). Some latent trait models and their use in inferring an examinee's ability. In F.M. Lord and M.R. Novick, *Statistical theories of mental test scores* (pp. 397 –545). Reading, Mass.: Addison-Wesley.

Guttman, L. (1950). The basis for scalogram analysis. In S.A. Stouffer, L. Guttman, E.A. Suchman, P.F. Lazarsfeld, S.A. Star and J.A. Clausen (Eds.), *Measurement and Prediction,* pp.60-90. New York: Wiley.

Harris, J. (1991). Consequences for social measurement of collapsing adjacent categories with three or more ordered categories. Unpublished Master of Education Dissertation, Murdoch University, Western Australia.

Jansen P.G.W. & Roskam, E.E. (1986). Latent trait models and dichotomization of graded responses. *Psychometrika, 51*(1), 69-91.

Luo, G. & Andrich, D. (2004). Estimation in the presence of null categories in the reparameterized Rasch model. *Journal of Applied Measurement*, Under review.

Masters, G.N. and Wright, B.D. (1997) The partial credit model. In W.J. van der Linden and R.K. Hambleton (Eds.) *Handbook of Item Response Theory.* (pp. 101– 121). New York. Springer.

Rasch, G. (1966). An individualistic approach to item analysis. In P.F. Lazarsfeld and N.W. Henry, (Eds.). *Readings in Mathematical Social Science* (pp.89-108). Chicago: Science Research Associates.

Wright, B.D. & Masters, G.N. (1982). *Rating Scale Analysis: Rasch Measurement.* Chicago: MESA Press.

Chapter 4

MONITORING MATHEMATICS ACHIEVEMENT OVER TIME

A SECONDARY ANALYSIS OF FIMS, SIMS and TIMS: A RASCH ANALYSIS

Tilahun Mengesha Afrassa
South Australian Department of Education and Children's Services

Abstract: This paper is concerned with the analysis and scaling of mathematics achievement data over time by applying the Rasch model using the QUEST (Adams & Khoo, 1993) computer program. The mathematics achievements of the students are brought to a common scale. This common scale is independent of both the samples of students tested and the samples of items employed. The scale is used to examine the changes in mathematics achievement of students in Australia over 30 years from 1964 to 1994. Conclusions are drawn as to the robustness of the common scale, and the changes in students' mathematics achievements over time in Australia.

Key words: Mathematics, achievement, measurement, Rasch analysis, change

1. FIMS, SIMS AND TIMS

Over the past five decades, researchers have shown considerable interest in the study of student achievement in mathematics at all levels across educational systems and over time. Many important conclusions can be drawn from various research studies about students' achievement in mathematics over time. Willett (1997, p.327) argued that by measuring change over time, it is possible to map phenomena at the heart of the educational enterprise. In addition, he argued that education seeks to enhance learning, and to develop change in achievement, attitudes and values. It is Willett's belief that 'only by measuring individual change is it possible to document each person's progress and, consequently, to evaluate the effectiveness of educational systems' (Willett, 1997, p. 327). Therefore,

S. Alagumalai et al. (eds.), Applied Rasch Measurement: A Book of Exemplars, 61–77.

the measurement of change in achievement over time is one of the most important tools for finding ways and means of improving the education system of a country.

Since Australia participated in the 1964, 1978 and 1994 International Association for the Evaluation of Educational Achievement (IEA) Mathematics Studies, it should be possible to examine the mathematics achievement differences over time across the 30-year time period. The IEA Mathematics Studies were conducted in Australia under the auspices of the IEA.

The First International Mathematics Study (FIMS) was the first large project of this kind (Keeves & Radford, 1969) and also included a detailed curriculum analysis (Keeves, 1968). Prior to FIMS, there was a lack of comparative international achievement data. For the last 50 years, however, the number and nature of the variables included in comparative studies of educational achievement have continued to expand.

The main purpose of FIMS was to investigate differences among different school systems and the interrelations between the achievement, attitudes and interests of 13-year-old students and final-year secondary school students (Husén, 1967; Keeves, 1968; Keeves & Radford, 1969; Rosier, 1980; Moss, 1982). Countries that participated in the FIMS are listed in Table 4-1.

School and students who participated in the FIMS study were selected using two-stage random sampling procedures, involving age and grade level samples. The age level sample included all 13-year-old students in Years 7, 8 and 9. The grade level sample involved Year 8 students, including 13-year-old students at that year level. All students in the samples were government school students. In the cluster sample, design schools were selected randomly at the first stage and students were selected randomly from within schools at the second stage. The results of the international analyses of the FIMS data are given in Husén (1967), Postlethwaite (1967) and summarised in Keeves (1995).

The Second International Mathematics Study (SIMS) was conducted in the late 1970s and early 1980s in 21 countries. The main purpose of SIMS 'was to produce an international portrait of mathematics education with a particular focus on the mathematics classroom' (Garden, 1987, p. 47). Countries that participated in SIMS are presented in Table 4-1.

The schools and students who participated in the SIMS study were selected using a two-stage sampling procedure. The students were all 13-year-olds and were from both government and non-government schools. The results of the analyses of SIMS data are reported by Rosier (1980), Moss (1982), Garden (1987), Robitaille and Travers (1992), and are summarised in Keeves (1995).

Table 4-1 Countries and number of students who participated in FIMS, SIMS and SIMS

	Number of students who participated in		
Country	**FIMS**[a]	**SIMS**[b]	**TIMS**[c]
Australia	4320	5120	5599
Austria	-	-	3013
Belgium (Flemish)	5900	1370	2768
Belgium (French)	-	1875	2292
Bulgaria	-	-	1798
Canada	-	6968	8219
Colombia	-	-	2655
Cyprus	-	-	2929
Czech Republic	-	-	3345
Denmark	-	-	2073
England	12740	2585	1803
Finland	7346	4394	-
France	3423	8215	3016
Germany	5767	-	2893
Greece	-	-	3931
Hong Kong	-	5548	3413
Hungary	-	1752	3066
Iceland	-	-	1957
Iran, Islamic Republic	-	-	3735
Israel	4509	3362	-
Japan	10257	8091	5130
Korea	-	-	2907
Kuwait	-	-	-
Latvia (LSS)	-	-	2567
Lithuania	-	-	2531
Luxembourg	-	2005	-
Netherlands	2510	5436	2097
New Zealand	-	5203	3184
Nigeria	-	1429	-
Norway	-	-	2469
Philippines	-	-	5852
Portugal	-	-	3362
Romania	-	-	3746
Russian Federation	-	-	4138
Scotland	17472	1356	2913
Singapore	-	-	3641
Slovak Republic	-	-	3600
Slovenia	-	-	2898
South Africa	-	-	5301
Spain	-	-	3741
Swaziland	-	899	-
Sweden	32704	3490	2831
Switzerland	-	-	4085
Thailand	-	3821	5845
United States	23063	6654	3886

Source: a= Husén (1967); b= Hanna (1989, p. 228); c=Beaton et al.(1996, p. A-16)

In 1994/1995 IEA conducted the Third International Mathematics and Science Study (TIMSS) that was the largest of its kind. It was a cross-national study of student achievement in mathematics and science that was administered at three levels of the school system (Martin, 1996). Table 4-1 shows countries that participated in FIMS, SIMS and TIMS.

The sampling procedure employed in FIMS and SIMS was a two-stage simple random sample. In the first stage schools and in the second stage students were selected from the schools chosen in Stage 1. However, in TIMSS, there were three stages of sampling (Foy, Rust & Schleicher, 1996). The first stage of sampling consisted of selecting schools using a probability proportional to size method. The second sampling stage involved selecting classrooms within the sampled schools by employing either equal probabilities or with probabilities proportional to their size. Meanwhile, the third stage involved selecting students from within the sampled classrooms. However, this sampling stage was optional (Foy, et al., 1996). The target populations at the lower secondary school level were students in the two adjacent grades containing the largest proportions of 13-year-olds at the time of testing. The results of the analyses of TIMSS data are presented in Beaton, Mulls, Martin, Gonzalez, Kelly & Smith (1996a) for mathematics and Beaton, Mulls, Martin, Gonzalez, Smith & Kelly (1996b) for science for Population 2.

A second round of TIMSS data collection was undertaken in 1998 and 1999 in the Southern Hemisphere and Northern Hemisphere respectively. This time the study was called TIMSS Repeat, because the first round of TIMSS was conducted in 1994/1995. According to Robitaille and Beaton (2002), 38 countries participated in TIMSS Repeat. Australia was one of these countries which participated in TIMSS Repeat (Zammit, Routitsky & Greenwood, 2002).

Since Australia participated in the 1964, 1978 and 1994 International Mathematics Studies, it is possible to examine the mathematics achievement differences over time across the 30-year period. Therefore, the purpose of this study is to investigate changes in achievement in mathematics of Australian lower secondary school students between 1964, 1978 and 1994.

[redacted]

[redacted]

2. SAMPLING PROCEDURE

Table 4-2 shows the target populations of the three mathematics studies included in the present analysis. In 1964 and 1978 the samples were age samples and included students from years 7, 8 and 9 in all participating states and territories, while in TIMS the samples were grade samples drawn from years 7 and 8 or years 8 and 9.

Therefore, in order to make meaningful comparisons of mathematics achievement over time by using the 1964, 1978 and 1994 data sets, the following steps were taken.

The 1978 students were chosen as an age sample and included students from both government and non-government schools. In order to make meaningful comparisons between the 1978 sample and the 1964 sample, students from non-government schools in all participating states and all students from South Australia and the Australian Capital Territory were excluded from the analyses presented in this paper.

Table 4-2 Target populations in FIMS, SIMS and TIMS

Target population	Label	Size	Sampling procedure	Primary unit	Secondary unit	Design effect	Effective sample size
13-year-old	FIMSA	2917	SRS	School	Student	7.66	380
Grade 8	FIMSB	3081	SRS	School	Student	11.82	261
Total	FIMS	4320	SRS	School	Student	11.11	389
13-year-old	SIMS	3038	PPS	School	Student	5.4	563
Year 8	TIMS	3786	PPS	School	Class	16.52	229

SRS = Stratified-random-sample of schools and students within schools
PPS = Probability-proportional-to-size sample of schools

Meanwhile, in TIMS the only common sample for all states and territories was the year 8 students. In order to make the TIMS samples comparable with the FIMS samples, only year 8 government school students in the five states that participated in FIMS are considered as the TIMS data set in this study. After excluding schools and the states and territories that did not participate in the 1964 study, two sub-populations of students were identified for comparison between occasions. The two groups were 13-year-old students in FIMSA and SIMS: all were 13-year-old students and were distributed across years 7, 8 and 9 on both occasions, whereas, for the comparison between FIMSB and TIMS, the other sub-populations consisted of 1964 and 1994 year 8 students. Students in both groups were at the same year level. Hence, the comparisons in this study are between 13-year-old students in FIMSA and SIMS on the one hand, and FIMSB and TIMS year 8 students on the other. Details of the sampling procedures employed in this study are presented in Tilahun (2002).

3. METHODS EMPLOYED IN THE STUDY

In this chapter the mathematics achievements of students over time is measured using the Rasch model.

The purpose of this analysis is to identify the differences in achievement in mathematics of Australian students between 1964, 1978 and 1994.

3.1 Use of the Rasch model

Since the beginning of the 20th century, research into the methods of test equating has been an ongoing process in order to examine change in the levels of student achievement over time. However, research has been intensified since the 1960s due to the development of Item Response Theory (IRT) and the availability of appropriate computer programs. Among the many test-equating procedures, the IRT techniques are generally considered the best. However, only the one parameter model, or Rasch model, has strong measurement properties. Therefore, in order to examine the achievement level of students over time, it is desirable to apply the Rasch model test equating procedures. Hence, in this study of the mathematics achievement of 13-year-old students over time, the horizontal test equating strategy with the concurrent, anchor item equating and common item equating techniques, using the Rasch model, are best applied.

3.2 Unidimensionality

Before the Rasch model could be used to analyse the mathematics test items in the present study, it was important to examine whether or not the items of each test were unidimensional, since the unidimensionality of the test items is one of the requirements for the use of the Rasch model (Hambleton & Cook, 1977; Anderson, 1994). Consequently, Tilahun (2002) employed confirmatory factor analysis to test the unidimensionality of the mathematics items in FIMS and SIMS. The results of the confirmatory factor analyses revealed that the nested model in which the mathematics items were assigned to three specific first-order factors (arithmetic, algebra and geometry), as well as a general higher order factor, which was labelled as mathematics provided the best fitting model.

In addition, using confirmatory factor analysis procedures, Tilahun also examined the factor structures of the mathematics items by categorising into three types of cognitive processes (namely, computation and verbal processes), lower and higher mental processes, and computation, knowledge, translation and analysis. The results of these analyses are reported in Tilahun (2002).

3.3 Developing a common mathematics scale

The calibration of the mathematics data permitted a scale to be constructed that extended across the three groups on the mathematics scale: namely, FIMS, SIMS and TIMS students. The fixed point of the scale was set at 500 with one logit, the natural metric of the scale, being set at 100 units. The choice of the fixed point of the scale, namely 500, was an arbitrary value, which was necessary to fix the scale, and which has been used by several authors in comparative studies (Keeves & Schleicher, 1992; Elley, 1994; Keeves & Kotte, 1996; Lietz, 1996; Tilahun, 1996; Tilahun & Keeves, 2001; Tilahun, 2002). The graphical representation of the mathematics scale constructed in this way for the different sample groups of students in FIMS, SIMS and TIMS is presented in Figure 4-1, with 100 scale units (centilogits) being equivalent to one logit.

3.4 Rasch analysis

Three groups of students, FIMS (4320), SIMS (3038) and TIMS (3786), were involved in the present analyses. The necessary requirement to calibrate a Rasch scale is that the items must fit the unidimensional scale. Items that do not fit the scale must be deleted in calibration. In order to examine whether or not the items fitted the scale, it was also important to evaluate both the item fit statistics and the person fit statistics. The results of these analyses are presented below.

3.4.1 Item fit statistics

One of the key item fit statistics is the infit mean square (INFIT MNSQ). The infit mean square measures the consistency of fit of the students to the item characteristic curve for each item with weighted consideration given to those persons close to the 0.5 probability level. The acceptable range of the infit mean squares statistic for each item in this study was taken to be from

0.77 to 1.30 (Adams & Khoo, 1993). Values outside this acceptable range that is above 1.30 indicate that these items do not discriminate well, and below 0.77 the items provide redundant information. Hence, consideration must be given to excluding those items that are outside the range. In calibration, items that do not fit the Rasch model and which are outside the acceptable range must be deleted from the analysis (Rentz & Bashaw, 1975; Wright & Stone, 1979; Kolen & Whitney, 1981; Smith & Kramer, 1992). Hence, in the FIMS data two items (Items 13 and 29), in SIMS data two items (Items 21 and 29) and in TIMS data one item [(Item T1b No. 148) with one item (No. 94) having been excluded from the international TIMSS analysis] were removed from the calibration analyses due to the misfitting of these items to the Rasch model. Consequently, 68 items for FIMS, 70 for SIMS and 156 for TIMS fitted the Rasch model.

3.4.2 The fit of case estimates

The other way of investigating the fit of the Rasch scale to data is to examine the estimates for each case. The case estimates express the performance level of each student on the total scale. In order to identify whether the cases fit the scale or not, it is important to examine the case OUTFIT mean square statistic (OUTFIT MNSQ) which measures the consistency of the fit of the items to the student characteristic curve for each student, with special consideration given to extreme items. In this study, the general guideline used for interpreting t as a sign of misfit is if $t>5$ (Wright & Stone, 1979, p. 169). That is, if the OUTFIT MNSQ value of a person has a t value >5, that person does not fit the scale and is deleted from the analysis. However, in this analysis, no person was deleted, because the t values for all cases were less than 5.

4. EQUATING OF MATHEMATICS ACHIEVEMENT BETWEEN OCCASIONS AND OVER TIME

The equating of the mathematics tests requires common items between occasions that are between FIMS (1964), SIMS (1978) and TIMS (1994).

In this study, the number of common items in the mathematics tests for FIMS and SIMS data sets was 65. These common items formed approximately 93 per cent of the items for FIMS, and 90 per cent for SIMS. Thus, the common items in the mathematics test for the two occasions were all above the percentage ranges proposed by Wright and Stone (1979), and Hambleton, Zaal & Peters (1991).

There were also some items, which were common for FIMS, SIMS and TIMS data sets. Garden and Orpwood (1996, p. 2-2) reported that achievement in TIMSS was intended to be linked with the results of the two earlier IEA studies. Thus, in the TIMS data set, there were nine items which were common to the other two occasions. Therefore, it was possible to claim that there were just sufficient numbers of common items to equate the mathematics tests on the three occasions.

Rasch model equating procedures were employed for equating the three data sets. Rentz and Bashaw (1975), Beard and Pettie (1979), Sontag (1984) and Wright (1995) have argued that Rasch model equating procedures are better than other procedures for equating achievement tests. The three types of Rasch model equating procedures, namely concurrent equating, anchor item equating and common item difference equating, were all used for equating the data sets in this study.

Concurrent equating was employed for equating the data sets from FIMS and SIMS. In this method, the data sets from FIMS and SIMS were combined into one data set. Hence, the analysis was done with a single data file. Only one misfitting item was deleted at a time so as to avoid dropping some items that might eventually prove to be good fitting items. The acceptable infit mean square values were between 0.77 and 1.30 (Adams & Khoo, 1993). The concurrent equating analyses revealed that, among the 65 common items, 64 items fitted the Rasch model. Therefore, the threshold values of these 64 items were used as anchor values in the anchor item equating procedures employed in the scoring of the FIMS and SIMS data sets separately. Among the 64 common items, nine were common to the FIMS, SIMS and TIMS data sets. The threshold values of these nine items generated in this analysis are presented in Table 4-3 and were used in equating the FIMS data set with the TIMS data set.

The design of TIMS was different from FIMS and SIMS in two ways. In the first place, only one mathematics test was administered in both FIMS and SIMS. However, in the 1994 study, the test included mathematics and science items and the study was named TIMSS (Third International Mathematics and Science Study). The other difference was that in the first two international studies, the test was designed as one booklet. Every participant used the same test booklet. Whereas in TIMSS, a rotated test design was employed. The test was designed in eight booklets. Garden and Orpwood (1996, p. 2-16) have explained the arrangement of the test in eight booklets as follows:

> This design called for items to be grouped into 'clusters', which were distributed (or 'rotated') through the test booklets so as to obtain eight booklets of approximately equal difficulty and equivalent content coverage. Some items (the core cluster) appeared in all

booklets, some (the focus cluster) in three or four booklets, some (the free-response clusters) in two booklets, and the remainder (the breadth clusters) in one booklet only. In addition, each booklet was designed to contain approximately equal numbers of mathematics and science items.

All in all, there were 286 (both mathematics and science) unique items that were distributed across eight booklets for Population 2 (Adams & Gonzalez, 1996, p. 3-2).

Garden and Orpwood (1996) also reported that the core cluster items (six items for mathematics) were common to all booklets. In addition, the focus cluster and free-response clusters were common to some booklets. Thus, it was possible to equate these eight booklets and report the achievement level in TIMS on a common scale. Hence, among the Rasch model test equating procedures, concurrent equating was chosen for equating these eight booklets. Consequently, the concurrent equating procedure was employed for the TIMS data set. The result of the Rasch analysis indicated that only one item was deleted from the analysis. Out of 157 items, 156 of the TIMS test items fitted the Rasch model well. The item which was deleted from the analysis was Item 148 (T1b), whose infit mean square value was below the critical value of 0.77. From this concurrent equating procedure, it was possible to obtain the threshold values of the nine common items in TIMS. These threshold values are shown in Table 4-3.

Table 4-3 Description of the common item difference equating procedure employed in FIMS, SIMS, and TIMS

FIMS and SIMS		**TIMS**		**TIMS - FIMS**
Item number	Thresholds	Item number	Thresholds	Thresholds
12	0.21	K4	0.87	0.66
26	0.21	J14	1.90	1.69
31	-2.38	A6	-0.84	1.54
32	-0.08	R9	1.45	1.53
33	-1.10	Q7	-0.38	0.72
36	-0.82	M7	-0.87	-0.05
38	0.28	G6	1.31	1.03
54	0.27	F7	1.67	1.40
67	0.26	G3	0.47	0.21
Sum				8.73
N				9
Mean				0.97

Notes

N = number of common items

Equating Constant = 0.970

Standard deviation of equating constant = 0.59

Standard error of equating constant = 0.197

The next step involved the equating of the FIMS data set with the TIMS data set using the common item difference equating procedure. In this method the threshold value of each common item from the concurrent equating run for the combined FIMS and SIMS mathematics test data set was first subtracted from the threshold value for the item in the TIMS test. Then the differences were summed up and divided by the number of anchor test items to obtain a mean difference. Subsequently, the mean difference was subtracted from the case estimated mean value on the second test to obtain the adjusted mean value. In addition, the standard deviation of the nine difference values and the standard error of the mean were calculated and are recorded in Table 4-3.

4.1 Comparisons between students on mathematics test

The comparisons of the performance of students on the mathematics test for the three occasions were undertaken for two different subgroups: namely, (a) year 8 students who participated in the study, and (b) 13-year-old students in both government and non-government schools, who participated in the study. Some of the FIMS students were 13-year-old students, while others were younger or older students who were in year 8. Therefore, for comparison purposes, FIMS students were divided into two groups: namely, (a) FIMSA, which involved all 13-year-old students, and (b) FIMSB, which involved all year 8 students, including 13-year-olds. Thus, FIMSA students' results could be compared with the SIMS government school students' results, because all students were 13-year-olds. In TIMS, a decision was made to include only year 8 government school students, because they were the only group of students who were common in all participating states in FIMS. Thus TIMS year 8 government school students could be compared with FIMSB students, because in the three groups students were at different age levels but at the same year level.

The Australian Capital Territory (ACT) and South Australia (SA) participated in the SIMS and TIMS, but not in FIMS, and the Northern Territory (NT) participated in the TIMS, but not in the FIMS and SIMS studies. Consequently, these two territories and South Australia were excluded from the comparisons between FIMS, SIMS and TIMS. Non-government schools were not involved in FIMS. However, they participated in SIMS and TIMS. Therefore, for comparability, the non-government school students who participated in the SIMS and TIMS were also excluded from the comparison between FIMSA and SIMS, and between FIMSB and TIMS.

This section considers two types of comparisons: the first section compares the 13-year-old students between FIMSA and SIMS, and the year 8 students between FIMSB and TIMS.

The first comparison is between FIMSA and SIMS students. Table 4-4 shows the descriptive statistics for students who participated in the mathematics studies on the three occasions.

4.1.1 Comparisons between FIMSA and SIMS students

When the mathematics test estimated mean scores of FIMSA (13-year-old students) and SIMS (13-year-old students excluding ACT and SA and all non-government school students in Australia) were compared, the FIMSA score was higher than the SIMS mean score (see Table 4-4 and Figure 4-1). The estimated mean score difference between the two occasions was 19 centilogits, the difference was in favour of the 1964 13-year-old Australian students. This revealed that the mathematics achievement of Australian students declined from 1964 to 1978. The differences in standard deviation and standard error values for the two groups were small, while the design effect was slightly larger in 1964 than in 1978. The effect size was small (0.19) and the t-value was 2.91. Hence, the mean difference was statistically significant at the 0.01 level (see Table 4-4, Figure 4-1). Thus in Australia the mathematics achievement level of the 13-year-old students declined over time, between 1964 and 1978, to an extent that represented approximately half of a year of mathematics learning.

Table 4-4 Descriptive statistics for mathematics achievement of students for the three occasions

	FIMSA	FIMSB	SIMS	TIMS		
Mean	460.0	451.0	441.0	427.0		
Standard deviation	96.0	82.0	102.0	124.0		
Standard error of the mean	4.9	5.1	3.9	7.6		
Design effect	7.7	11.8	5.7	17.3		
Sample size	2917	3081	3989	4648		
	Mean differences		**Effect size**		**t-value**	**Significance level**
FIMSA vs SIMS	19.0		0.19		2.91	<0.01
FIMSB vs TIMS	25.0		0.24		1.13	NS
Alternative estimation of equating error						
FIMSB vs TIMS	31		0.29		-2.16	<0.05

Notes
NS = not significant

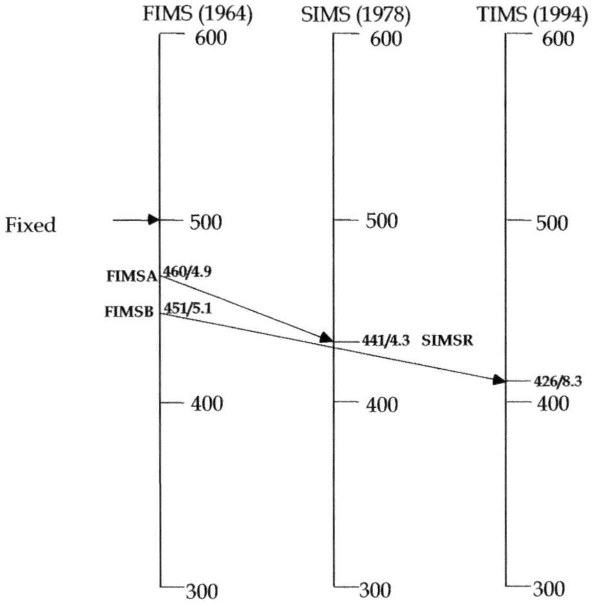

Fixed point - 500, metric - 100 = 1 logit, 1 unit = 1 centilogit
Values indicated for each occasion are Rasch estimated scores and standard errors of the mean respectively.

Figure 4-1. The mathematics test scale of government school students in FIMSA, FIMSB, SIMS and TIMS

4.1.2 Comparisons between FIMSB and TIMS students

The next comparison was between FIMSB and TIMS students. The estimated mean score of the 1964 Australian year 8 students was 451, while it was 426 in 1994 for the TIMS sample. The difference was 25 centilogits in favour of the 1964 students (see Table 4-4 and Figure 4-1). This difference revealed that the mathematics achievement level of Australian year 8 students has declined over the last 30 years. The standard deviation, standard error and the design effects, were markedly larger in 1994 than in 1964. The effect size was small (0.24) and the t-value was 1.13. While the effect size difference between FIMSB and TIMS was approximately three-quarters of a year of school learning, this difference was not statistically significant as a consequence of the large standard error of the equating constant shown in Table 4-3 and considered to be about 19.7 centilogits. Because of this extremely large standard error for the equating constant, which arose from the use of only nine common items, it was considered desirable to undertake alternative procedures to estimate the equating constant and its standard errors. Tilahun and Keeves (1997) used the five state subsamples and the nine common items to provide more accurate estimation. With these alternative procedures, a mean difference of 31.0 with an effect size of 0.29 (see Table 4-4), or nearly a full year of mathematics learning, was obtained which was found to be statistically significant at the five per cent level of significance.

4.2 Summary

The investigation using Rasch modelling revealed that the mathematics achievement of Australian students declined significantly over time at the 13-year-old level. Moreover, there was not a statistically significant decline at the year 8 student level.

5. CONCLUSION

In this chapter, Rasch analysis was employed to investigate differences in mathematics achievement between the 1964, 1978 and 1994 Australian students. The findings of the study are summarised as follows:

1. The achievement level of Australian 13-year-old students declined between 1964 and 1978.
2. Moreover, there was a decline of uncertain significance statistically but of clear significance in practical terms at the year 8 level between 1964 and 1994.

The findings in both comparisons between FIMSA and SIMS, FIMSB and TIMS showed that the achievement level of Australian students had declined over time. However, the decline between FIMS and TIMS was of uncertain significance because of the large error in the calculation of the equating constant. This arose from the relatively fewer common items that were employed in the tests on the two occasions. These findings indicated that there is a need to investigate differences in conditions of learning over the three occasions. Carroll (1963) has identified five factors that influence school learning. One of the factors identified by Carroll was students' perseverance (motivation and attitude) towards the subject they are learning. Therefore, it is important to examine FIMS, SIMS and TIMS students' views and attitudes towards mathematics and schooling.

6. REFERENCES

Adams, R. J. & Khoo, S.T. (1993). *Quest- The Iinteractive Test Analysis System*. Hawthorn, Victoria: ACER.

Adams, R. J. & Gonzalez, E. J. (1996). The TIMSS test design. In M.O. Martin & D.L. Kelly (eds), *Third International Mathematics and Science Study Technical Report vol. 1*, Boston: IEA, pp. 3-1 - 3-26.

Anderson, L. W. (1994). Attitude Measures. In T. Husén (ed), *The International Encyclopedia of Education, vol. 1, (second ed.)*, Oxford: Pergamon, pp. 380-390.

Beaton, A. E., Mulls, I. V. S., Martin, M. O, Gonzalez, E. J., Kelly, D. L. and Smith, T. A. (1996a). *Mathematics Achievement in the Middle School Years: IEA's Third International Mathematics and Science Study*. Boston: IEA.

Beaton, A. E., Martin, M. O, Mulls, I. V. S., Gonzalez, E. J., Smith, T. A. & Kelly, D. L. (1996b). *Science Achievement in the Middle School Years: IEA's Third International Mathematics and Science Study*. Boston: IEA.

Beard, J. G. & Pettie, A. L. (1979). A comparison of Linear and Rasch Equating results for basic skills assessment Tests. Florida State University, Florida: ERIC.

Brick, J. M., Broene, P., James, P. & Severynse, J. (1997). *A user's guide to WesVarPC. (Version 2.11)*. Boulevard, MD: Westat, Inc.

Elley, W. B. (1994). *The IEA Study of Reading Literacy: Achievement and Instruction in Thirty-Two School Systems*. Oxford: Pergamon Press.

Foy, R., Rust, K. & Schleicher, A. (1996). Sample design. In M O Martin & D L Kelly (eds), *Third International Mathematics and Science Study: Technical Report Vol 1: Design and Development*, Boston: IEA, pp. 4-1 to 4-17.

Garden, R. A. (1987). The second IEA mathematics study. *Comparative Education Review, 31* (1), 47-68.

Garden , R. A. & Orpwood, G. (1996). Development of the TIMSS achievement tests. In M 0. Martin & D L Kelly (eds), *Third International Mathematics and Science Study Technical Report Volume 1: Design and Development*, Boston: IEA, pp. 2-1 to 2-19.

Hambleton, R. K.& Cook, L. L. (1977). Latent trait models and their use in the analysis of educational test data. *Journal of educational measurement*, 14 (2), 75-96.

Hambleton, R. K., Zaal, J. N.& Pieters, J. P. M. (1991). Computerized adaptive testing: theory, applications, and standards. In R.K Hambleton & J.N. Zaal (eds), *Advances in Educational and Psychological Testing*, Boston, Mass.: Kluwer Academic Publishers, pp. 341-366.

Hanna, G. (1989). Mathematics achievement of girls and boys in grade eight: Results from twenty countries. *Educational Studies in Mathematics, 20* (2), 225-232.

Husén, T. (ed.), (1967). *International Study of Achievement in Mathematics (vols 1 & 2)*. Stockholm: Almquist & Wiksell.

Keeves, J. P. (1995). *The World of School Learning: Selected Key Findings from 35 Years of IEA Research*. The Hague, The Netherlands: The International Association for the Evaluation of Education.

Keeves, J. P. (1968). V*ariation in Mathematics Education in Australia*: Some Interstate *Differences in the Organization, Courses of Instruction, Provision for and Outcomes of Mathematics Education in Australia*. Hawthorn, Victoria: ACER.

Keeves, J. P. & Kotte, D. (1996). The Measurement and reporting of key competencies. In *Teaching and Learning the Key Competencies in the Vocational Education and Training sector*, Adelaide: Flinders Institute for the Study of Teaching, pp. 139-168.

Keeves, J. P. & Schleicher, A. (1992). Changes in Science Achievement: 1970-84. In J P Keeves (ed.), *The IEA Study of Science III : Changes in Science Education and Achievement: 1970 to 1984*, Oxford: Pergamon Press, pp. 141-151.

Keeves, J. P. & Radford, W. C. (1969). *Some Aspects of Performance in Mathematics in Australian schools.* Hawthorn, Victoria: Australian Council for Educational Research.

Kolen, M. J. & Whitney, D. R. (1981). Comparison of four procedures for equating the tests of general educational development. Paper presented at the annual meeting of thee American Educational Research Association. Los Angeles, California.

Lietz, p. (1996). *Changes in Reading Comprehension across Culture and Over Time.* German: Waxman, Minister.

Martin, M. O. (1996). Third international mathematics and science study: An overview. In M O Martin & D L Kelly (eds), *Third International Mathematics and Science Study: Technical Report vol 1: Design and Development*, Boston: IEA, pp. 1.1-1.19.

Moss, J. D. (1982). *Towards Equality: Progress by Girls in Mathematics in Australian Secondary Schools.* Hawthorn, Victoria: ACER.

Norusis, M. J.(1993). *SPSS for Windows: Base System User's Guide: Release 6.0.* Chicago: SPSS Inc.

Postlethwaite, T. N. (1967). *School Organization and Student Achievement a Study Based on Achievement in Mathematics in Twelve Countries.* Stockholm: Almqvist & Wiksell.

Rentz, R. R. & Bashaw, W. L. (1975*). Equating Reading tests with the Rasch model, Vol. I Final Report.* Athens, Georgia: University of Georgia: Educational Research Laboratory, College of Education.

Robitaille, D. F. (1990). Achievement comparisons between the first and second IEA studies of mathematics. *Educational Studies in Mathematics, 21* (5), 395-414.

Robitaille, D.F. and Beaton, A.E. (2002). TIMSS: A Brief Overview of the Study. In D.F. Robitaille and A.E. Beaton (ed.), *Secondary Analysis of the TIMSS Data*, Dordrecht, KLUWER, pp. 11-180.

Robitaille, D. F. & Travers, K. J. (1992). International studies of achievement in mathematics. In D A Grouws, (ed.), *Hand book of Research on Mathematics Teaching and Learning,* New York: Macmillan, pp. 687-709.

Rosier, M. J. (1980). *Changes in Secondary School Mathematics in Australia.* Hawthorn, Victoria: ACER.

Smith, R. M. and Kramer, G. A. (1992). A comparison of two methods of test equating in the Rasch model. *Educational and Psychological Measurement,* 52 (4), 835-846.

Sontag, L. M. (1984). Vertical equating methods: A comparative study of their efficacy. DAI, 45-03B, page 1000.

Tilahun Mengesha Afrassa. (2002). *Changes in Mathematics Achievement Overtime in Australia and Ethiopia.* Flinders University Institute of International Education, Research Collection, No 4, Adelaide.

Tilahun Mengesha Afrassa (1996). *Students' Attitudes towards Mathematics and Schooling over time: A Rasch analysis.* Paper presented in the Joint Conference of Educational Research Association, Singapore (ERA) and Australian Association for Research in Education (AARE), Singapore Polytechnic, Singapore, 25 - 29 November 1996.

Tilahun Mengesha Afrassa and Keeves, J. P. (2001). Change in differences between the sexes in mathematics achievement at the lower secondary school level in Australia: Over Time. *International Education Journal, 2(2),* 96-107.

Tilahun Mengesha Afrassa. and Keeves, J. P.(1997). *Changes in Students' Mathematics Achievement in Australian Lower Secondary Schools over Time: A Rasch analysis.* Paper presented in 1997 Annual Conference of the Australian Association for Research in Education (AARE). Hilton Hotel, Brisbane, 30 November to 4 December 1997.

Wright, B. D. (1995). 3PL or Rasch? *Rasch Measurement Transactions*, 9 (1), 408-409.

Wright, B. D., and Stone, M. H. (1979). Best Test Design: Rasch Measurement. Chicago: Mesa Press.

Willett, J. B. (1997). Change, Measurement of. In J.P. Keeves (ed.), *Educational Research, Methodology, and Measurement: An International Handbook, (second ed.),* Oxford: Pergamon, pp. 327-334.

Zammit, S.A, Routitsky, A. and Greenwood, L. (2002). *Mathematics and Science Achievement of Junior Secondary School Students in Australia.* Camberwell, Victoria: ACER.

Chapter 5

MANUAL AND AUTOMATIC ESTIMATES OF GROWTH AND GAIN ACROSS YEAR LEVELS: HOW CLOSE IS CLOSE?

Petra Lietz
International University, Bremen, Germany

Dieter Kotte
Causal Impact, Germany

Abstract: Users of statistical software are frequently unaware of the calculations underlying the routines that they use. Indeed, users—particularly in the social sciences—are often somewhat adverse towards the underlying mathematics. Yet, in order to appreciate the thrust of certain routines, it is beneficial to understand the way in which a program arrives at a particular solution. Based on data from the Economic Literacy Study conducted at year 11 and 12 level across Queensland in 1998, this article renders explicit the steps involved in calculating growth and gain estimates in student performance. To this end, the first part of the article describes the Omanual calculation of such estimates using the Rasch estimates of item thresholds of common items at the different year levels produced by Quest (Adams & Khoo, 1993) as a starting point for the subsequent calibrating, scoring and equating. In the second part of the chapter, we explore the extent to which estimates of change in performance across year levels that are calculated with ConQuest (Wu, Adams & Wilson, 1997).

The article shows that the manual and automatic way of calculating growth and gain estimates produce nearly identical results. This is not only reassuring from a technical point of view but also from an educational point of view as this means that the reader of the non-mathematical discussion of the manual calculation procedure will develop a better understanding of the processes involved in calculating growth and gain estimates.

Key words: unidimensional latent regression, gain, calibrate, scoring, equating across year levels, test performance, economic literacy

S. Alagumalai et al. (eds.), Applied Rasch Measurement: A Book of Exemplars, 79–96.

For a number of years, ConQuest (Wu, Adams & Wilson, 1997) has offered the possibility of calculating estimates of growth and gain—for example of student performance between year levels—automatically using unidimensional latent regression.

Prior to that, one way of obtaining estimates of growth and gain was to calculate these estimates 'manually' using the Rasch estimates of item thresholds of common items at the different year levels produced by Quest (Adams & Khoo, 1993) as a starting point for the subsequent calibrating, scoring and equating.

It should be noted that in this context, 'growth' refers to the increase in performance that occurs just as a result of development from one year to the next while 'gain' refers to the yield as an outcome of educational efforts.

The focus of this chapter is twofold. Firstly, the description of the 'manual' calculation is aimed at illustrating the process underlying the automatised calculation of growth and gain estimates. Secondly, we explore the extent to which estimates of change in performance across year levels are calculated 'manually' differs from estimates that are produced 'automatically' by ConQuest.

To this end, student achievement data of a study of economic literacy in year 11 and 12 in Queensland, Australia, in 1998, are first analysed using the program Quest (Adams & Khoo, 1993), which produces Rasch estimates of student performance separately for each year level. The subsequent calibrating, scoring and equating across year levels, are done manually.

The same data are then analysed using ConQuest (Wu, Adams & Wilson, 1997), which automatically calculates estimates of change in performance between year 11 and 12. Results of the two ways of proceeding are then compared.

In order to locate these analyses within the context of the larger endeavour, a brief summary of the study of economic literacy in Queensland in 1998 is given before proceeding with the analyses and arriving at an evaluation of the extent of the difference in manually and automatically produced estimates of change across year levels.

1. THE STUDY OF ECONOMIC LITERACY IN QUEENSLAND IN 1998

Economic literacy refers to an understanding of basic economic concepts which is necessary for members of society to make informed decisions not only about personal finance or private business strategies but also about the relative importance of differing political arguments. Walstad (1988, p. 327) operationalises economic literacy as involving economic concepts that are mentioned in the daily national media including 'tariffs and trade, economic growth and investment, inflation and unemployment; supply and demand, the federal budget deficit; and the like'.

The Test of Economic Literacy (TEL, Walstad & Soper, 1987), which was developed originally in the United States to assess economic literacy, has also been employed in the United Kingdom (Whitehead & Halil, 1991), China (Shen & Shen, 1993) as well as Austria, Germany and Switzerland (Beck & Krumm, 1989, 1990, 1991). In Australia, however, despite the fact that economics is an elective subject in the last two years of secondary schooling in all states, no standardised instrument has been developed to allow comparisons across schools and states.

As a first step towards developing such an instrument, a content analysis of the curricula of the eight Australian states and territories was undertaken and compared with the coverage of the TEL. Results showed not only a large degree of overlap between the eight curricula, but also between the curricula and the TEL. Only a few concepts which were covered in the curricula of some Australian states were not covered by the TEL. These included environmental economics and primary industry, minerals and energy, reflecting the importance of agriculture and mining for the economy of particular states.

In addition to the content analysis, six economics teachers attended a session to rate the appropriateness of the proposed test items for economics students. An electronic facility, called the Group Support System (GSS), was used to streamline this process. The GSS generated a summary of the teachers' ratings for each item and distractor, and enabled discussions about contentious items or phrases. The rating process was undertaken twice, once for each year level.

As a result of the curricular content analysis and the teachers' ratings of the original TEL, 42 items were included in the year 11 test (AUSTEL-11) and 52 items in the year 12 test component (AUSTEL-12). Thirty items were common to the two test forms allowing the equating of student performance across year levels. Test items covered four content areas: namely,

1. fundamental economic concepts
2. microeconomics
3. macroeconomics and
4. international economics.

It should be noted that items assessing international economics were mainly incorporated in the year 12 test as this content area as the curricular analyses had shown that this content area was hardly taught in year 11.

As a second step towards developing an instrument assessing economic literacy in Australia, a pilot study was conducted in 1997 (Lietz & Kotte,

1997). The adapted test was administered to a total of 246 students enrolled in economics at years 11 and 12 in 18 schools in Central Queensland (Capricornia school district). Testing was undertaken in the last two weeks of term 3 of the 1997 school year. This time was selected so that students would have had the opportunity to learn a majority of the intended subject content for the year and before the early release of year 12 students in their final school year.

The pilot study also served to check the suitability of item format, background questionnaires (addressed to students and teachers) and test administration. Observation during testing by the researchers, as well as feed-back from students and teachers, did not reveal any difficulties in respect to the multiple-choice format or the logistics of test administration. With few exceptions, students needed less than the maximum amount of time available (that is, 50 minutes) to complete the test. Hence, the test was not a speeded but a power test as had also been the case in the United States (Soper & Walstad, 1987, p. 10).

Achievement data in the pilot study were obtained by means of paper and pencil testing, a format with which students, as well as teachers, were rather comfortable. Two new means of data collection, using PCs and the internet, were also pretested in a few schools prior to the main Queensland-wide study in September 1998. Some minor adjustments, such as web-page design to suit different monitor sizes, were made as a result of the piloting.

2. ECONOMIC LITERACY LEVELS IN QUEENSLAND

Before presenting details of the different procedures of calculating estimates of growth and gain between the two year levels, a brief overview of the levels of economic literacy in the Queensland study of year 11 and year 12 students is given below.

2.1 Economic literacy levels at year 11

The year 11 test comprised 42 items of which 13 were assigned to the fundamental concepts sub-scale, 12 to the microeconomics sub-scale and 14 to the macroeconomics sub-scale. Four Rasch scores were calculated for the overall test performance as well as for the subsets of items for fundamental concepts, microeconomics and macroeconomics. Scores were calculated on a scale ranging from 0 (zero ability) to 1000 (maximum ability) and a midpoint of 500 in line with common notations (Keeves & Kotte, 1996; Lokan, Ford & Greenwood, 1996, 1997).

Table 5-1 Rasch scores and their standard deviations (in brackets) for the overall sample as well as for selected sub-samples, year 11

	All test items	Fundamental concepts	Micro-economics	Macro-economics
All QLD (N=884)	521 (86)	527 (102)	514 (104)	529 (115)
All females (N=408)	511 (72)	517 (93)	500 (93)	523 (102)
All males (N=416)	532 (96)	539 (109)	529 (114)	536 (127)
State schools (N=306)	500 (73)	508 (89)	492 (88)	506 (105)
Independent schools (N=379)	542 (89)	547 (111)	537 (107)	549 (116)
Catholic schools (N=199)	515 (88)	518 (97)	506 (112)	528 (122)

Results presented in Table 5-1 indicate that year 11 students across Queensland achieve the highest level of competence in macroeconomics which covers concepts such as gross national product, inflation and deflation as well as monetary and fiscal policy. In contrast, students show the lowest performance level in microeconomics, which covers concepts such as markets and prices, supply and demand, as well as competition and market structure. An examination of the variation of scores reveals a considerably greater range of performance levels between students on the macroeconomics sub-scale than on the microeconomics sub-scale—an observation which applies to all three types of schools.

Table 5-1 also shows differences in performance levels of male and female students. Thus, scores on the total scale, as well as all sub-scales, are considerably higher for male students than for female students. At the same time, results of percentile analyses presented elsewhere (Lietz & Kotte, 2000) indicate that the range of achievement levels between the highest and the lowest achievers is greater for boys than it is for girls. On the Macroeconomics sub-scale, for example, the lowest performing boys are as low performing as the lowest female performers. However, in macroeconomics, 75 per cent of boys achieve at a level at which only the highest female performers can be found.

A t-test of the mean differences demonstrates that all differences would be significant, except for the Macroeconomics sub-scale, if the data stemmed from a simple random sample. However, as in most educational research, the sample in this study was not a simple random sample. Instead, intact classes were tested within schools. Thus, it should be sufficient to state that the gender differences in economics achievement in this study were considerable.

While gender differences in economics achievement have been reported in some studies (Walstad & Robson, 1997), no significant differences between male and female students had emerged in the 1997 pilot study at the total score, sub-score or item levels (Kotte & Lietz, 1998). Moreover, gender differences have frequently been shown to be mediated by other factors such as homework effort or type and extent of out-of-school activities (Keeves, 1992; Kotte, 1992). Hence, it will be of interest to look beyond a bivariate relationship between gender and achievement and examine the emerging gender differences in a larger model of factors influencing economics performance. This, however, is beyond the scope of this chapter.

Table 5-1 also reveals differences between achievement levels of different school types. Thus, students enrolled in independent schools show the highest level of performance across all scales, followed by Catholic and state schools. State schools exhibit the lowest spread of scores for the total scale and the sub-scale measuring fundamental concepts in economics. In contrast, the spread of scores for the fundamental economics sub-scale displayed for Catholic schools is considerable. Here the lowest achievers perform below the lowest performers in state schools while the upper 25th percentile of students in Catholic schools achieves at a higher level than the highest achievers in the state schools on the fundamental economics sub-scale (Kotte & Lietz, 2000).

For the microeconomics sub-scale, independent schools demonstrate a lower range of ability levels than state schools while the range reported for macroeconomics is similar for the two school types. Again, Table 5-1 illustrates that the differences between high achieving and low achieving students is greater in Catholic schools than in state or independent schools.

It should be noted, though, that such comparisons of mean performance levels—as already pointed out for the gender differences reported above—are frequently misleading in that they fail to consider important variables which lead to these differences in achievement. With respect to school types, for example, it is frequently argued (Postlethwaite & Wiley, 1991; Kotte, 1992; Keeves, 1996) that not the school type itself but the associated differences in resource levels contribute to differences in student performance. Evidence supporting this assumption emerged from the analysis of a more sophisticated model of factors influencing economics achievement to estimate the effects of different variables on achievement while holding the effects of other important variables constant (Lietz & Kotte, 2000).

2.2 Economic literacy levels at year 12

The year 12 economic literacy test, AUSTEL-12, was comprised of 52 items of which 14 were assigned to the fundamental concepts sub-scale, 15 to the Microeconomics sub-scale, 13 to the Macroeconomics sub-scale and 10 to the sub-scale measuring International economics. The latter scale was only added at year 12 level as a teacher rating of the test items in the Central Queensland Pilot Study in 1997 had shown that the content required to answer these items was not taught until year 12.

Hence, five Rasch scores were calculated for this year level: namely, for the overall test performance as well as for the subsets of items for fundamental concepts, microeconomics, macroeconomics and international economics.

Table 5-2 Rasch scores and their standard deviations (in brackets) for the overall sample as well as for selected sub-samples, year 12

	All test items	Fundamental concepts	Micro-economics	Macro-economics	International economics
All QLD (N=583)	568 (86)	562 (103)	594 (101)	559 (117)	558 (124)
All females (N=266)	560 (76)	556 (91)	590 (97)	551 (108)	544 (110)
All males (N=268)	578 (95)	567 (114)	603 (105)	569 (125)	574 (135)
State schools (N=210)	550 (79)	546 (98)	577 (92)	541 (111)	538 (118)
Independent schools (N=227)	589 (92)	580 (114)	615 (111)	583 (115)	583 (126)
Catholic schools (N=146)	562 (81)	557 (88)	585 (91)	549 (121)	548 (125)

Table 5-2 shows that year 12 students achieved the highest level of competence in microeconomics. This is in contrast to the findings for year 11 students who exhibited the lowest performance on that sub-scale. This is likely to reflect the shift in the content focus from year 11 to year 12. The high performance in microeconomics (594) is followed by the mean achievement on the fundamental concepts sub-scale (562), which is closely followed by the mean score for macroeconomics (559) and international economics (558). Like the year 11 data, year 12 results show that differences between the highest and lowest achievers are greatest for the macroeconomics sub-scale.

The scores presented in Table 5-2 are consistently higher for male students than for female students across the total, as well as for the four sub-scales. However, a t-test of the mean achievement levels reveals that these differences are only significant for the total and the international economics sub-score. Again, this test should only be regarded as an indicator. The same cautionary note applies regarding the application of tests, which assumes simple random samples to data from resulting from different sampling designs put forward in the previous section. As is the case for year 11, boys display a greater range in performance than girls.

A finding at the year 11 level which also emerges in the year 12 data is that students from independent schools show the highest performance across all scales, followed by students from Catholic and state schools. At the same time, an examination of the spread of scores provides evidence that independent schools are also confronted with the greatest differences between high and low achievers. Only for the international economics sub-scale are differences between the highest and lowest achievers greatest in Catholic schools.

In summary, students at year 12 across all schools perform well above average (568). However, a number of noticeable differences are found when comparing independent, Catholic and state schools. Though this is not necessarily surprising—and in line with findings relevant for other subjects (Lokan, Ford & Greenwood, 1996, 1997)—students enrolled in independent schools perform, on average, better than other students. A possible explanation might be the better teaching facilities and resources available in independent schools, as well as the greater emphasis given to economics as an elective.

3. GROWTH AND GAIN FROM YEAR 11 TO YEAR 12

Economics is studied in Queensland only at the upper secondary school level. Hence, it is of interest to educators to examine whether the potential yield between year 11 and year 12 is realised in actual gain between the two year levels.

In the previous sections, the mean performance level of year 11 students was reported to be slightly above average (521; midpoint being 500). In comparison, the year 12 students were found to perform well above average (568; midpoint, again, being 500). While one might be tempted to proclaim an increase in performance levels between the two year levels, the two estimates are not directly comparable as they involve different samples and different test items.

In order to obtain estimates of the potential learning that took place between year 11 and year 12, provisions had been made in the test design to incorporate a number of bridging items that were common to both the

AUSTEL-11 and the AUSTEL-12 forms. The two ways in which estimates of growth and gain were calculated, namely the 'manual' and the 'automatic' calculation, are described below.

3.1 'Manual' calculation of growth and gain from year 11 to year 12

The manual calculation of estimates for growth and gain involves three steps: namely, calibration, scoring and equating. While calibration refers to the calculation of item difficulty levels or thresholds, scoring denotes the estimation of scores taking into account the difficulty levels of the items answered by a student, and equating is the last step of arriving at the estimate of gain between year 11 and year 12.

These steps are described in detail below:

Calibration

1. A Rasch analysis using Quest (Adams & Khoo, 1993) was based on the responses of only those year 11 students who had attempted all items. The use of only those students who responded to all items was intended to minimise the potential of bias introduced by inappropriate handling or ignoring missing data as a result of differences in student test-taking behaviour or differences in actual testing conditions.
2. Year 11 item threshold values for those 30 items that were common to the year 11 and year 12 test were recorded (see Table 5-3).
3. A Rasch analysis was performed using only the responses those year 12 students who had attempted all items.
4. Year 12 item threshold values for those 30 items that were common to the year 11 and year 12 test were recorded (see Table 5-3).
5. An examination of the Year 11 and Year 12 item threshold values revealed that the common items were more difficult in the context of the year 12 test. In other words, the test as a whole was easier at the Year 12 level which made the common items relatively more difficult.
6. Differences were calculated between the year 11 and year 12 threshold values of common items.
7. The sum of all differences was divided by the number of common items (that is, 30). The resulting mean difference was 26 points (that is, 0.26 logits).

Scoring

8. Rasch scores were calculated for all year 11 students using the threshold values for all items obtained in step 1.
9. Rasch scores were calculated for all year 12 students using the threshold values for all items obtained in step 3.
10. In order to undertake the actual equating, the mean Rasch score for the year 11 students (that is, 521) was subtracted from the mean Rasch score for the year 12 students (that is, 568). This raw difference of 47 points was adjusted for the expected higher performance of year 12 students by subtracting the mean difference of 26 points. Thus, the gain from year 11 to year 12 was calculated to be 21 Rasch scale points.

Keeves (1992, p. 8) states that, generally, an increase of 33 score points equates to one year of schooling. Hence, the resulting yield of 21 Rasch score points is equivalent to approximately two-thirds of a full year of schooling. At this point, it can only be speculated as to why the gain between year 11 and year 12 falls short of that of a full year of schooling. Thus, it might be a consequence of the fact that students in the final year of schooling in Queensland spend a considerable amount of time preparing for the final school-leaving exams.

3.2 'Automatic' calculation of growth and gain from year 11 to year 12

The Rasch-scaling software, ConQuest, developed by ACER (Wu, Adams & Wilson, 1997), is an enhancement of the first Rasch-scaling software, QUEST, released by ACER in the early 1990s (see Adams & Khoo, 1993). ConQuest employs a number of additional modules and options carrying the application of Rasch-scaling considerably further (Adams, Wilson & Wu, 1997; Wang, Wilson & Adams, 1997; Loehlin, 1998).

One particular enhancement of ConQuest—the earliest version released in 1996—is the capability to estimate directly unidimensional regression models (see Wu, Adams & Wilson, 1996, pp. 55 – 69). This approach is used when comparing achievement differences across year levels. Appendix 1 specifies the syntax that is required to obtain growth and gain estimates 'automatically'.

The input syntax illustrates that 30 items common to AUSTEL-11 and AUSTEL-12 were used to estimate the latent regression. The underlying ASCII data file contained the responses of all students attempting all items (RITEM1 to RITEM42) plus a student identification variable (here called: ID) and a variable containing the student's grade (labelled: year).

For the analyses with ConQuest, only those students who attempted to answer all items of the AUSTEL were selected. Thus, 298 students who answered all 42 items at year 11 and 494 students who responded to all 52 items at year 12—which amounts to a total of 792 students—were included in the latent regression estimation.

Results of the analysis presented in Output 5-1 show that the constant—in other words the gain between year 11 and year 12—is 0.217 while the 'year'—in other words the growth occurring from one year to the next which has to be taken into account—is estimated to be 0.257 logits. Thus, the results produced by ConQuest largely coincide with the results produced by manual calculation of these values as shown in the previous section, since they only differ in the third decimal place. Therefore, in answer to the question 'how close is close?' it can be argued that such a result is 'very close', hence close enough.

Table 5-3 Rasch estimates of item thresholds for 30 AUSTEL items common to year 11 and year 12

Common item number	Item thresholds year 12	Item thresholds year 11	Difference year 12 – year 11
1	-1.33	-1.54	0.21
2	-1.38	-1.53	0.15
3	-0.94	-1.32	0.38
4	-1.13	-1.28	0.15
5	-1.18	-1.35	0.17
6	-1.65	-1.83	0.18
7	-0.48	-0.73	0.25
8	-0.11	-0.29	0.18
9	0.26	-0.12	0.38
10	0.43	0.14	0.29
11	-0.28	-0.46	0.18
12	-0.10	-0.41	0.31
13	-1.17	-1.16	0.01
14	0.53	0.58	0.05
15	0.63	0.61	0.02
16	0.27	-0.14	0.41
17	-0.40	0.06	0.46
18	-0.01	0.00	0.01
19	0.62	0.55	0.07
20	0.62	0.40	0.22
21	-0.43	-0.77	0.34
22	0.83	0.61	0.22
23	0.47	0.57	0.10
24	0.54	0.23	0.31

25	1.58	1.42	0.16
26	1.55	1.11	0.44
27	1.58	1.12	0.46
28	1.05	0.60	0.45
29	0.52	0.83	0.31
30	2.79	2.13	0.66

Notes

Total difference	7.53
Average difference (growth)	0.26
Year 12 mean total Rasch score	568
Year 11 mean total Rasch score	521
Raw difference (year 12 – year 11)	47
Gain from year 11 to year 12, adjusted for growth (i.e. 0.26)	21

4. SUMMARY

In this chapter, two ways of calculating growth and gain estimates between student performance in economics in year 11 and year 12 were presented. On the one hand, the estimates were produced using Rasch item thresholds of common items as a starting point with subsequent 'manual' calculations during calibrating, scoring and equating. On the other hand, the estimates were calculated automatically using ConQuest.

In addition, the background of the study of economic literacy in Queensland, Australia, was outlined and results of the performance levels of year 11 and year 12 students were presented.

Students at the upper secondary school level in Queensland who participated in the Economic Literacy Survey in 1998 showed, in general, satisfactory performance in the test of economic literacy as adapted to the Australian context. The Rasch scores for both year 11 and year 12 students were above the theoretical average of 500 points.

The estimates of gain indicated that learning had occurred in the subject of economics between year 11 and year 12. However, the observed gain between the two year groups appeared to be less than that of a full year of subject exposure. The observation that year 12 students spent a considerable amount of instructional time preparing for their school-leaving examinations at the end of the school year, leaving less time for in-depth treatment of topics in elective subjects, such as economics, was put forward as a possible explanation.

In respect to the comparison of the manual and automatic way of calculating growth and gain, the fact that both procedures resulted in nearly identical estimates was reassuring, for—in answer to the question in this chapter's heading—differences in just the third decimal place were considered to be 'sufficiently close' not to invalidate results produced by either procedure.

5. REFERENCES

Adams RJ & Khoo SK 1993 Quest - The interactive test analysis system. Hawthorn, Vic.: Australian Council for Educational Research.

Adams RJ, Wilson M & Wu M 1997 Multilevel item response models: An approach to errors in variables regression. Journal of Educational and Behavioral Statistics, 22(1), pp. 47-76.

Australian Bureau of Statistics (ABS) 1999 Census Update. http://www.abs.gov.au/websitedbs/D3110129.NSF.

Beck K & Krumm V 1989 Economic literacy in German speaking countries and the United States. First steps to a comparative study. Paper presented at the annual meeting of AERA, San Francisco.

Elley WB 1992 How in the world do students read? The Hague: IEA.

Harmon M, Smith TA, Martin MO, Kelly DL, Beaton AE, Mullis IVS, Gonzalez EJ & Orpwood G 1997 Performance Assessment in IEA's Third International Mathematics and Science Study. Chestnut Hill, MA: Center for the Study of Testing, Evaluation, and Educational Policy, Boston College & Amsterdam: IEA.

Keeves JP & Kotte D 1996 The measurement and reporting of key competencies. The Flinders University of South Australia, Adelaide.

Keeves JP 1992 Learning Science in a Changing World. Cross-national studies of Science Achievement: 1970 to 1984. The Hague: IEA.

Keeves JP 1996 The world of school learning. Selected key findings from 35 years of IEA research. The Hague: IEA.

Kotte D & Lietz P 1998 Welche Faktoren beeinflussen die Leistung in Wirtschaftskunde? Zeitschrift für Berufs- und Wirtschaftspädagogik, Vol. 94, No. X, pp. 421-434.

Lietz P & Kotte D 1997 Economic literacy in Central Queensland: Results of a pilot study. Paper presented at the Australian Association for Research in Education (AARE) annual meeting, Brisbane, 1 - 4 December, 1997.

Lietz P 1996 Reading comprehension across cultures and over time. Münster/New York: Waxmann.

Loehlin JC 1998 Latent variable models (3rd Ed.). Mahwah, NJ: Erlbaum.

Lokan J, Ford P & Greenwood L 1996 Maths & Science on the Line: Australian junior secondary students' performance in the Third International Mathematics and Science Study. Melbourne: Australian Council for Educational Research.

Lokan J, Ford P & Greenwood L 1997 Maths & Science on the Line: Australian middle primary students' performance in the Third International Mathematics and Science Study. Melbourne: Australian Council for Educational Research.

Martin MO & Kelly DA (eds) 1996 Third International Mathematics and Science Study Technical Report, Volume II: Design and Development. Primary and Middle School Years. Chestnut Hill, MA: Center for the Study of Testing, Evaluation, and Educational Policy, Boston College & Amsterdam: International Association for the Evaluation of Educational Achievement (IEA).

OECD 1998 The PISA Assessment Framework - An Overview. September 1998: Draft of the PISA Project Consortium. Paris: OECD.

Postlethwaite TN & Ross KN 1992 Effective schools in reading. Implications for educational planners. The Hague: IEA.
Postlethwaite TN & Wiley DE 1991 Science Achievement in Twenty-Three Countries. Oxford: Pergamon Press.
Shen R & Shen TY 1993 Economic thinking in China: Economic knowledge and attitudes of high school students. Journal of Economic Education, Vol. 24, pp. 70-84.
Soper JC & Walstad WB 1987 Test of economic literacy. Examiner's manual 2nd ed. New York: Joint Council on Economic Education (now the National Council on Economic Education).
Walstad WB & Robson D 1997 Differential item functioning and male-female differences on multiple-choice tests in economics. Journal of Economic Education, Spring, pp. 155-171.
Walstad WB & Soper JC 1987 A report card on the economic literacy of U.S. High school students. American Economic Review, Vol. 78, pp. 251-256.
Wang W, Wilson M & Adams RJ 1997 Rasch models for multidimensionality between and within items. In: Wilson M, Engelhard G & Draney K (eds), Objective measurement IV: Theory into practice. Norwood, NJ: Ablex.
Whitehead DJ & Halil T 1991 Economic literacy in the United Kingdom and the United States: A comparative study. Journal of Economic Education, Spring, pp. 101-110.
Wu M, Adams RJ & Wilson MR 1996 ConQuest: Generalised Item Response Modelling Software. Draft Version 1. Camberwell: ACER.
Wu M, Adams RJ & Wilson MR 1997 ConQuest: Generalised Item Response Modelling Software. Camberwell: ACER.

6. OUTPUT 5-1

ConQuest input and output files

This input syntax was used with *ConQuest* Rasch-scaling software released by ACER to estimate the latent regression between year 11 and year 12 students in the AUSTEL.

The input syntax had to be kept in ASCII format and followed the syntax specifications given in the user manual of ConQuest (Wu, Adams & Wilson 1996):

```
==================================================================
datafile yr1112a.dat;
title   EcoLit 1998 equating 30 common items Yr11 & Yr12;
format  ID 1-8 year 10 responses 12-38, 40-42;
labels  << labels30.txt;
key  111111111111111111111111111111  ! 1;
regression year;
model item;
estimate ! fit=no;
show ! tables=1:2:3:4:5:6 >>eco_04.out;
quit;
==================================================================
```

The GUI-based version of ConQuest produces an on-screen protocol of the different iteration and estimation steps (called E-step and M-step; not shown here). However, the requested results (keyword: SHOW; options: TABLES) are listed in the actual output file on the following pages (contents of the ASCII file 'eco_04.out').

```
===================================================================
```

6. OUTPUT 5-2

```
EcoLit 1998 equating 30 common items Yr11 & Yr12 Wed Jan 08 23:05:54
SUMMARY OF THE ESTIMATION
===================================================================


The Data File was: yr1112a.dat
The format was:   ID 1-8 year 10 responses 12-38, 40-42
The model was: item

Sample size was: 792
Deviance was: 27847.78842
Total number of estimated parameters was: 32

The number of iterations was: 65

Iterations terminated because the convergence criteria were reached
===================================================================
EcoLit 1998 equating 30 common items Yr11 & Yr12 Wed Jan 08 23:05:54
TABLES OF POPULATION MODEL PARAMETER ESTIMATES
===================================================================
REGRESSION COEFFICIENTS


Regression Variable

CONSTANT                    0.217 ( 0.032)
year                        0.257 ( 0.053)
----------------------------------------------
An asterisk next to a parameter estimate indicates that it is
constrained
==============================================
COVARIANCE/CORRELATION MATRIX


Dimension

    1
-------------------------------------------------------------------
Variance                    0.517
-------------------------------------------------------------------
```

```
===================================================================
EcoLit 1998 equating 30 common items Yr11 & Yr12 Wed Jan 08 23:05:54
TABLES OF RESPONSE MODEL PARAMETER ESTIMATES
===================================================================
TERM 1: item
-------------------------------------------------------------------
   VARIABLES                           UNWGHTED FIT    WGHTED FIT
---------------                        -------------  -------------
     item          ESTIMATE  ERROR      MNSQ     T       MNSQ    T
-------------------------------------------------------------------
 1 RITEM01           -1.531   0.072
 2 RITEM02           -1.520   0.072
 3 RITEM03           -1.091   0.068
 4 RITEM04           -1.292   0.070
 5 RITEM05           -1.205   0.069
 6 RITEM06           -1.655   0.074
 7 RITEM07           -0.607   0.064
 8 RITEM08           -0.192   0.062
 9 RITEM09           -0.063   0.061
 10 RITEM10           0.256   0.061
 11 RITEM11          -0.441   0.063
 12 RITEM12          -0.264   0.062
 13 RITEM13          -1.146   0.068
 14 RITEM14           0.556   0.061
 15 RITEM15           0.556   0.061
 16 RITEM17           0.057   0.061
 17 RITEM19          -0.064   0.061
 18 RITEM21           0.080   0.061
 19 RITEM23           0.510   0.061
 20 RITEM25           0.481   0.061
 21 RITEM27          -0.596   0.064
 22 RITEM28           0.641   0.061
 23 RITEM30           0.503   0.061
 24 RITEM32           0.412   0.061
 25 RITEM34           1.334   0.065
 26 RITEM36           1.293   0.064
 27 RITEM38           1.313   0.065
 28 RITEM40           0.704   0.061
 29 RITEM41           0.728   0.062
 30 RITEM42           2.242*
-------------------------------------------------------------------
Separation Reliability =  0.995
Chi-square test of parameter equality = 4899.212,  df = 29,  Sig Level
= 0.000
===================================================================
```

```
=====================================================================
EcoLit 1998 equating 30 common items Yr11 & Yr12 Wed Jan 08 23:05:54
MAP OF LATENT DISTRIBUTIONS AND RESPONSE MODEL PARAMETER ESTIMATES
=====================================================================
                                              Terms in the Model Statement
                         +item
---------------------------------------------------------------------
  3                      |                            |
            Case         |                            |
          estimates      |                            |
             not         |                            |
          requested      |                            |
                         |                            |
                         |30                          |
                         |                            |
  2                      |                            |
                         |                            |
                         |                            |
                         |                            |
                         |25                          |
                         |26 27                       |
                         |                            |
  1                      |                            |
                         |                            |
                         |28 29                       |
                         |14 15 22                    |
                         |19 20 23 24                 |
                         |                            |
                         |10                          |
                         |16 18                       |
  0                      |9 17                        |
                         |8 12                        |
                         |                            |
                         |11                          |
                         |7 21                        |
                         |                            |
                         |                            |
 -1                      |                            |
                         |3 13                        |
                         |4 5                         |
                         |                            |
                         |1 2                         |
                         |6                           |
                         |                            |
                         |                            |
 -2                      |                            |
                         |                            |
=====================================================================
```

```
==================================================================
EcoLit 1998 equating 30 common items Yr11 & Yr12 Wed Jan 08 23:05:54
MAP OF LATENT DISTRIBUTIONS AND THRESHOLDS
==================================================================
                              Generalised-Item Thresholds
------------------------------------------------------------------
   3                |
         Case       |
       estimates    |
          not       |
       requested    |
                    |
                    |30.1
                    |
   2                |
                    |
                    |
                    |
                    |25.1
                    |26.1 27.1
                    |
   1                |
                    |
                    |28.1 29.1
                    |14.1 15.1 22.1
                    |19.1 20.1 23.1 24.1
                    |
                    |10.1
                    |16.1 18.1
   0                |9.1 17.1
                    |8.1 12.1
                    |
                    |11.1
                    |7.1 21.1
                    |
                    |
  -1                |
                    |3.1 13.1
                    |4.1 5.1
                    |
                    |1.1 2.1
                    |6.1
                    |
                    |
  -2                |
                    |
==================================================================
The labels for thresholds show the levels of item, and step,
respectively
==================================================================
```

Chapter 6

JAPANESE LANGUAGE LEARNING AND THE RASCH MODEL

Kazuyo Taguchi
University of Adelaide; Flinders University

Abstract: This study attempted to evaluate outcome of foreign language teaching by measuring the reading and writing proficiency achieved by students studying Japanese as a second language in six different year levels from year 8 to the first of university in the classroom setting. In order to measure linguistic gains across six years, it was necessary, firstly, to define operationally what reading and writing proficiency was; and secondly, to create measuring instruments, and, thirdly, to identify suitable statistical analysis procedures.

The study sought to answer the following research questions: Can reading and writing performance in Japanese as a foreign language be measured?; and Does reading and writing performance in Japanese form a single dimension on a scale?

The participants of this project were drawn from one independent school and two universities, while the instruments used were the routine tests produced and marked by the teachers. The estimated test scores of the students calculated indicated that the answers to all research questions are in the affirmative. In spite of some unresolved issues and limitations the results of the study indicated a possible direction and methods to commence an evaluation phase of foreign language teaching. The study also identified the Rasch model as not only robust measuring tools but also as capable of identifying grave pedagogical issues that should not be ignored.

Key words: linguistic performance, learning outcomes, person estimates, item estimates, measures of growth, pedagogical implications

S. Alagumalai et al. (eds.), Applied Rasch Measurement: A Book of Exemplars, 97–113.

1. INTRODUCTION

In order to enhance Australia's economic performance, a number of reports have highlighted the need for Australians to be proficient in the languages of Asia and to develop greater awareness of Asian culture (Lo Bianco, 1987; Asian Studies Council, 1988; Leal, 1991; Rudd, 1994). This is the result of recognition within Australian economic circles that the Asian market is vital and thus employers are searching for an Asia-literate work force to compete in it.

However, after a substantial expenditure in funding for the teaching of Asian languages for nearly three decades, the outcomes that have been reaped from the past endeavour are not clear. Several questions need to be asked: What levels of linguistic proficiency do the students reach after one year of language instruction in the school setting, and after five years? What ultimate level of proficiency can be expected, or should be expected? Are the demands of employers and of the nation being met to a greater extent by producing Asia-literate school leavers and university graduates?

Measuring learning outcomes is essential as a basis for any discussion of the process and products of our educational endeavour. It is overdue, therefore, to examine how much learning has been taking place with the curriculum that has been in use.

Although a great number of studies have been carried out in the area of reading and writing proficiency in both the first language (L1) and the second language (L2) to date in a variety of languages (Gordon & Braun, 1982; Eckhoff, 1983; Krashen, 1984) using various instruments and statistical analyses, they are mostly qualitative in nature (Scarino, 1995). In order to add a quantitative dimension to the description of linguistic growth, this study departed from a very fundamental point, that is, whether linguistic performance in Japanese is measurable, especially with significant diversity in learners' proficiency levels spanning six years from year 8 to year 12 and university. The positive findings to this question that linguistic proficiency growth is measurable by deploying the Rasch analysis (Rasch, 1960; Keeves & Alagumalai, 1999) have laid a foundation on which future research can be built. Without it, the measurement of linguistic growth must continue to rely on a vague, non-specific interpretation of terms which are typically used to describe learning outcomes to date. This study also investigated whether it is possible to set up a scale for reading and writing proficiency independent of both the difficulty level of the test items used as well as the people whose test scores are calibrated. By providing evidence that such a scale can be set up using software (Adams & Khoo, 1993) based on the Rasch model, the findings of this study offered a tool which can be used with confidence for future research in various areas of second language acquisition where learning outcomes form part of their investigations.

Although limited in its scale, due to the small sample size and non-longitudinal nature, this study added a basic piece of information to the world of knowledge: that is, identification of the Rasch model as a useful and robust means for measuring foreign language learning outcomes. Furthermore, the Rasch analysis made important pedagogical issues explicit to the attention of researchers and practitioners in unmistakable terms. These issues, that is, missing data, misfitting items, and local independence, might have been sounding a warning signal in the subconscious level of the test producers and the teachers have not yet decisively surfaced to demand serious attention.

2. METHODS

2.1 Samples

The participants of this project consisted of students of two educational sectors who were taking Japanese language as one of their academic subjects. They were: years 8 to 12 students (216 in total) of a co-educational independent secondary school, and students from two universities.

The tertiary participants (69 in total) were taking the university second year course of Japanese. With five or more years of exposure to Japanese language at their secondary schools (and at primary schools for some), students who obtained a score or 15 out of 20 or above in their matriculation examination enrolled directly into the second year level course (identified as university 1 group). They were in the same classes as those who had done only one year of the Japanese beginners' course at the university (labelled as university 2 group). The university 2 group was included in the study but was not the main focus of the study.

The obvious limitation of this study is its inability to generalise its findings to a wider population due to the small size and homogeneous nature of the participants. However, the beneficial outcome of the very limitation in this study is the fact that numerous unknown institution-related, students-related as well as teacher-related variables, such as attitudes and learning aptitude, have been controlled.

2.2 Measuring instruments

Two types of testing materials were used in the study: that is, routine tests and common items tests. The results of tests which would have been administered to the students as part of their assessment procedures, even if this research had not been conducted with them, were collected as measures of reading and writing proficiency.

In order to equate different tests which were not created as equal in difficulty level, it was necessary to have common test items (McNamara, 1996; Keeves & Alagumalai, 1999) which were administered to students of adjacent grade levels. These tests of 10 to 15 items were produced by the teachers and given to the students as warm-up exercises. Counting the results of both routine tests and anchor items tests towards their final grades ensured that students took these tests seriously. Although 10 to 15 anchor items were produced by the teachers, due to statistical non-fitting and over-fitting nature, some are deleted and, as a consequence, the valid number of anchor items was smaller (see Figure 6-1 and Figure 6-2). Since scholars such as Umar (1987) and Wingersky and Lord (1984) claim that the minimum number of common items necessary for Rasch analysis is as few as five, equating using these test items in this study is considered valid.

Marking and scoring of the tests and examinations were the responsibility of the class teachers. These were double-checked by the researcher.

For statistical calculation, the omitted items to which a student did not respond were treated as wrong, while not-reached items were ignored in calibration. Non-reached items are the first item to which a student did not respond, plus all the non-responded items that appeared after that particular item in the test. Obviously, this decision is a cause for concern and can be counted as one of the limitations of this study.

2.3 Measures of reading and writing proficiency

Based on the following two principles, the constructs called ‘reading proficiency’ and ‘writing proficiency’ were operationally defined for this project.

Principle 1: The results of the tests and examinations produced and marked by the classroom teachers were defined in this study as proficiency in reading and writing.

Principle 2: In years 8 and 9, reading or writing of a single vocabulary item was justifiably defined as part of proficiency. This was judged acceptable in the light of the reported research evidence which indicated the established high correlation between reading ability and vocabulary knowledge (Davis, 1944, 1968; Thorndike, 1973; Becker, 1981; Anderson &

Freebody, 1985). This was also believed to be justifiable due to the non-alphabetical nature of the Japanese language in which learners were required to master two sets of syllabaries consisting of 46 letters each, as well as the third set of writing system called kanji, ideographic characters which originated from the Chinese language. Thus, mastery of the orthography of the Japanese language is demanding and sine qua non to become literate in Japanese. Shaw and Li (1977) offer a theoretical rationale for this decision. That is, the importance placed on different aspects of language which language users need to access in order to either read or write, moves from letter-sound correspondences -> syllables-> morphemes-> words -> sentences -> linguistic context all the way to pragmatic context.

2.4 Data analysis

The following procedures were followed for data analysis:

Firstly, the test results are calibrated and scored using Rasch scaling for reading and writing separately as well as combined. The concurrent equating of the test scores that assessed performance of each of the six grade levels is carried out. Secondly, estimated proficiency scores are plotted on a graph both for writing and reading in order to address research questions.

3. RESULTS

Figure 6-1 below shows both person and item estimates on one scale. The average is set at zero and the greater the value the higher the ability of the person and difficulty level of an item.

It is to be noted that the letter ‘r’ indicates reading items and ‘w’ writing items. The most difficult item is r005.2 (.2 after the item name indicates partial credit containing two components in this particular case to gain a full mark), while the easiest item is Item r124. For the three least able students identified by three x’s in the left bottom of the figure, all the items except for Item r124 are difficult, while for the five students represented by five x’s on 5.0 level, all the items are easy except for the ones above this level which are Items r005.2, r042.2, r075.2 and r075.3.

4. PERFORMANCE IN READING AND WRITING

The visual display of these results is shown as Figures 6.2 – 6.7 below where the vertical axis indicates mean performance and the horizontal axis shows the year levels. The dotted lines are regression lines.

```
--------------------------------------------------------------------------------------------------
Item Estimates (Thresholds)                                       18/ 3/2001 12: 4
all on jap (N = 278 L = 146 Probability Level=0.50)
--------------------------------------------------------------------------------------------------
                   | r005.2
                   |
                   |
                   | r042.2
 7.0               |
                   |
                   |
                   | w042.2
                   |
                   |
                   |
 6.0               |
                   |
                   |
                   |
             XXXX  |
                   |
                   |
 5.0               |
                X  |
                   |
                   | w035.2
               XX  |
             XXXX  |
                   | r003  r021.2
 4.0          XXX  |
             XXXX  | r020  w043.2
             XXXX  | r012.2
               XX  | w012.2
                X  | r014.3 r018
             XXXX  | r017.2 r028  r029
              XXX  | w003.5 w004.3
 3.0     XXXXXXXX  | r017.1 w006.5 w028.2
         XXXXXXXX  | r035
          XXXXXXX  | r014.2 r031  w005.4 w008.2
       XXXXXXXXXX  | r023  w001.4 w013  w064.2
             XXXX  | w002.5 w007.4 w009.5 w010.2
      XXXXXXXXXXX  |
             XXXX  | r016.2 r019  r024  r030  w003.4 w006.4
 2.0       XXXXXX  | r016.1 r036  w002.4 w006.1 w006.2 w006.3
          XXXXXXX  | r025  w002.1 w002.2 w002.3 w007.3 w009.3 w009.4 w019  w027.2
               XX  | w004.2 w007.1 w007.2 w009.1 w009.2 w017  w018  w035.1 w092
               XX  | r021.1 r022  r098
                X  | r015  r081
             XXXX  | r004.2 r048.2 r097  w044.2 w045  w071
               XX  | r005.1 r006  r103  w003.3 w005.3 w008.1 w020  w064.1
 1.0           XX  | r012.1 r052  w001.3 w003.2 w014  w067.2
                X  | r009  r014.1 r032  w012.1 w016
             XXXX  | r033  r034  r087  w001.1 w001.2 w003.1 w005.1 w005.2 w028.1
               XX  | r027  w010.1
                X  | r007  r072  w047  w066.2
             XXXX  | r048.1 r053  r054  w023.2
               XX  | r001  r004.1 r083  w027.1 w032  w037  w043.1 w062  w065.2
           XXXXXX  | r066.2 r068.3
 0.0          XXX  | r041  r050  r074  r099  w042.1 w050  w089.2
         XXXXXXXX  | r011  r013  r077  w023.1 w044.1 w066.1
               XX  | r071  r095  r096  w090.2
            XXXXX  | r046  r063  r068.2 r104  w038  w046  w061  w067.1
            XXXXX  | r110.2 w004.1 w011  w015  w072
           XXXXXX  | r086  w065.1
          XXXXXXX  | w091.2
-1.0      XXXXXXX  | r068.1 r102  r171  w069
                X  | r084  r101  w048  w070
        XXXXXXXXX  | r042.1 r045  w088.2
          XXXXXXX  | r066.1 r079  r093  w049
              XXX  | r010  r094  w081  w089.1
               XX  | r172  w090.1 w091.1
               XX  | r085
-2.0               |
              XXX  | r044  w068  w087
              XXX  | r100  r110.1
              XXX  | r082  r178  w084
              XXX  | r180.3
               XX  | r091
            XXXXX  | w088.1
-3.0        XXXXX  |
              XXX  | r179.3
          XXXXXXX  | r176.4 r180.2
                   | r122
               XX  | r174  r176.3
               XX  | r176.2 w082  w083
               XX  | r176.1 r180.1
-4.0          XXX  |
            XXXXX  | r179.2
                   |
               XX  | r177  r179.1 w086
                   | r175
               XX  | r173  w080
               XX  |
-5.0               |
                   |
                   | r121
                X  | r123
                   | r120
                   |
                   |
-6.0               |
                   | r124
                   |
--------------------------------------------------------------------------------------------------
 Each X represents  1 students
==================================================================================================
```

Figure 6-1. Person and item estimates

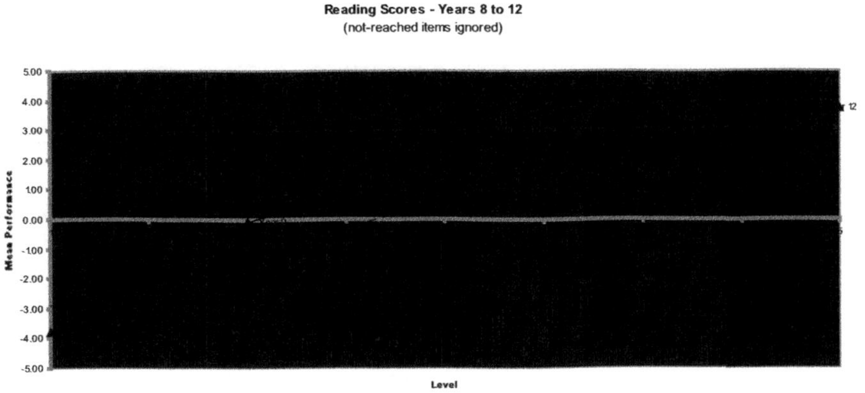

Figure 6-2. Reading scores (years 8 to 12)

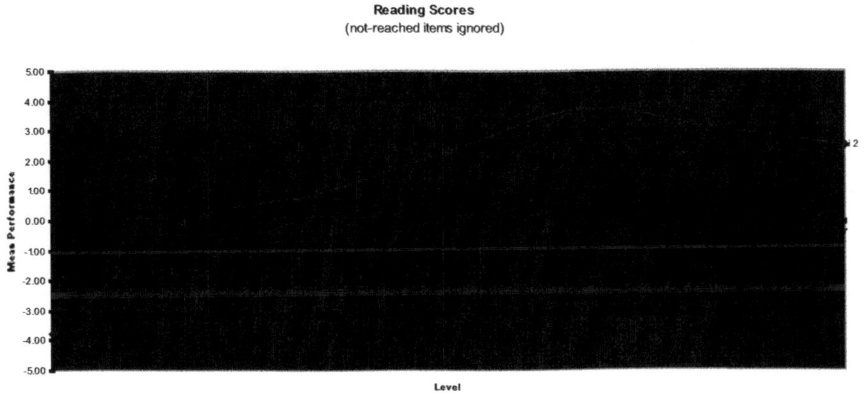

Figure 6-3. Reading scores (year 8 to university group 2)

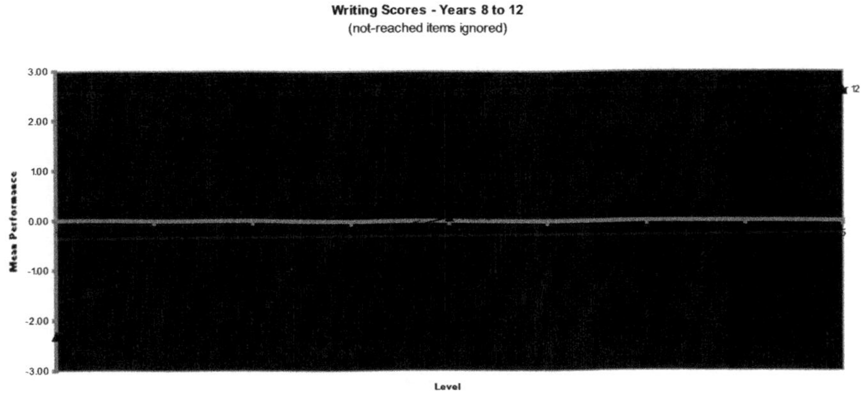

Figure 6-4. Writing scores (years 8 to 12)

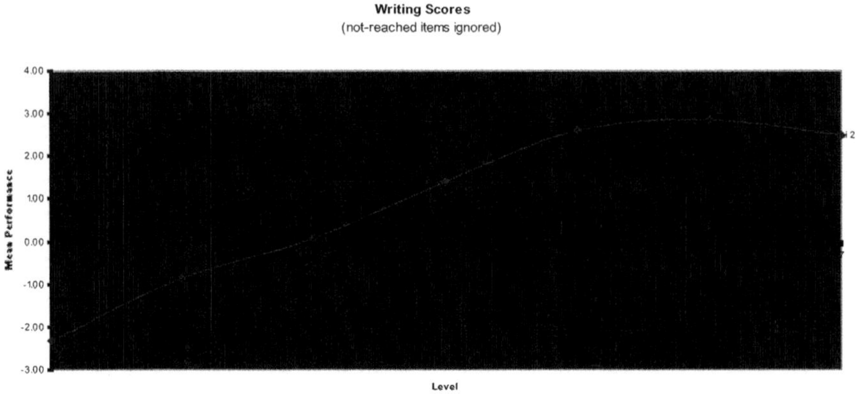

Figure 6-5. Writing scores (year 8 to university group 2)

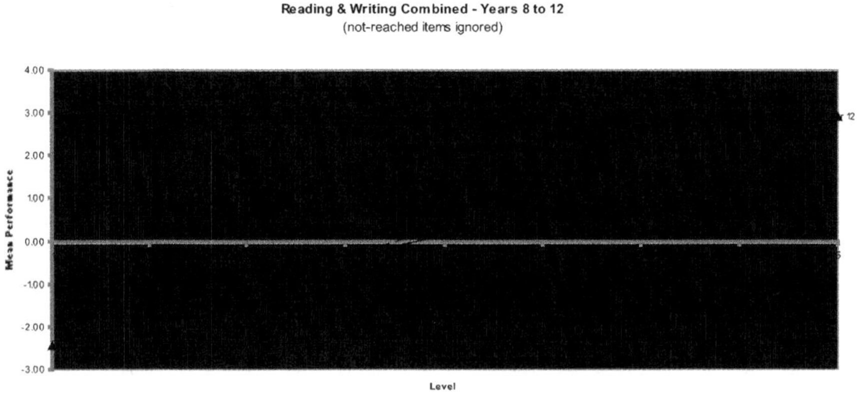

Figure 6-6. Combined scores of reading and writing (years 8 to 12)

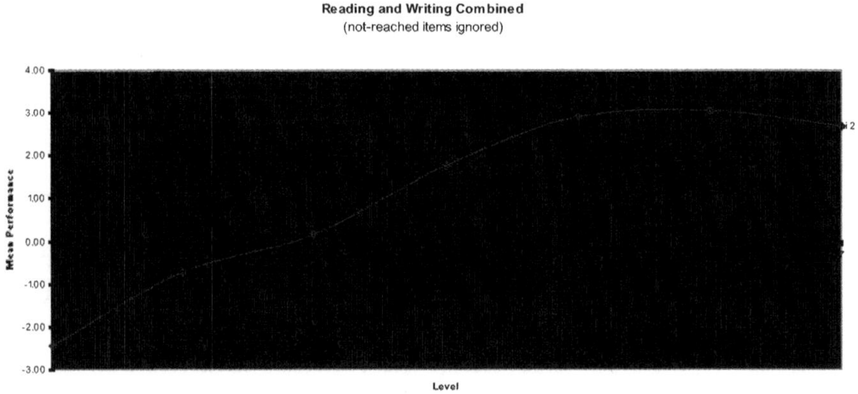

Figure 6-7. Combined scores of reading and writing (year 8 to university group 2)

As evident from the figures above, the performance of reading and writing in Japanese has been proved to be measurable in this study. The answer to research question 1 is: 'Yes, reading and writing performance can be measured'. The answer to research question 2 is also in the affirmative: that is, the result of this study indicated that a scale independent of the sample whose test scores were used in calibration and the difficulty level of the test items could be set up. It should, however, be noted that the zero for the scale, without loss of generality, is set at the mean average difficulty level of the items.

5. THE RELATIONSHIP BETWEEN READING AND WRITING PERFORMANCE

The answer to research question 3 (Do reading and writing performance in Japanese form a single dimension on a scale?) is also 'Yes', as seen in Figure 6-8 below.

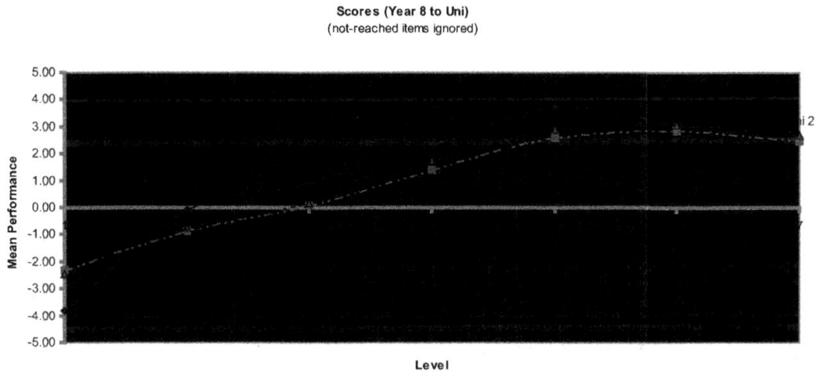

Figure 6-8. Reading and writing scores (year 8 to university)

The graph indicates that reading performance is much more varied or diverse than writing performance. That the curve representing writing performance is closer to the centre (zero line) indicates a smaller spread compared to the line representing reading performance.

As a general trend of the performance, two characteristics are evident. Firstly, reading proficiency would appear to increase more rapidly than writing. Secondly, the absolute score of the lowest level of year 8 in reading

(-3.8) is lower than writing (-2.4) while the absolute highest level of year 12 in reading (3.7) is higher than in writing (2.7). Despite these two characteristics, performance in reading and writing can be fitted to a single scale as shown in Figure 6-8. This indicates that, although they may be measuring different psychological processes, they function in unison: that is, the performance on reading and writing is affected by the same process, and, therefore, is unidimensional (Bejar, 1983, p. 31).

6. DISCUSSION

The measures of growth were examined in the following ways.

6.1 Examinations of measures of growth recorded in this study

Figures 6-2 to 6-8 suggest that the lines indicating reading and writing ability growth recorded by the secondary school students are almost disturbance-free and form straight lines. This, in turn, means that the test items (statistically 'fitting' ones) and the statistical procedures employed were appropriate to serve the purpose of this study: namely, to examine growth in reading and writing proficiency across six year levels. Not only did the results indicate the appropriateness of the instrument, but they also indicated its sensitivity and validity: that is, the usefulness of the measure as explained by Kaplan (1964:116):

One measuring operation or instrument is more sensitive than another if it can deal with smaller differences in the magnitudes. One is more reliable than another if repetitions of the measures it yields are closer to one another. Accuracy combines both sensitivity and reliability. An accurate measure is without significance if it does not allow for any inferences about the magnitudes save that they result from just such and such operations. The usefulness of the measure for other inferences, especially those presupposed or hypothesised in the given inquiry, is its validity.

The Rasch model is deemed sensitive since it employs an interval scale unlike the majority of extant proficiency tests that use scales of five or seven levels. The usefulness of the measure for this study is the indication of unidimensionality of reading and writing ability. By using Kaplan's yardstick to judge, the results suggested a strong case for inferencing that reading and writing performance are unidimensional as hypothesised by research question 3.

6.2 The issues the Rasch analysis has identified

In addition to its sensitivity and validity, the Rasch model has highlighted several issues in the course of the current study. Of them the following three have been identified by the researcher as being significant and are discussed below. They are: (a) misfitting items, (b) treatment of missing data, and (c) local independence. The paper does not attempt to resolve these issues but rather merely reports them as issues made explicit by Rasch analysis procedures. First, misfitting items are discussed below.

Rasch analysis identified 23 reading and eight writing items as misfitting: that is, these items are not measuring the same latent traits as the rest of the items in the test (McNamara, 1996). Pedagogical implication of these items (if included in the test) is that the test as a whole no longer can be considered valid. That is, it is not measuring what it is supposed to measure.

The second issue discussed is missing data. Missing data (= non-responded) in this study were classified into two categories: namely, either (a) non-reached, or (b) wrong. That is, although no response was given by the test taker, these items were treated as identical to the situation where a wrong response was given. The rationale for these decisions is based on the assumption that the candidate did not attempt to respond to non-reached items: that is, they might have arrived at the correct responses if the items had been attempted. Some candidates' responses indicate that it is questionable to use this classification.

The third issue highlighted by the Rasch analysis is local independence. Weiss et al. (1992) define the term 'local independence' as the probability that a correct response of an examinee to an item is unaffected by responses to other items in the test and it is one of the assumptions of Item Response Theory. In the Rasch model, one of the causes for an item being overfitting is its violation of local independence (McNamara, 1996), which is of concern for two different reasons. Firstly, as a valid part of data in a study such as this, these items are of no value since they add no new information which other items have already given (McNamara, 1996). The second concern is more practical and pedagogical.

One of the frequently sighted forms in foreign language tests is to pose questions in the target language which require answers in the target language as well. How well a student performs in a reading item influences the performance. If the comprehension of the question were not possible, it would be impossible to give any response. Or if comprehension were partial or wrong, an irrelevant and/or wrong response would result.

The pedagogical implications of locally dependent items such as these are: (1) students may be deprived of an opportunity to respond to the item, and (2) a wrong/partial answer may be penalised twice.

In addition to the three issues which have been brought to the attention of the researcher, in the course of present investigation, old unresolved problems confronted the researcher as well. Again, they are not resolved but two of them are reported here as problems yet to be investigated. They are: (1) allocating weight to test items, and (2) marker inferences.

One of the test writers' perpetual tasks is the valid allocation of the weight assigned to each of the test items that should indicate the relative difficulty level in comparison to other items in the test. One way to refine an observable performance in order to assign a number to a particular ability is to itemise discrete knowledge and skills of which the performance to be measured is made up. In assigning numbers to various reading and writing abilities in this study, an attempt has been made to refine the abilities measured to an extent that only the minimum inferences were necessary by the marker (see Output 6-1). In spite of the attempt, however, some items needed inferences.

The second problem confronted the researcher is marker inferences. Regardless of the nature of data, either quantitative or qualitative, in marking human performance in education, it is inevitable that instances arise where the markers must resort to their power of inferences no matter how refined the characteristics that are being observed (Brossell, 1983; Wilkinson, 1983; Bachman, 1990; Scarino, 1995; Bachman & Palmer, 1996). Every allocation of a number to a performance demands some degree of abstraction; therefore, the abilities that are being measured must be refined. However, in research such as this study which investigates human behaviour, there is a limit to that refinement and the judgment relies on the marker's inferences.

Another issue brought to the surface by the Rasch model is the identification of items that violate local independence and this is discussed below.

The last section of this paper discusses various implications of the findings, the implication for theories, teaching, teacher education and future research.

6.3 Implications for theories

The findings of this study suggest that performance in reading and writing is unidimensional. This adds further evidence to the long debated nature of linguistic competence by providing evidence that some underlying common skills are in force in the process of learning the different aspects of reading and writing a L2. The unidimensionality of reading and writing, which this project suggests, may imply that the comprehensible input hypothesis (Krashen, 1982) is fully supported, although many applied linguists like Swain (1985) and Shanahan (1984) believe that comprehensible output or the explicit instruction on linguistic production is necessary for the transfer of skills to take place.

As scholars in linguistics and psychology state, theories of reading and writing are still inconclusive (Krashen, 1984, p. 41; Clarke 1988; Hamp-Lyons, 1990; Silva, 1990, p. 8). The present study suggests added urgency for the construction of theories.

6.4 Implications for teaching

The performance recorded by the university group 2 students, who had studied Japanese for only one year compared to year 12 students who had five years of language study, indicated that approximately equal proficiency could be reached in one year of intense study. As concluded by Carroll (1975), commencing foreign language studies earlier seems to have little advantage, except for the fact that these students have longer total hours of study to reach a higher level. The results of this study suggest two possible implications. Firstly, secondary school learners may possess much more potential to acquire competence in five years of their language study than presently required since one year of tertiary study could take the learners to almost the same level.

Secondly, the total necessary hours of language study need to be considered. For an Asian language like Japanese, to reach a functional level, it is suggested by the United States Foreign Services Institute that 2000 to 2500 hours are needed. If the educational authorities are serious about producing Asian literate school leavers and university graduates, the class hours for foreign language learning must be reconsidered on the basis of evidence produced by such an investigation as this.

6.5 Implications for teacher education

The quality of language teachers, in terms of their own linguistic proficiency, background linguistic knowledge, and awareness of learning in general, is highlighted in the literature (Nunan, 1988, 1991; Leal, 1991; Nicholas, 1993; Elder & Iwashita, 1994; Language Teachers, 1996; Iwashita & Elder, 1997). The teachers themselves believe in the urgent need for improvement in these areas. Therefore, teacher education must be considered seriously in setting up a set of more stringent criteria for qualification as a language teacher, especially in proficiency in the target language and knowledge in linguistics.

6.6 Suggestions for future research

While progress in language learning appears to be almost linear, growth achieved by the year 9 students is an exception. In order to discover the reasons for unexpected linguistic growth achieved by these students, further research is necessary.

The university 2 group students' linguistic performance has reached a level quite close to that of the year 12 students, in spite of their limited period of exposure to the language. A further research project involving the addition of one more group of students who are in their third year of a university course could indicate whether, by the end of their third year, these students would reach the same level as the other students who, by then, would have had seven years of exposure to the language. If this were the case, further implications are possible on the commencement of language studies and the ultimate linguistic level that is to be expected and achieved by the secondary and tertiary students. Including primary school language learners in a project similar to the present one would indicate what results in language teaching are being achieved across the whole educational spectrum.

Aside from including one more year level, namely, the third year of university students, a similar study that extended its horizon to a higher level of learning, say to the end of intermediate level, would contribute further to the body of knowledge. The present study focused on only the beginner's level, and consequently limited its thinking in terms of transfer, linguistic proficiency and their implications. A future study focused on later stages of learning could reveal the relationship between the linguistic threshold level which, it is suggested, plays a role for transfer to take place, and the role which higher-order cognitive skills play. These include pragmatic competence, reasoning, content knowledge and strategic competence (Shaw & Li, 1997).

Another possible future project would be, now that a powerful and robust statistical model for measuring linguistic gains has been identified, to investigate the effectiveness of different teaching methods (Nunan, 1988). As Silva (1990, p. 18) stated 'research on relative effectiveness of different approaches applied in the classroom is nonexistent'.

7. CONCLUSION

To date, the outcomes of students' foreign language learning are unknown. It is overdue, in order to plan for the future, to examine the results of substantial public and private resources directed to the foreign language education, not to mention the time and effort spent by the students and teachers.

This study, in quite a limited scale, suggested a possible direction in order to measure linguistic gains achieved by the students whose proficiency varied greatly from the very beginning level to the intermediate level. The capabilities and possible application of the Rasch model demonstrated in this study added confidence in the use of extant softwares for educational research agenda. The Rasch model deployed in this study has proven to be not only appropriate, but also powerful in measuring linguistic growth achieved by students across six different year levels. By using a computer software QUEST (Adams & Khoo, 1993), the tests that were measuring different difficulty levels were successfully equated by using common test items contained in the tests of adjacent year levels. Rasch analysis also examined the test items routinely to check whether they measure the same traits as the rest of the test items and deleted those that did not. The results of the study imply that the same procedures could confidently be applied to measure learning outcomes, not limited to the studies of languages, but in other areas of learning. Furthermore, the pedagogical issues which need consideration and which have not yet received much attention in testing were made explicit by the Rasch model. This study may be considered as groundbreaking work in terms of establishing the basic direction such as identifying the instruments to measure proficiency as well as being a tool for the statistical analysis.

It is hoped that the appraisal of foreign language teaching practices commences as a matter of urgency in order to reap the maximum result from the daily effort of teachers and learners in the classrooms.

8. REFERENCES

Adams, R. & S-T Khoo (1993) QUEST: The Interactive test analysis system. Melbourne: ACER.

Asian languages and Australia's economic future. A report prepared for COAG on a proposed national Asian languages/studies strategies for Australian schools. [Rudd Report] Canberra: AGPS (1994).

Bachman, L. & Palmer, A.S. (1996) Language testing in practice: Oxford: Oxford University Press.

Bachman, L. (1990) Fundamental considerations in language testing. Oxford: Oxford University Press.

Bejar, I.I. (1983) Achievement testing: Recent advances. Beverly Hills, California: Sage Publication.

Brossell, G. (1983) Rhetorical specification in essay examination topics. College English, (45) 165-174.

Carroll, J.B. (1975) The teaching of French as a foreign language in eight countries. International studies in evaluation V. Stockholm:Almqvist & Wiksell International.

Clarke, M.A. (1988) The short circuit hypothesis of ESL reading – or when language competence interferes with reading performance. In P. Carrell, J. Devine & D. Eskey (Eds.).

Eckhoff, B. (1983) How reading affects childrens' writing. Language Arts, (60) 607-616.

Elder, C. & Iwashita, N. (1994) Proficiency Testing: a benchmark for language teacher education. Babel, (29) No. 2.

Gordon, C.J., & Braun, G. (1982) Story schemata: Metatextual aid to reading and writing. In J.A. Niles & L.A. Harris (Eds.). New inquiries in reading research and instruction. Rochester, N. Y.: National Reading Conference.

Hamp-Lyons, L. (1989) Raters respond to rhetoric in writing. In H. Dechert & G. Raupach. (Eds.). Interlingual processes. Tubingen: Gunter Narr Verlag.

Iwashita, N. and C. Elder (1997) Expert feedback: Assessing the role of test-taker reactions to a proficiency test for teachers of Japanese. In Melbourne papers in Language Testing, (6)1. Melbourne: NLLIA Language Testing Research Centre.

Kaplan, A. (1964) The Conduct of inquiry. San Francisco, California: Chandler.

Keeves, J. & Alagumalai, S. (1999) New approaches to measurement. In G. Masters, & J. Keeves. (Eds.).

Keeves, J. (Ed.) (1997) (2nd edt.) Educational research, methodology, and measurement: An international handbook. Oxford: Pergamon.

Krashen, S. (1982) Principles and practice in second language acquisition. Oxford: Pergamon.

Language teachers: The pivot of policy: The supply and quality of teachers of languages other than English. 1996. The Australian Language and Literacy Council (ALLC). National Board of Employment, Education and Training. Canberra: AGPS.

Leal, R. (1991) Widening our horizons. (Volumes One and Two). Canberra: AGPS.

McNamara, T. (1996) Measuring second language performance. London: Longman.

Nicholas, H. (1993) Languages at the crossroads: The report of the national inquiry into the employment and supply of teachers of languages other than English. Melbourne: The National Languages & Literacy Institute of Australia.

Nunan, D. (1988) The learner-centred curriculum. Cambridge: Cambridge University Press.

Rasch, G. (1960) Probabilistic models for some intelligence and attainment tests. Copenhagen: Danmarks Paedagogiske Institut.

Rudd, K.M. (Chairperson) (1994) Asian languages and Australian economic future. A report prepared for the Council of Australian Governments on a proposed national Asian languages/ studies strategies for Australian schools. Queensland: Government Printer.

Scarino, A. (1995) Language scales and language tests: development in LOTE. In Melbourne papers in language testing, (4) No. 2, 30-42. Melbourne: NLLIA.

Shaw, P & Li, E.T. (1997) What develops in the development of second – language writing? Applied Linguistics, 225 –253.

Silva. T. (1990) Second language composition instruction: developments, issues, and directions in ESL. In Kroll (Ed.). (1990).

Swain, M. (1985) 'Communicative competence: some roles of comprehensible input and comprehensible output in its development'. In S. Gass, S. & C. Madden (Eds.). Input in second language acquisition. Cambridge: Newbury House.

Taguchi, K. (2002) The linguistic gains across seven grade levels in learning Japanese as a foreign language. Unpublished EdD desertation, Flinders University: South Australia.

Umar, J. (1987) Robustness of the simple linking procedure in item banking using the Rasch model. (Doctorial dissertation, University of California: Los Angeles).

Weiss, D. J. & Yoes, M.E. (1991) Item response theory. In R. Hambleton, & J. Zaal. (Eds.). Advances in educational and psychological testing: Theory and applications. London: Kluwer Academic Publishers.

Wilkinson, A. (1983) Assessing language development: The Credition Project. In A. Freedman, I. Pringle, & J. Yalden (Eds.). Learning to write: First language/ second language. New York: Longman.

Wingersky, M. S., & Lord, F. (1984) An investigation of methods for reducing sampling error in certain IRT procedures. Applied Psychology Measurement, (8) 347-64.

Chapter 7

CHINESE LANGUAGE LEARNING AND THE RASCH MODEL

Measurement of students' achievement in learning Chinese

Ruilan Yuan
Oxley College, Victoria

Abstract: The Rasch model is employed to measure students' achievement in learning Chinese as a second language in an Australian school. Comparison between occasions and between year levels were examined. The performance in Chinese achievement tests and English word knowledge tests are discussed. The chapter highlights the challenges of equating multiple tests across levels and occasions.

Key words: Chinese language, Rasch scaling, achievement

1. INTRODUCTION

After World War II, and especially since the middle of the 1960s, when Australia's involvement in business affairs with some Asian countries in the Asian region started to occur, more and more Australian school students started to learn Asian languages. The Chinese language is one of the four major Asian languages taught in Australian schools. The other three Asian languages are Indonesian, Japanese and Korean. In the last 30 years, like other school subjects, some of the students who learned the Chinese language in schools achieved high scores in learning the Chinese language, and others were poor achievers. Some students continued learning the language to year 12, while most dropped out at different year levels. Therefore, it is considered worth investigating what factors influence student achievement in the Chinese language. The factors might be many or various, such as school factors, factors related to teachers, classes and peers. This

S. Alagumalai et al. (eds.), Applied Rasch Measurement: A Book of Exemplars, 115–137.

study, however, only examines student-level factors that influence achievement in learning the Chinese language.

The Chinese language program has been introduced into Australian school systems since the 1960s. Several research studies have noted factors influencing students' continuing with the learning of the Chinese language as a school subject in Australian schools (Murray & Lundberg, 1976; Fairbank & Pegalo, 1983; Tuffin & Wilson, 1990; Smith et al., 1993). Although various reasons are reported to influence continuing with the study of Chinese, such as attitudinal roles, peer and family pressure, gender difference, lack of interest in languages, the measurement of students' achievement growth in learning the Chinese language across year levels and over time, and the investigation of factors influencing such achievement growth have not been attempted. Indeed, measuring students' achievement across year levels and over time is important in the teaching of the Chinese language in Australian school systems because it may provide greater understanding of the actual learning that occurs and enable comparisons to be made between year levels and the teaching methods employed at different year levels.

2. DESIGN OF THE STUDY

The subjects for this study were 945 students who learned the Chinese language as a school subject in a private college of South Australia in 1999. The instruments employed for data collection were student background questionnaires and attitude questionnaires, four Chinese language tests, and three English word knowledge tests. All the data were collected during the period of one full school year in 1999.

3. PURPOSE OF THE STUDY

The purpose of this chapter is to examine students' achievement in learning Chinese between and within year levels on four different occasions; and to investigate whether proficiency of English word knowledge influences the achievement level of Chinese language.

In the present study, in order to measure students' achievement in learning the Chinese language across years and over time, a series of Chinese tests were designed and administered to each year from year 4 to year 12 as well as over school terms from term 1 to term 4 in the 1999 school year. It was necessary to examine carefully the characteristics of the test items before the test scores for each student who participated in the study could be calculated in appropriate ways, because meaningful scores were essential for the subsequent analyses of the data collected in the study.

In this chapter the procedures of analysing the data sets collected from the Chinese language achievement tests and English word knowledge tests are discussed, and the results of the Rasch analyses of these tests are presented. It is necessary to note that the English word knowledge tests were administered to students who participated in the study in order to examine whether the level of achievement in learning the Chinese language was associated with proficiency in English word knowledge and the student's underlying verbal ability in English (Thorndike, 1973a).

This chapter comprises four sections. The first section presents the methods used for the calculation of scores. The second section considers the equating procedures, while the third section examines the differences in scores between year levels, and between term occasions. The last section summarises the findings from the examination of the Chinese achievement tests and English word knowledge tests obtained from the Rasch scaling procedures. It should be noted that the data obtained from the student questionnaires are examined in Chapter 8.

4. THE METHODS EMPLOYED FOR THE CALCULATION OF SCORES

It was considered desirable to use the Rasch measurement procedures in this study in order to calculate scores and to provide an appropriate data set for subsequent analyses through the equating of the Chinese achievement tests, English word knowledge tests and attitude scales across years and over time. In this way it would be possible to generate the outcome measures for the subsequent analyses using PLS and multilevel analysis procedures. Lietz (1995) has argued that the calculation of scores using the Rasch model makes it possible to increase the homogeneity of the scales across years and over occasions so that scoring bias can be minimised.

4.1 Use of Rasch scaling

The Rasch analyses were employed in this study to measure (a) the Chinese language achievement of students across eight years and over four term occasions, (b) English word knowledge tests across years, and (c) attitude scales between years and across two occasions. The examination of the attitude scales is undertaken in the next chapter. The estimation of the scores received from these data sets using the Rasch model involved two different procedures, namely, calibration and scoring, which are discussed below.

The raw scores on a test for each student were obtained by adding the number of points received for correct answers to each individual item in the test, and were entered into the SPSS file. In respect to the context of the current study, the calculation of these raw scores did not permit the different tests to be readily equated. In addition, the difficulty levels of the items were not estimated on an interval scale. Hence, the Rasch model was employed to calculate appropriate scores to estimate accurately the difficulty levels of the items on a scale that operated across year levels and across occasions. In the Chinese language achievement tests, and English word knowledge tests, omissions of the items were considered as wrong.

4.2 Calibration and equating of tests

This study used vertical equating procedures so that achievement of students in learning the Chinese language at different year levels could be measured on the same scale. The horizontal equating approach was also employed to measure student achievement across the four term occasions. In addition, two different types of Rasch model equating, namely, anchor item equating and concurrent equating, were employed at different stages in the equating processes. The equating of the Chinese achievement tests requires common items between years and across terms. The equating of the English word knowledge tests requires common items between the three tests: namely, tests 1V, 2V and 3V. The following section reports the results of calibrating and equating of the Chinese tests between years and across the four occasions as well as the equating of English word knowledge tests between years.

4.2.1 Calibration and scoring of tests

There were eight year level groups of students who participated in this study (year 4 to year 12). A calibration procedure was employed in this study in order to estimate the difficulty levels (that is, threshold values) of the items in the tests, and to develop a common scale for each data set. In the calibration of the Chinese achievement test data and English word knowledge test data in this study, three decisions were made. Firstly, the calibration was done with data for all students who participated in the study. Secondly, missing items or omitted items were treated as wrong in the Chinese achievement test and the English word knowledge test data in the calibration. Finally, only those items that fitted the Rasch scale were employed for calibration and scoring. This means that, in general, the items whose infit mean square values were outside an acceptable range were deleted from the calibration and scoring process. Information on item fit estimates and individual person fit estimates are reported below.

4.2.2 Item fit estimates

It is argued that Rasch analysis estimates the degree of fit of particular items to an underlying or latent scale, and that the acceptable range of item fit taken in this study for each item in the three types of instruments, in general, was between 0.77 and 1.30. The items whose values were below 0.77 or above 1.30 were generally considered outside the acceptable range. The values of overfitting items are commonly below 0.77, while the values of misfitting items are generally over 1.30. In general, the misfitting items were excluded from the calibration analysis in this study, while in some cases it was considered necessary and desirable for overfitting items to remain in the calibrated scales.

It should be noted that the essay writing items in the Chinese achievement tests for level 1 and upward that were scored out of 10 marks or more were split into sub-items with scores between zero to five. For example, if a student scored 23 on one writing item, the extended sub-item scores for the student were 5, 5, 5, 4, and 4. The overfitting items in the Chinese achievement tests were commonly those subdivided items whose patterns of response were too predictable from the general patterns of response to other items.

Table 7-1 presents the results of the Rasch calibration of Chinese achievement tests across years and over the four terms. The table shows the total number of all the items for each year level and each term, the numbers of items deleted, the anchor items across terms, the bridge items between year levels and the number of items retained for analysis.

The figures in Table A in Output 7-1 show that 17 items (6.9% of the total items) did not fit the Rasch model and were removed from the term 1 data file. There were 46 items (13%) that were excluded from the term 2 data file, while 33 items (13%) were deleted from the term 3 data file. A larger number of 68 deleted items was seen in the term 4 data file (22%). This might be associated with the difficulty levels of the anchor items across terms and the bridge items between year levels because students were likely to forget what had been learned previously after learning new content. In addition, there were some items that all students who attempted these items answered correctly, and such items had to be deleted because they provided no information for the calibration analysis.

As a result of the removal of the misfitting items from the data files for the four terms (the items outside the accep5 range of 0.77 and 1.30, 237 items for the term 1 tests; 317 items for the term 2 tests; 215 items for the term 3 tests; and 257 items for the term 4 tests) fitted the Rasch scale. They were therefore retained for the four separate calibration analyses. There was, however, some evidence that the essay type items fitted the Rasch model less well at the upper year levels. Table 7-1 provides the details of the numbers of both anchor items and bridge items that satisfied the Rasch scaling requirement after deletion of misfitting items. The figures show that 33 out of 40 anchor items fitted the Rasch model for term 1 and were linked to the term 2 tests. Out of 70 anchor items in term 2, 64 anchor items were retained, among which 33 items were linked to the term 1 tests, and 31 items were linked to the term 3 tests. Of 58 anchor items in the term 3 data file, 31 items were linked to the term 2 tests, and 27 items were linked to the term 4 tests.

The last column in Table 7-2 provides the number of bridge items between year levels for all occasions. There were 20 items for years 4 and 5; 43 items between years 5 and 6; 32 items between year 6 and level 1; 31 items between levels 2 and 2; 30 items between levels 2 and 3; 42 items between levels 3 and 4; and 26 items between levels 4 and 5. The relatively small numbers of items linking between particular occasions and particular year levels were offset by the complex system of links employed in the equating procedures used.

Table 7-1 Final number of anchor and bridge items for analysis

Level	Term 1		Term 2		Term 3		Term 4		Total
	A	B	A	B	A	B	A	B	B
Year 4	4	5	4	5	14	5	4	5	20
Year 5	5	18	10	10	10	10	5	5	43
Year 6	2	7	7	10	10	10	5	5	32
Level 1	5	6	10	10	15	10	10	5	31
Level 2	5	8	8	9	3	8	0	5	30
Level 3	5	18	8	8	4	8	1	8	42
Level 4	4	10	4	8	2	5	2	3	26
Level 5	3	10	13	4	-	-	-	-	14
Total	**33**	-	**64**	-	**58**	-	**27**	-	**238**

Notes:
A = anchor items
B = bridge items

In the analysis for the calibration and equating of the tests, the items for each term were first calibrated using concurrent equating across the years and the threshold values of the anchor items for equating across occasions were estimated. Thus, the items from term 1 were anchored in the calibration of the term 2 analysis, and the items from term 2 were anchored in the term 3 analysis, and the items from term 3 were anchored in the term 4 analysis. This procedure is discussed further in a later section of this chapter.

Table 7-2 summarises the fit statistics of item estimates and case estimates in the process of equating the Chinese achievement tests using anchor items across the four terms. The first panel shows the summary of item estimates and item fit statistics, including infit mean square, standard deviation and infit t, as well as outfit mean square, standard deviation and outfit t. The bottom panel displays the summary of case estimates and case fit statistics as well as infit and outfit results.

Table 7-2 Summary of fit statistics between terms on Chinese tests using anchor items

Statistics	Terms 1/2	Terms 2/3	Terms 3/4
Summary of item estimates and fit statistics			
Mean	0.34	1.62	1.51
SD	1.47	1.92	1.87
Reliability of estimate	0.89	0.93	0.93
Infit mean square			
Mean	1.06	1.03	1.01
SD	0.37	0.25	0.23
Outfit mean square			
Mean	1.10	1.08	1.10
SD	0.69	0.56	1.04
Summary of case estimates and fit statistics			
Mean	0.80	1.70	1.47
SD	1.71	1.79	1.81
SD (adjusted)	1.62	1.71	1.73
Reliability of estimate	0.90	0.91	0.92
Infit mean square			
Mean	1.05	1.03	1.00
SD	0.6	0.30	0.34
Infit t			
Mean	0.20	0.13	0.03
SD	1.01	1.06	1.31
Outfit mean square			
Mean	1.11	1.12	1.11
SD	0.57	0.88	1.01
Outfit t			
Mean	0.28	0.24	0.18
SD	0.81	0.84	1.08

4.2.3 Person fit estimates

Apart from the examination of item fit statistics, the Rasch model also permits the investigation of person statistics for fit to the Rasch model. The item response pattern of those persons who exhibit large outfit mean square values and t values should be carefully examined. If erratic behaviour were detected, those persons should be excluded from the analyses for the calibration of the items on the Rasch model (Keeves & Alagumalai, 1999). In the data set of the Chinese achievement tests, 27 out of 945 cases were deleted from term 3 data files because they did not fit the Rasch scale. The high level of satisfactory response from the students tested resulted from the fact that in general the tests were administered as part of the school's normal testing program, and scores assigned were clearly related to school years awarded. Moreover, the HLM computer program was able to compensate appropriately for this small amount of missing data.

4.2.4 Calculation of zero and perfect scores

Zero scores received by a student on a test indicate that the student answered all the items incorrectly, while perfect scores indicate that a student answered all the items correctly. Since students with perfect scores or zero scores are considered not to provide useful information for the calibration analysis, the QUEST computer program (Adams & Khoo, 1993) does not include such cases in the calibration process. In order to provide scores for the students with perfect or zero scores and so to calculate the mean and standard deviation for the Chinese achievement tests and the English word knowledge tests for each student who participated in the study, it was necessary to estimate the values of the perfect and zero scores.

In this study, the values of perfect and zero scores in the Chinese achievement and English word knowledge tests were calculated from the logit tables generated by the QUEST computer program. Afrassa (1998) used the same method to calculate the values of the perfect and zero scores of the mathematics achievement tests. The values of the perfect scores were calculated by selecting the three top raw scores close to the highest possible score. For example, if the highest raw score was 48, the three top raw scores chosen were 47, 46 and 45. After the three top raw scores were chosen, the second highest value of logit (2.66) was subtracted from the first highest logit value (3.22) to obtain the first entry (0.56). Then the third highest logit value (2.33) was subtracted from the second highest logit value (2.66) to gain the second entry (0.33). The next step was to subtract the second entry (0.33) from the first entry (0.56) to obtain the difference between the two entries (0.23). The last step was to add the first highest logit value (3.22) and the first entry (0.56) and the difference between the two entries (0.23) so that the highest score value of 4.01 was estimated. Table 7-3 shows the procedures used for calculating perfect scores.

Table 7-3 Estimation of perfect scores

Scores	Estimate (logits)		Entries		Difference		Perfect score value
47	3.22						
46	2.66		0.56				
45	2.33		0.33		0.23		
MAX = 48	**3.22**	+	**0.56**	+	**0.23**	=	**4.01**

The same procedure was employed to calculate zero scores except that the three lowest raw scores and logit values closest to zero were chosen (that is, 1, 2 and 3) and subtractions were conducted from the bottom. Table 7-4 presents the data and the estimated zero score value using this procedure. The entry -1.06 was estimated by subtracting -5.35 from -6.41, and the entry -0.67 was obtained by subtracting -4.68 from -5.35. The difference -0.39 was estimated by subtracting -0.67 from -1.06, while the zero score value of-7.86 was estimated by adding -6.41 and -1.06 and -0.39.

Table 7-4 Estimation of zero scores

Scores	**Estimate (logits)**	**Entries**	**Difference**	**Zero score value**
3	-4.68			
2	-5.35	-0.67		
1	-6.41	-1.06	-0.39	
MIN 0	**-6.41** +	**-1.06** +	**-0.39**	**-7.86**

The above section discusses the procedures for calculating scores of the Chinese achievement and English word knowledge tests using the Rasch model. The main purposes of calculating these scores are to: (a) examine the mean levels of all students' achievement in learning the Chinese language between year levels and across term occasions, (b) provide data on the measures for individual students' achievement in learning the Chinese language between terms for estimating individual students' growth in learning the Chinese language over time, and (c) test the hypothesised models of student-level factors and class-level factors influencing student achievement in learning the Chinese language. The following section considers the procedures for equating the Chinese achievement tests between years and across terms, as well as the English word knowledge tests across years.

4.3 Equating of the Chinese achievement tests between terms

Table A in Output 7-1 shows the number of anchor items across terms and bridge items between years as well as the total number and the number of deleted items. The anchor items were required in order to examine the achievement growth of the same group of students over time, while the bridge items were developed so that the achievement growth between years could be estimated. It should be noted that the number of anchor items was greater in terms 2 and 3 than in terms 1 and 4. This was because the anchor items in term 2 included common items for both term 1 and term 3, and the anchor items in term 3 included common items for both term 2 and term 4, whereas term 1 only provided common items for term 2, and term 4 only had common items from term 3. Nevertheless, the relatively large number of linking items employed relatively small numbers involved in particular links overall.

The location of the bridge items in a test remained the same as their location in the lower year level tests for the same term. For example, items 28 to 32 were bridge items between year 5 and year 6 in the term 1 tests, and their numbers were the same in the tests at both levels. The raw responses of the bridge items were entered under the same item numbers in the SPSS data file, regardless of different year levels and terms. However, the anchor items were numbered in accordance with the items in particular year levels and different terms. This is to say that the anchor items in year 6 for term 2 were numbered 10 to 14, while in term 3 test they might be numbered 12 to 16, depending upon the design of term 3 test. It can be seen in Table A that the number of bridge items varied slightly. In general, the bridge items at one year level were common to the two adjacent year levels. For example, there were 10 bridge items in year 5 for the term 2 test. Out of the 10 items, five were from the year 4 test, and the other five were linked to the year 6 test. Year 4 only had five bridge items each term because it only provided common items for year 5.

In order to compare students' Chinese language achievement across year levels and over terms, the anchor item equating method was employed to equate the test data sets of terms 1, 2, 3 and 4. This is done by initially estimating the item threshold values for the anchor items in the term 1 tests. These threshold values were then fixed for these anchor items in the term 2 tests. Thus, the term 1 and term 2 data sets were first equated, followed by equating the terms 2 and 3 data files by fixing the threshold values of their common anchor items.

Finally terms 3 and 4 data were equated. In this method, the anchor items in term 1 were equated using anchor item equating in order to obtain appropriate thresholds for all items in term 2 on the scale that had been defined for term 1. In this way the anchor items in term 2 were able to be anchored at the thresholds of those corresponding anchor items in term 1. The same procedures were employed to equate terms 2 and 3 tests, as well as terms 3 and 4 tests. In other words, the threshold values of anchor items in the previous term scores were estimated for equating all the items in the subsequent term. It is clear that the tests for terms 2, 3, and 4 are fixed to the zero point of the term 1 tests. Zero point is defined to be the average difficulty level of the term 1 items used in calibration of the term 1 data set. Tables 7-6 to 7-7 present the anchor item thresholds used in equating procedures between terms 1, 2, 3 and 4. In Table 7-5, the first column shows the number of anchor items in the term 2 data set, the second column displays the number of the corresponding anchor items in the term 1 data, and the third column presents the threshold value of each anchor item in the term 1 data file. It is necessary to note that level 5 data were not available for terms 3 and 4 because the students at this level were preparing for year 12 SACE examinations. As a consequence, the level 5 data were not included in the data analyses for term 3 and term 4. The items at level 2 misfitted the Rasch model and were therefore deleted. Level 5 tests were not available for terms 3 and 4.

Table 7-5 Description of anchor item equating between terms 1 and 2

Term 2 items	Term 1 items	Thresholds
Year 4		
Item 2	3	anchored at -1.96
Item 3	4	anchored at -1.24
Item 4	5	anchored at -2.61
Item 5	6	anchored at -1.50
Year 5		
Item 16	18	anchored at 0.46
Item 17	2	anchored at -2.25
Item 18	19	anchored at -0.34
Item 19	20	anchored at 1.04
Item 20	6	anchored at -1.50
Year 6		
Item 29	22	anchored at -0.72
Item 30	23	anchored at 0.20
Level 1		
Item 46	47	anchored at 0.36
Item 47	48	anchored at -1.59
Item 48	49	anchored at 1.13
Item 49	50	anchored at 0.22
Item 50	51	anchored at 0.34
Level 2		
Item 95	66	anchored at -3.22
Item 96	70	anchored at -1.03
Item 97	79	anchored at -1.86
Item 98	78	anchored at -0.97
Item 99	76	anchored at -1.15
Level 3		
Item 175	126	anchored at -1.40
Item 176	127	anchored at 1.37
Item 177	128	anchored at -0.44
Item 178	129	anchored at -0.86
Item 179	130	anchored at 1.83
Level 4		
Item 240	142	anchored at -1.38
Item 241	143	anchored at 0.25
Item 243	144	anchored at -0.50
Item 244	145	anchored at -0.50
Level 5		
Item 302	137	anchored at 0.53
Item 304	139	anchored at 1.80
Item 306	141	anchored at 2.44
Total 33 items		

Note:
probability level = 0.50

5. EQUATING OF ENGLISH WORD KNOWLEDGE TESTS

Concurrent equating was employed to equate the three English language tests: namely, tests 1V, 2V and 3V. Test 1V was admitted to students at years 4 to 6 and level 1, test 2V was admitted to students at levels 2 and 3, and test 3V was completed by levels 4 and 5 students. In the process of equating, the data from the three tests were combined into a single file so that the analysis was conducted with one data set.

In the analyses of tests 1V and 3V, item 11 and item 95 misfitted the Rasch scale. However, when the three tests were combined by common items and analysed in one single file, both items fitted the Rasch scale. Consequently, no item was deleted from the calibration analysis.

Table 7-6 Description of anchor item equating between terms 2 and 3

Term 3 items	Term 2 items	Thresholds
Year 4		
Item 2	2	anchored at -2.25
Item 3	3	anchored at -1.96
Item 4	4	anchored at -1.24
Item 5	5	anchored at -2.61
Item 6	6	anchored at -1.50
Item 7	7	anchored at -4.00
Item 8	8	anchored at -3.95
Item 9	9	anchored at -3.58
Item 10	10	anchored at -1.55
Item 11	11	anchored at -0.25
Item 12	12	anchored at -1.91
Year 5		
Item 18	21	anchored at -0.34
Item 19	22	anchored at 1.04
Item 20	23	anchored at 0.34
Item 21	24	anchored at -0.31
Item 22	25	anchored at 1.11
Year 6		
Item 38	31	anchored at -1.49
Item 39	34	anchored at 0.64
Item 40	37	anchored at -0.47
Item 41	39	anchored at -0.35
Item 42	40	anchored at -0.30
Level 1		
Item 48	41	anchored at 0.45
Item 49	42	anchored at -0.26
Item 50	43	anchored at 0.76
Item 51	44	anchored at 0.61
Item 52	45	anchored at 0.69
Level 2		
Item 107	100	anchored at 0.28
Item 108	101	anchored at 2.43
Item 111	104	anchored at -0.67
Level 3		
Item 158	187	anchored at 1.14
Item 159	188	anchored at 1.14
Total 31 items		

Notes:

probability level = 0.50

Items at levels 4 and 5 misfitted the Rasch model and were therefore deleted.

Table 7-7 Description of anchor item equating between terms 3 and 4

Term 4 items	Term 3 items	Thresholds
Year 4		
Item 2	3	anchored at -1.96
Item 3	4	anchored at -1.24
Item 4	5	anchored at -2.61
Item 5	6	anchored at -1.50
Year 5		
Item 26	28	anchored at 0.70
Item 27	29	anchored at 1.12
Item 28	30	anchored at 0.25
Item 29	31	anchored at 1.27
Item 30	32	anchored at 1.17
Year 6		
Item 31	30	anchored at 0.25
Item 32	31	anchored at 1.27
Item 33	32	anchored at 1.17
Item 34	43	anchored at -0.48
Item 35	44	anchored at 1.30
Level 1		
Item 41	58	anchored at 2.03
Item 42	59	anchored at 2.60
Item 43	60	anchored at 1.79
Item 44	61	anchored at 2.37
Item 45	62	anchored at 1.60
Item 46	63	anchored at 0.38
Item 47	64	anchored at 0.84
Item 48	65	anchored at 0.80
Item 49	66	anchored at 0.48
Item 50	67	anchored at 0.59
Level 3		
Item 185	165	anchored at 5.46
Level 4		
Item 234	188	anchored at 2.38
Item 236	190	anchored at 3.41
Total 27 items		

Note:
probability level = 0.50

There were 34 common items between the three tests, of which 13 items were common between tests 1V and 2V, whereas 21 items were common between tests 2V and 3V. Furthermore, all the three test data files shared two of the 34 common items. The thresholds of the 34 items obtained during the calibration were used as anchor values for equating the three test data files and for calculating the Rasch scores for each student. Therefore, the 120 items became 86 items after the three tests were combined into one data file.

In the above sections, the calibration, equating and calculation of scores of both the Chinese language achievement tests and English word knowledge tests are discussed. The section that follows presents the comparisons of students' achievement in learning the Chinese language across year levels and over the four school terms, as well as the comparisons of the English word knowledge results across year levels.

6. DIFFERENCES IN THE SCORES ON THE CHINESE LANGUAGE ACHIEVEMENT TESTS

The comparisons of students' achievement in learning the Chinese language were examined in the following three ways: (a) comparisons over the four occasions, (b) comparisons between year levels, and (c) comparisons within year levels. English word knowledge tests results were only compared across year levels because the tests were administered on only one occasion.

6.1 Comparisons between occasions

Table 7-8 shows the scores achieved by students on the four term occasions, and Figure 7-1 shows the achievement level by occasions graphically. It is interesting to notice that the figures indicate general growth in student achievement mean score between terms 1 and 2 (by 0.53), terms 2 and 3 (by 0.84), whereas an obvious drop in the achievement mean score is seen between terms 3 and 4 (by 0.17). The drop of achievement level in term 4 might result from the fact that some students had decided to drop out from learning the Chinese language in the next year; thus they ceased to put an effort into the learning of the Chinese language.

Table 7-8 Average Rasch scores on Chinese achievement tests by term

Term 1	**Term 2**	**Term 3**	**Term 4**
0.43	**0.96**	**1.80**	**1.63**
N=781	**N=804**	**N=762**	**N=762**

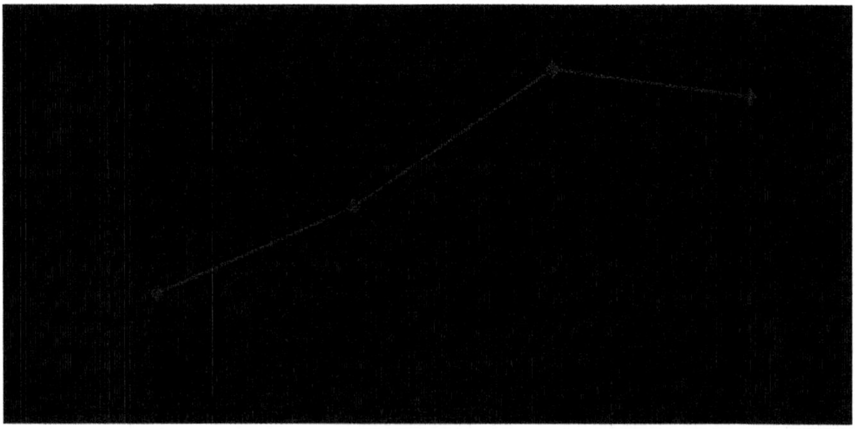

Figure 7-1. Chinese achievement level by four term occasions

6.2 Comparisons between year levels on four occasions

This comparison was made between year levels on the four different occasions. After scoring, the mean score for each year was calculated for each occasion. Table 7-9 presents the mean scores for the students at year 4 to year 6, and level 1 to level 5, and shows increased achievement levels between the first three terms. However, the achievement level decreases in term 4 for year 4, year 5, level 1, level 3, and level 4. The highest level of achievement for these years is, in general, on the term 3 tests. The achievement level for students in year 6 is higher for term 1 than for term 2. However, sound growth is observed between term 2 and term 3, and term 3 and term 4.

It is of interest to note that the students at level 2 achieved a marked growth between term 1 and term 2: namely, from -0.07 to 2.27. The highest achievement level for this year is at term 4 with a mean score of 2.88. Students at level 4 are observed to have achieved their highest level in term 3. The lowest achievement level for this year is at term 2. Because of the inadequate information provided for level 5 group, it was not considered possible to summarise the achievement level for that year. Figure 6.2 presents the differences in the achievement levels between year levels on four occasions based on the scores for each year as well as for each term in Table 7-9. Figure 7-2 below presents the mean differences in the achievement levels between years over the four occasions.

Table 7-9 Average Rasch scores on Chinese tests by term and by year level

TERM LEVEL	Term 1	Term 2	Term 3	Term 4	Mean
Year 4	-0.93	-0.15	0.35	0.09	-0.17
Year 5	0.20	0.52	1.86	1.47	1.01
Year 6	0.81	0.69	0.85	1.06	0.85
Level 1	0.73	0.95	2.33	1.77	1.45
Level 2	-0.07	2.27	2.13	2.88	1.80
Level 3	1.71	2.65	4.62	4.30	3.32
Level 4	1.35	0.31	4.64	1.86	2.04
Level 5	2.43	1.58	-	-	-
No. of cases	**N=782**	**N=804**	**N N=762**	**N=762**	-

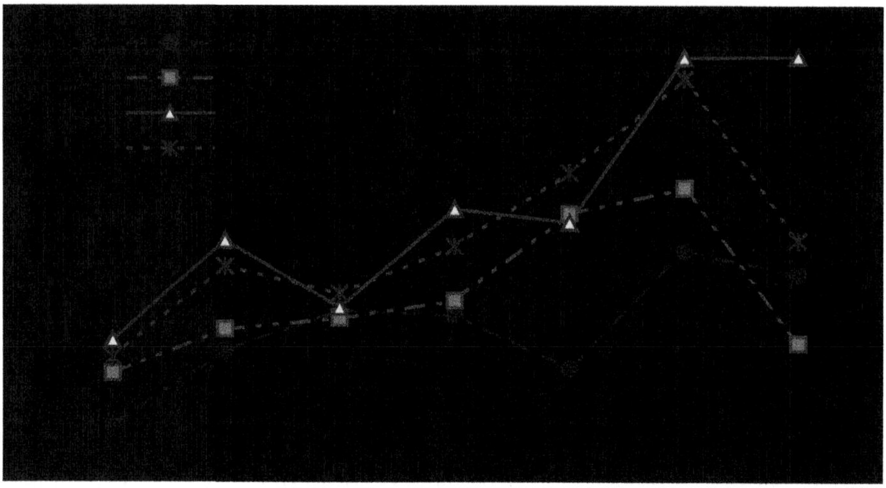

Figure 7-2. Description of achievement level between years on four occasions

Figures 7-1, 7-2 and 7-3 present graphically the achievement levels for each year for the four terms. Figure 7-1 provides a picture of students' achievement level on different occasions, while Figure 7-2 shows that there is a marked variability in the achievement level across terms between and within years. However, the general trend of a positive slope is seen for term 1 in Figure 7-2. A positive slope is also seen for performance at term 2 despite the noticeable drop at level 4. The slope of the graph for term 3 can be best described as erratic because a large decline occurs at year 6 and a slight decrease occurs at level 2. It is important to note that the trend line for term 4 reveals a considerable growth in the achievement level although it declines markedly at level 4.

Figure 7-3 presents the comparisons of the means, which illustrated the differences in student achievement levels between years. It is of importance to note that students at level 3 achieved the highest level among the seven year levels, followed by level 4, while students at year 4 were the lowest achievers as might be expected. This might be explained by the fact that four of the six year 4 classes learned the Chinese language only for two terms, namely, terms 1 and 2 in the 1999 school year. They learned French in terms 3 and 4.

Figure 7-3. Comparison of means in Chinese achievement between year levels

6.3 Comparisons within years on different occasions

This section compares the achievement level within each year. By and large, an increased trend is observed for each year level from term 1 to term 4 (see Figures 7-1, 7-2 and 7-4, and Table 7-10). Year 4 students achieve at a markedly higher level between terms 1, 2 and 3. The increase is 0.78 between term 1 and term 2, and 0.20 between term 2 and term 3. However, the decline between term 3 and term 4 is 0.26. Year 5 is observed to show a similar trend in the achievement level as year 4. The growth difference is 0.32 between term 1 and term 2. A highly dramatic growth difference is seen of 1.34 between term 2 and term 3. Although a decline of 0.39 is observed between term 3 and term 4, the achievement level in term 4 is still considered high in comparison with terms 1 and 2.

The tables and graphs above show a consistent growth in achievement level for year 6 except for a slight drop in term 2. The figures for achievement at level 1 reveal a striking progress in term 3 followed by term 4, and a consistent growth is shown between terms 1 and 2. At level 2 while a poor level of achievement is indicated in term 1, considerably higher levels are achieved for the subsequent terms. The students at level 3 achieve a remarkable level of performance across all terms even though a slight decline is observed in term 4. The achievement level at level 4 appears unstable because a markedly low level and extremely high level are achieved in term 2 and term 3, respectively. Figure 4 shows the achievement levels within year levels on the four occasions.

Despite the variability in the achievement levels on the different occasions at each year level, a common trend is revealed with a decline in achievement that occurs in term 4. This decline might have resulted from the fact that some students had decided to drop out from the learning of the Chinese language for the next year, and therefore ceased to put in effort in term 4. With respect to the differences in the achievement level between and within years on different occasions, it is more appropriate to withhold comment until further results from subsequent analyses of other relevant data sets are available.

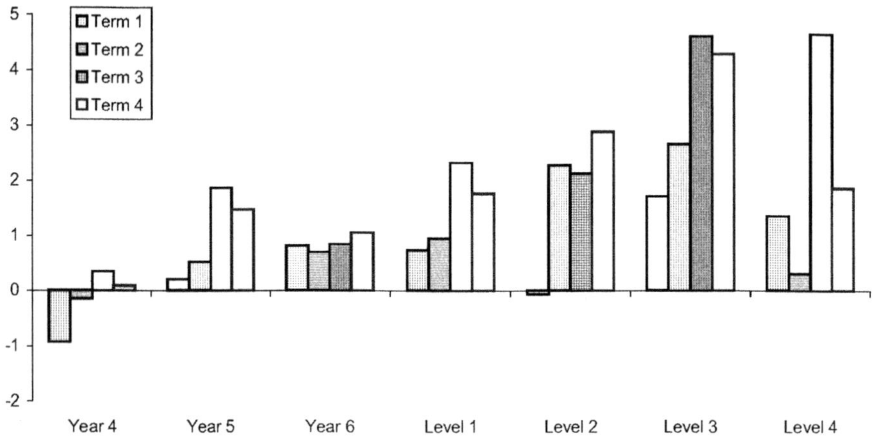

Figure 7-4. Description of achievement level within year levels on four occasions

7. DIFFERENCE IN ENGLISH WORD KNOWLEDGE TESTS BETWEEN YEAR LEVELS

Three English word knowledge tests were administered to students who participated in this study in order to investigate the relationship between the Chinese language achievement level and proficiency in the English word knowledge tests. Table 7-10 provides the results of the scores by each year level, and Figure 7-5 presents graphically the trend in English word knowledge proficiency between year levels using the scores recorded in Table 7-10.

Table 7-10 Average Rasch score on English word knowledge tests by year level

Level	Number of students (N)	Scores
Year 4	154	-0.20
Year 5	167	0.39
Year 6	168	0.63
Level 1	158	0.70
Level 2	105	1.13
Level 3	46	1.33
Level 4	22	1.36
Level 5	22	2.07
Total	**842 (103 cases missing)**	**Mean = 0.93**

Table 7-10 presents the mean Rasch scores on the combined English word knowledge tests for the eight year levels. It is of interest to note the general improvement in English word knowledge proficiency between years. The difference is 0.59 between years 4 and 5; 0.24 between years 5 and 6; 0.07 between year 6 and level 1; 0.43 between levels 1 and 2; 0.20 between levels 2 and 3; a small difference of 0.03 between levels 3 and 4; and a large increase between levels 4 and 5.

It is also of interest to notice the marked development in the English word knowledge proficiency between year levels. Large differences occur between years 4 and 5, as well as between levels 4 and 5. Medium or slight differences occur between other years: namely, between years 5 and 6; year 6 and level 1; levels 1 and 2; levels 2 and 3; and levels 3 and 4. The differences between year levels, whether large or small, are to be expected because, as students grow older and move up a year, they learn more words and develop their English vocabulary and thus may be considered to advance in verbal ability.

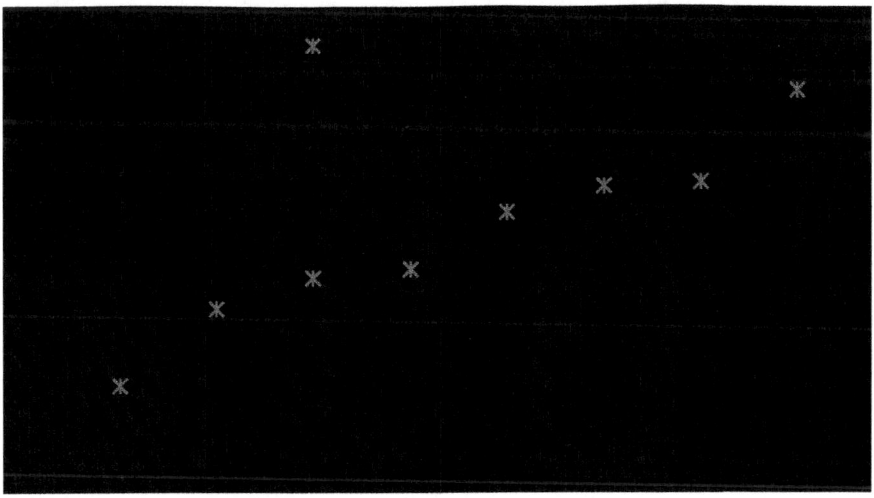

Figure 7-5. Graph of scores on English word knowledge tests across year levels

In order to examine whether the development in the Chinese language is associated with the development in English word knowledge proficiency as well as to investigate the interrelationship between the two languages, students' achievement level in the Chinese language and development of English word knowledge (see Figures 7-3 and 7-5) are combined to produce Figure 7-6. The combined lines indicate that the development of the two languages is by an large interrelated except for the drops at year 6 from year 5 and at level 4 from level 3 in the level of achievement in learning the Chinese language. This suggests that both students' achievement in the Chinese language and development in English word knowledge proficiency generally increase across year levels.

It should be noted that both sets of scores are recorded on logit scales in which the average levels of difficulty of the items within the scales determine the zero or fixed point of the scales. It is thus fortuitous that the scale scores for the students in year 4 for English word knowledge proficiency and performance on the tests of achievement in learning the Chinese language are so similar.

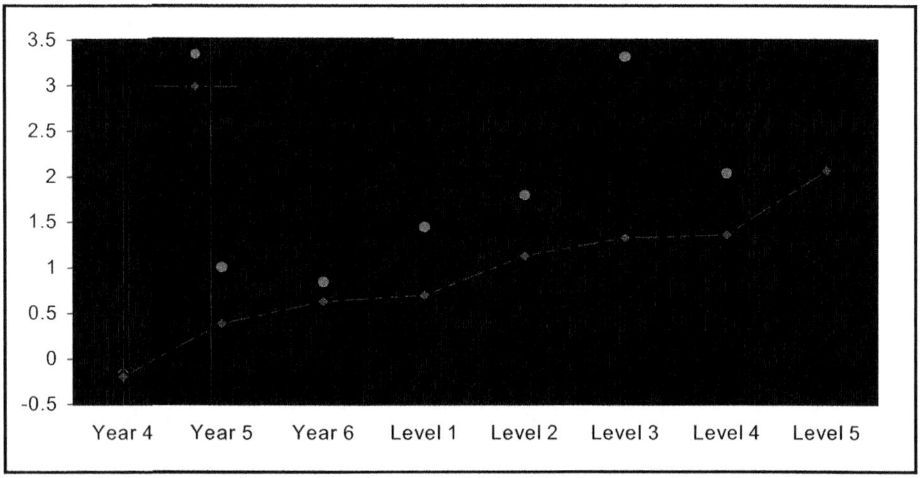

Figure 7-6. Comparison between Chinese and English scores by year levels

8. CONCLUSION

In this chapter the procedures of scaling the Chinese achievement tests and English word knowledge tests are discussed, and followed by the presentation of the results of the analyses of both the Chinese achievement data and English word knowledge test data. The findings from the two types of tests are summarised in this section.

Firstly, the students' achievement level in the Chinese language generally improves between occasions though there is a slight decline in term 4. Secondly, overall, the higher the year level in the Chinese language, the higher the achievement level. Thirdly, within year achievement level in learning the Chinese language for each year level indicates a consistent improvement across the first three terms but shows a decline in term 4. The decline in performance for term 4, particularly of students at level 4, may have been a consequence of the misfitting essay type items that were included in the tests at this level, which resulted in an underestimate of the students' scores. Finally, the achievement level in learning the Chinese language appears to be associated with the development of English word knowledge: namely, the students at higher year levels are higher achievers in both English word knowledge and in the Chinese language tests.

Although a common trend is observed in the findings for both the Chinese language achievement and English word knowledge development, differences still exist within and between year levels as well as across terms. It is considered necessary to investigate what factors gave rise to such differences. Therefore, the chapter that follows focuses on the analysis of the student questionnaires in order to identify whether students' attitudes towards the learning of the Chinese language and schoolwork might play a role in learning the Chinese language and in the achievement levels recorded.

Nevertheless, the work of calibrating and equating so many tests across so many different levels and so many occasions is to some extent a hazardous task. There is the clear possibility of errors in equating, particularly in situations where relatively few anchor items and bridge items are being employed. However, stability and strength are provided by the requirement that items must fit the Rasch model, which in general they do well.

9. REFERENCES

Adams, R. and Khoo, S-T. (1993). *Quest: The Interactive Test Analysis System*, Melbourne: ACER.

Afrassa, T. M. (1998). *Mathematics achievement at the lower secondary school stage in Australia and Ethiopia: A comparative study of standards of achievement and student level factors influencing achievement.* Unpublished Doctoral Thesis. School of Education, The Flinders University of South Australia, Adelaide.

Anderson, L.W. (1992). Attitudes and their measurement. In J.P.Keeves (ed.), *Methodology and Measurement in International Educational Surveys: The IEA Technical Handbook.* The Netherlands: the Hague, pp.189-200.

Andrich, D. (1988). *Rasch Models for Measurement.* Series: Quantitative applications in the social sciences. Newbury Park, CA: Sage Publications.

Andrich, D. and Masters, G. N. (1985). Rating scale analysis. In T. Husén and T. N. Postlethwaite (eds.), *The International Encyclopedia of Education.* Oxford: Pergamon Press, pp. 418-4187.

Angoff, W. H. (1982). Summary and derivation of equating methods used at ETS. In P. W. Holland and D. B. Rubin (eds.), *Test Equating.* New York: Academic Press, pp. 55-69.

Auchmuty, J. J. (Chairman) (1970). *Teaching of Asian Languages and Cultures in Australia.* Report to the Minister for Education. Canberra: Australian Government Publishing Service (AGPS).

Australian Education Council (1994). *Languages other than English: A Curriculum Profile for Australian Schools.* A joint project of the States, Territories and the Commonwealth of Australia initiated by the Australian Education Council. Canberra: Curriculum Corporation.

Baker, F. B. and Al-Karni (1991). A comparison of two procedures for computing IRT equating coefficients. *Journal of Educational Measurement*, 28 (2), 147-162.

Baldauf, Jr., R. B. and Rainbow, P. (1995). *Gender Bias and Differential Motivation in LOTE Learning and Retention Rates: A Case Study of Problems and Materials.* Canberra: DEET

Birnbaum, A. (1968). Some latent trait models and their use in inferring an examinee's ability. In F. Lord and M.Novick, *Statistical Theories of Mental Test Scores.* Reading MA: Addison-Wesley, pp.397-472.

Bourke, S. F. and Keeves, J. P. (1977). *Australian Studies in School Performance: Volume III, the Mastery of Literacy and Numeracy, Final Report.* Canberra: AGPS.

Buckby, M. and Green, P. S. (1994). Foreign language education: Secondary school programs. In T. Husén and T.N. Postlethwaite (eds.), *The International Encyclopedia of Education (2nd. edn.)* Oxford: Pergamon Press, pp. 2351-2357.

Carroll, J. B. (1963a). A model of school learning. *Teachers College Record*, 64, 723-733.

Carroll, J. B. (1963b). Research on teaching foreign languages. In N. L. Gage (ed.), *Handbook of Research on Teaching.* Chicago: Rand McNally, pp. 1060-1100.

Carroll, J. B. (1967). *The Foreign Language Attainments of Language Majors in the Senior Year: A Survey Conducted in U.S. Colleges and Universities.* Cambridge, Mass: Laboratory for Research in Instruction, Graduate School of Education, Harvard University.

Fairbank, K. and Pegalo, C. (1983). *Foreign Languages in Secondary Schools.* Queensland: Queensland Department of Education.

Murray, D. and Lundberg, K. (1976). *A Register of Modern Language Teaching in South Australia.* INTERIM REPORT, Document No. 50/76, Adelaide.

Keeves, J. & Alagumalai, S. (1999) *New approaches to measurement.* In G. Masters, & J. Keeves. (Eds.). *Advances in Measurement in Educational Research and Assessment.* Amsterdam: Pergamon.

Smith, D., Chin, N. B., Louie, K., and Mackerras. C. (1993). *Unlocking Australia's Language Potential: Profiles of 9 Key Languages in Australia, Vol. 2: Chinese.* Canberra: Commonwealth of Australia and NLLIA.

Thorndike, R. L. (1973a). *Reading Comprehension Education in Fifteen Countries.* International Studies in Evaluation III. Stockholm, Sweden: Almqvist & Wiksell.

Thorndike, R.L. (1982). *Applied Psychometrics.* Houghton Mifflin Company: Boston.

Tuffin, P. and Wilson, J. (1990). *Report of an Investigation into Disincentives to Language Learning at the Senior Secondary Level.* Commissioned by the Asian Studies Council, Adelaide.

Chapter 8

EMPLOYING THE RASCH MODEL TO DETECT BIASED ITEMS

Njora Hungi
Flinders University

Abstract: In this study, two common techniques for detecting biased items based on Rasch measurement procedures are demonstrated. One technique involves an examination of differences in threshold values of items among groups and the other technique involves an examination of fit of item in different groups.

Key words: Item bias, DIF, gender differences, Rasch model, IRT

1. INTRODUCTION

There are cases in which some items in a test have been known to be biased against a particular subgroup of the general group being tested and this fact has become a matter of considerable concern to users of test results (Hambleton & Swaminathan, 1985; Cole & Moss, 1989; Hambleton, 1989). This concern is regardless of whether the test results are intended for placement or selection or are just as indicators of achievement in the particular subject. The reason for this is apparent, especially considering that test results are generally taken to be a good indicator of a person's ability level and performance in a particular subject (Tittle, 1988). Under these circumstances it is clearly necessary to apply item bias detection procedures to 'determine whether the individual items on an examination function in the same way for two groups of examinees' (Scheuneman & Bleistein, 1994, p. 3043). Tittle (1994) notes that the examination of a test for bias towards groups is an important part in the evaluation of the overall instrument as it influences not only testing decisions, but also the use of the test results. Furthermore, Lord and Stocking (1988) argue that it is important to detect

S. Alagumalai et al. (eds.), Applied Rasch Measurement: A Book of Exemplars, 139–157.

biased items as they may not measure the same trait in all the subgroups of the population to which the test is administered.

Thorndike (1982, p. 228) proposes that 'bias is potentially involved whenever the group with which a test is used brings to test a cultural background noticeably different from that of the group for which the test was primarily developed and on which it was standardised'. Since diversity in the population is unavoidable, it is logical that those concerned with ability measurements should develop tests that would not be affected by an individual's culture, gender or race. It would be expected that, in such a test, individuals having the same underlying level of ability would have equal probability of getting an item correct, regardless of their subgroup membership.

In this study, real test data are used to demonstrate two simple techniques for detecting biased items based on Rasch measurement procedures. One technique involves examination of differences in threshold values of items among subgroups (to be called "item threshold approach') and the other technique involves an examination of infit mean square values (INFT MNSQ) of the item in different subgroups (to be called 'item fit approach').

The data for this study were collected as part of the South Australian Basic Skills Testing Program (BSTP) in 1995, which involved 10 283 year 3 pupils and 10 735 year 5 pupils assessed in two subjects; literacy and numeracy. However, for the purposes of this study, a decision was made to use only data from the pupils who answered all the items in the 1995 BSTP (that is, 3792 and 3601 years 3 and 5 pupils respectively). This decision was based on findings from a study carried out by Hungi (1997), which showed that the amount of missing data in the 1995 BSTP varied considerably from item to item at both year levels and that there was a clear tendency for pupils to omit certain items. Consequently, Hungi concluded that item parameters taken considering all the students who participated in these tests were likely to contain more errors compared to those taken considering only those students who answered all items.

The instruments used to collect data in the BSTP consisted of a student questionnaire and two tests (a numeracy test and a literacy test). The student questionnaire sought to gather information regarding background characteristics of students (for example, gender, race, English spoken at home and age). The numeracy test consisted of items that covered three areas (number, measurement and space), while the literacy test consisted of two sub-tests (language and reading). Hungi (1997) examined the factor structure of the BSTP instruments and found strong evidence to support the existence of (a) a numeracy factor and not clearly separate number, measurement, and space factors, and (b) a literacy factor and clearly separate language and reading factors. Hence, in this study, the three aspects of numeracy are considered together and the two separate sub-tests of literacy are considered separately.

This study seeks to examine the issues of item bias in the 1995 BSTP sub-tests (that is, numeracy, reading and language) for years 3 and 5. For purposes of parsimony, the analyses described in this study focus on detection of items that exhibited gender bias. A summary of the number of students who took part in the 1995 BSTP, as well as those who answered all the items in the tests by the gender groups, is given in Table 8-1.

Table 8-1. Year 3 and year 5 students' genders

	Year 3				Year 5			
	All cases		Completed cases		All cases		Completed cases	
	N	%	N	%	N	%	N	%
Boys	5158	50.2	1836	48.4	5425	50.5	1685	46.8
Girls	5125	49.8	1956	51.6	5310	49.5	1916	53.2
Total cases	10 283		3792		10 735		3601	

Notes:
There were no missing responses to the question.
All cases—All the students who participated in the BSTP in South Australia
Completed cases—Only those students who answered all the items in the tests

2. MEANING OF BIAS

Osterlind (1983) argues that the term 'bias' when used to describe achievement tests has a different meaning from the concept of fairness, equality, prejudice, preference or any other connotations sometimes associated with its use in popular speech. Osterlind states:

> Bias is defined as systematic error in the measurement process. It affects all measurements in the same way, changing measurement - sometimes increasing it and other times decreasing it. ... Bias, then, is a technical term and denotes nothing more or less than consistent distortion of statistics. (Osterlind, 1983, p. 10)

Osterlind notes that in some literature the terms 'differential item performance' (DIP) or 'differential item functioning' (DIF) are used instead of item bias. These alternative terms suggest that the item function differently for different groups of students and this is the appropriate meaning attached to the term 'bias' in this study.

Another suitable definition based on item response theory is the one given by Hambleton (1989, p. 189): 'a test is unbiased if the item characteristic curves across different groups are identical'. Equally suitable is the definition provided by Kelderman (1989):

A test item is biased if individuals with the same ability level from different groups have a different probability of a right response: that is, the item has different difficulties in different subgroups (Kelderman, 1989, p. 681).

3. GENERAL METHODS

An important requirement before carrying out the analysis to detect biased items in a test is the assumption that all the items in the test conform to a unidimensional model (Vijver & Poortinga, 1991). Thus, ensuring that the test items measure a single attribute of the examinees. Osterlind (1983, p. 13) notes that 'without the assumption of unidimensionality, the interpretation of item response is profoundly complex'. Vijver and Poortinga (1991) say that, in order to overcome this unidimensionality problem, it is common for most bias detection studies to assume the existence of a common scale, rather than demonstrate it. For this study, the tests being analysed were shown to conform to the unidimensionality requirement in a study carried out by Hungi (1997).

There are numerous techniques available for detecting biased items. Interested readers are referred to Osterlind (1983), and Cole and Moss (1989) for an extensive discussion of the methods for detecting biased item. Generally, the majority of the methods for detecting biased items fall either under (a) the classical test theory (CTT), or under (b) the item response theory (IRT). Outlines of the popular methods are presented in the next two sub-sections.

3.1 Classical tests theory methods

Adams (1992) has provided a summary of the methods used to detect biased items within the classical test theory framework. He reports that among the common methods are (a) the ANOVA method, (b) transformed item difficulties (TID) method, and (c) the Chi-square method. Another popular classical test theory-based technique is the Mantel-Haenszel (MH) procedure (Hambleton & Rogers, 1989; Klieme & Stumpf, 1991; Dorans & Holland, 1992; Ackerman & Evans, 1994; Allen & Donoghue, 1995; Parshall & Miller, 1995).

Previous studies indicate that the usefulness of the classical theory approaches in detecting items that are biased cannot be underestimated (Narayanan & Swaminathan, 1994; Spray & Miller, 1994; Chang, 1995; Mazor, 1995). However, several authors, including Osterlind, have noted that 'several problems for biased item work persist in procedures based on classical test models' (Osterlind, 1983, p. 55). Osterlind indicates that the main problem is the fact that a vast majority of the indices used for detection of biased items are dependent on the sample of students under study. In addition, Hambleton and Swaminathan (1985) argue that classical item approaches to the study of item bias have been unsuccessful because they fail to handle adequately true ability differences among groups of interest.

3.2 Item response theory methods

Several research findings have indicated support to employment of IRT approaches in detection of item bias (Pashley, 1992; Lautenschlager, 1994; Tang, 1994; Zwick, 1994; Kino, 1995; Potenza & Dorans, 1995). Osterlind (1983, p.15) describes the IRT test item bias approaches as 'the most elegant of all the models discussed to tease out test item bias'. He notes that the assumption of IRT concerning item response function (IRF) makes it suitable for identifying biased items in a test. He argues that, since item response theory is based on the use of items to measure a given dominant latent trait, items in an unidimensional test must measure the same trait in all subgroups of the population. A similar argument is provided by Lord and Stocking (1988):

> Items in a test that measures a single trait must measure the same trait in all subgroups of the population to which the test is administered. Items that fail to do so are biased for or against a particular subgroup. Since item response functions in theory do not depend upon the group used for calibration, item response theory provides a natural method for detecting item bias. (Stocking, 1997, p. 839)

Kelderman (1989, p.681) relates the strength of IRT models in the detection of test item bias to their 'clear separation of person ability and item difficulty'.

In summary, there seems to be a general agreement amongst researchers that IRT approaches have an advantage over CTT approaches in the detection of item bias. However, it is common for researchers to apply CTT approaches and IRT approaches in the same study to detect biased items either for comparison purposes or to enhance the precision of detecting the biased items (Cohen & Kim, 1993; Rogers & Swaminathan, 1993; Zwick, 1994).

Within the IRT framework, Osterlind (1983) reports that, to detect a biased item in a test taken by two subgroups, the item response function for a particular item is estimated separately for each group. The two curves are then compared. Those items that are biased would have curves that would be significantly different. For example, one subgroup's ICC could be clearly higher when compared to the other subgroup. In such a case, the members of the subgroup with an ICC that is higher would stand a greater chance of getting the item correct at the same ability level. Osterlind notes that in practice the size of the area between curves is considered as the measure of bias because it is common to get variable probability differences across the ability levels, which result in inter-locking ICCs.

However, Pashley (1992) argued that the use of the area between the curves as a measure of bias considers only the overall item level DIF information, and does not indicate the location and magnitude of DIF along the ability continuum. He consequently proposed a method for producing simultaneous confidence bands for the difference between item response curves, which he termed as 'graphical IRT-based DIF analyses'. He also argued that, after these bands had been plotted, the size and regions of DIF were easily identified.

For studies (such as the current one) whose primary aim is to identify items that exhibit an overall bias regardless of the ability level under consideration, it seems sufficient to regard the area between the curves as an appropriate measure of bias. The Pashley technique needs to be applied where additional information about the location and magnitude of bias along the ability continuum is required.

Whereas the measure of bias using IRT is the area between the ICCs, the aspects actually examined to judge whether an item is biased or not are (a) item discrimination, (b) item difficulty, and (c) guessing parameters of the item (Osterlind, 1983). These three parameters are read from the ICCs of the item under analysis for the two groups being compared. If any of the three parameters differ considerably between the two groups under comparison, the item is said to be biased because the difference between these parameters can be translated as indicating differences in probability of responding correctly to the item between the two groups.

4. METHODS BASED ON THE RASCH MODEL

Within the one-parameter IRT (Rasch) procedures, Scheuneman and Bleistein (1994) report that the two most common procedures used for evaluating item bias examine either the differences in item difficulty (threshold values) between groups or item discrimination (INFT MNSQ values) in each group. This is because the guessing aspect mentioned above is not examined when dealing with the Rasch model. The proponents of the Rasch model argue that guessing is a characteristic of individuals and not the items.

4.1 Item threshold approach

Probably the easiest index to employ in the detection of a biased item is the difference between the threshold values (difficulty levels) of the item in the two groups. If the difference in the item threshold values is noticeably large, it implies that the item is particularly difficult for members of one of the groups being compared, not because of their different levels of achievement, but due to other factors probably related to being members of that group.

There is no doubt that the major interest in the detection of item bias is the difference in the item's difficulty levels between two subgroups of a population. However, as Scheuneman (1979) suggests, it takes more than the simple difference in item difficulty to infer bias in a particular item. Thorndike (1982) agrees with Scheuneman that:

> In order to compare the difficulty of the items in a pool of items for two (or more) groups, it is first necessary to convert the raw percentage of correct answers for each item to a difficulty scale in which the units are approximately equal. The simplest procedure is probably to calculate the Rasch difficulty scale values separately for each group. If the set of items is the same for each group, the Rasch procedure has the effect of setting the mean scale value at zero within each group, and then differences in scale value for any item become immediately apparent. Those items with largest differences in a scale value are the suspect items. (Thorndike, 1982, p. 232)

Through use of the above procedure, items that are unexpectedly hard, as well as those unexpectedly easy for a particular subgroup, can be identified.

4.2 Item fit approach

Through use of the Rasch model, all the items are assumed to have equal discriminating power as that of the ideal ICC. Therefore, all items should have infit mean square (INFT MNSQ) values equal to unity or within a predetermined range, regardless of the groups of students used. However, some items may record INFT MNSQ values outside the predetermined range, depending on the subgroup of the general population being tested. Such items are considered to be biased as they do not discriminate equally for all subgroups of the general population being tested.

The main problem with the employment of an item fit approach in identification of biased items is the difficulty in the determination of the possible bias. With the item threshold approach, an item found to be more difficult for a group than the other items in a test is biased against that group. When, however, the item's fit in the two groups is compared, such straightforward interpretation of bias cannot be made (see Cole and Moss, 1989, pp.211–212).

5. BIAS CRITERIA EMPLOYED

The main problem in detection of item bias within the IRT framework, as noted by Osterlind (1983), is the complex computations that require the use of computers. This is equally true for item bias detection approaches based on the CTT. The problem is especially critical for analysis involving large data sets such as the current study. Consequently, several computer programs have been developed to handle the detection of item bias. The main computer software employed in item bias analysis in this study is QUEST (Adams & Khoo, 1993).

The Rasch model item bias methods available using QUEST involve (a) the comparison of item threshold levels between any two groups being compared, and (b) the examination of the item's fit to the Rasch model in any two groups being compared.

In this study, the biased items are identified as those that satisfy the following requirements.

5.1 For item threshold approach

1. Items whose difference in threshold values between two groups are outside a pre-established range. Two major studies carried out by Hungi (1997, 2003) found that the growth in literacy and numeracy achievement between years 3 and 5 in South Australia is about 0.50 logits per year. Consequently, a difference of ±0.50 logits in item threshold values between two groups should be considered substantial because it represents a difference of one year of school learning between the groups: that is,

$$d_1 - d_2 > \pm 0.50 \qquad (1)$$

where:
d_1 = the item's threshold value in group 1, and
d_2 = the item's threshold value in group 2.

2. Items whose differences in standardised item threshold between any of the groups fall outside a predefined range. Adams and Khoo (1993) have employed the range -2.00 to +2.00: that is,

$$\text{st (d1 - d2)} > \pm 2.00 \qquad (2)$$

where:
st = standardised

For large samples (greater than 400 cases), it is necessary to adjust the standardised item threshold difference. The adjusted standardised item threshold difference can be calculated by using the formula below:

$$\text{Adjusted standardised difference} = \text{st(d1 - d2)} \div [N/400]^{0.5} \qquad (3)$$

where:
N = pooled number of cases in the two groups,

The purpose of dividing by the parameter $[N/400]^{0.5}$ is to adjust the standardised item threshold difference to reflect the level it would have taken were the sample size approximately 400. For this study, the cutoff values (calculated using Formula 3 above) for the adjusted standardised item threshold difference for the year 3 as well as the year 5 data are presented in Table 8-2.

Table 8-2. Cut off values for the adjusted standardised item threshold difference

	Number of cases	Cut off values	
		Lower limit	Upper limit
Year 3	3,792	-6.16	6.16
Year 5	3,601	-6.00	6.00

It is necessary to discard all the items that do not conform to the model employed before identifying biased items (Vijver & Poortinga, 1991). Consequently, items outside a predefined INFT MNSQ value would need to be discarded when employing the item difficulty technique to identify biased items within the Rasch model framework.

5.2 For item fit approach

Through the use of QUEST, the misfitting (and therefore biased) items are identified as those items whose INFT MNSQ values are outside a predefined range for a particular group of students. Based on extensive experience, Adams and Khoo (1993), as well as McNamara (1996), advocated for INFT MNSQ values in the range of approximately 0.77 to 1.30, the range employed in this study.

The success of identifying biased items using the criterion of an item's fit relies on the requirement that all the items in the test being analysed have adequate fit when the two groups being compared are considered together. In other words, the item should be identified as misfitting only when used in a particular subgroup and not when used in the general population being tested. Hence, items that do not have adequate fit to the Rasch model when used in the general population should be dropped before proceeding with the detection of biased items.

In this study, all the items recorded INFT MNSQ values within the desired range (0.77–1.30) when data from both gender groups were analysed together and, therefore, all the items were involved in the item bias detection analysis.

6. RESULTS

Tables 8-3 and 8-4 present examples of results of the gender comparison analyses carried out using QUEST for years 3 and 5 numeracy tests. In these tables, starting from the left, the item being examined is identified, followed by its INFT MNSQ values in 'All' (boys and girls combined). The next two columns record the INFT MNSQ of the item in boys only and girls only. The next set of columns list information about the items' threshold values, starting with the;

1. the items' threshold value for boys (d_1);
2. the items' threshold value for girls (d_2);
3. the difference between the threshold value of the item for boys and the threshold value of the item for girls (d_1-d_2); and
4. the standardised item threshold differences {st(d_1-d_2)}.

The tables also provide the rank order correlation coefficients (ρ) between the rank orders of the item threshold values for boys and for girls.

Pictorial representation of the information presented in the Tables 8-3 and 8-4 is provided in Figure 8-1 and Figure 8-2. The figures are plots of the standardised differences generated by QUEST for comparison of the performance of the boys and girls in the Basic Skills Tests items for years 3 and 5 numeracy tests.

Osterlind (1983), as well as Adams and Rowe (1988), have described the use of rank order correlation coefficient as an indicator of item bias. However, they have termed the technique as 'quick but incomplete' and it is only useful as an initial indicator of item bias. Osterlind says that:

> For correlations of this kind one would look for rank order correlation coefficients of .90 or higher to judge for similarity in ranking of item difficulty values between groups. (Osterlind, 1983, p. 17)

The observed rank order correlation coefficients were ≥0.95 for all the sub-tests (that is, numeracy, language and reading) in the year 3 test, as well as in the year 5 test. These results indicated that there were no substantial changes in the order of the items according to their threshold values when considering boys compared to the order when considering girls. Osterlind (1983) argues that such high correlation coefficients should reduce the suspicion of the existence of items that might be biased. Thus, using this strategy, it would appear that gender bias was not an issue in any of the sub-tests of the 1995 Basic Skills Tests at either year level.

Table 8-3. Year 3 numeracy test (item bias results)

Item	INFT MNSQ approach			Threshold approach			
	All	Boys	Girls	Boys	Girls		
	(INFT MNSQ)	(INFT MNSQ)	(INFT MNSQ)	(d_1)	(d_2)	d_1-d_2	st(d_1-d_2)
y3n01	1.01	1.01	1.01	-0.68	-0.60	-0.09	-0.74
y3n02	0.98	0.96	1.00	-1.67	-1.43	-0.24	-1.43
y3n03	1.00	0.96	1.01	0.32	1.10	-0.78 [a]	-9.28 [b]
y3n04	0.98	0.99	0.97	-1.34	-1.24	-0.10	-0.66
y3n05	0.93	0.93	0.93	0.88	0.75	0.14	1.71
y3n06	1.02	1.00	1.05	2.35	2.35	0.01	0.08
y3n07	0.96	0.97	0.96	-0.64	-0.30	-0.34	-3.03
y3n08	1.03	1.02	1.04	-0.59	-0.28	-0.30	-2.72
y3n09	0.94	0.95	0.93	-0.43	-0.81	0.38	3.19
y3n10	1.07	1.10	1.05	0.13	-0.15	0.28	2.86
y3n11	0.93	0.92	0.93	0.13	-0.02	0.14	1.51
y3n12	1.07	1.07	1.07	-1.54	-1.60	0.06	0.34
y3n13	1.09	1.11	1.07	0.85	0.79	0.06	0.74
y3n14	0.99	0.97	0.99	-1.63	-1.23	-0.40	-3.71
y3n15	1.00	1.00	0.99	-1.05	-0.75	-0.30	-2.27
y3n16	0.97	0.98	0.97	-1.11	-1.30	0.19	1.29
y3n17	0.93	0.93	0.92	0.65	1.05	-0.40	-4.94
y3n18	1.02	1.00	1.03	1.24	1.19	0.05	0.64
y3n19	0.92	0.91	0.93	2.66	2.68	-0.02	-0.26
y3n20	0.98	0.98	0.99	-1.41	-1.64	0.23	1.39
y3n21	1.13	1.11	1.15	2.12	2.17	-0.05	-0.69
y3n22	0.89	0.88	0.90	-1.03	-1.28	0.25	1.76
y3n23	1.02	1.06	0.98	-0.05	0.15	-0.20	-2.10
y3n24	1.03	1.04	1.03	0.14	0.20	-0.07	-0.71
y3n25	1.01	1.00	1.01	0.27	0.05	0.21	2.29
y3n26	0.98	1.00	0.96	-1.20	-1.49	0.30	1.91
y3n27	1.05	1.06	1.03	-1.14	-1.63	0.48	3.05
y3n28	1.01	1.01	1.00	0.18	-0.02	0.20	2.07
y3n29	0.96	0.96	0.96	2.46	2.30	0.16	2.08
y3n30	0.91	0.90	0.92	0.86	0.73	0.13	1.62
y3n31	1.04	1.02	1.06	-0.53	-0.70	0.16	1.38
y3n32	0.99	1.01	0.97	0.94	0.87	0.07	0.83
				$\rho =$	0.95		

Notes:
All items had INFT MNSQ value within the range 0.77 - 1.30
a difference in item difficulty outside the range ±0.50
b adjusted st(d1-d2) outside the range ±6.16
ρ rank order correlation
All All Students who answered all items (N = 3792)
Boys (N = 1836)
Girls (N = 1956)

Table 8-4. Year 5 numeracy test (item bias results)

	INFT MNSQ approach			Threshold approach			
	All	Boys	Girls	Boys	Girls		
	(INFT MNSQ)	(INFT MNSQ)	(INFT MNSQ)	(d_1)	(d_2)	d_1-d_2	st(d_1-d_2)
y5n01	0.98	0.97	0.99	-2.22	-2.25	0.02	0.09
y5n02	0.99	0.99	0.99	-2.03	-2.38	0.36	1.37
y5n03	0.99	0.99	0.99	-0.35	-0.78	0.43	5.09
y5n04	1.03	1.05	1.01	1.16	1.47	-0.31	-4.00
y5n05	0.98	0.98	0.98	0.11	0.10	0.01	0.15
y5n06	1.11	1.11	1.10	1.89	1.70	0.19	2.62
y5n07	1.06	1.07	1.06	1.63	1.52	0.10	1.41
y5n08	0.92	0.91	0.93	-1.57	-1.12	-0.45	-2.51
y5n09	1.00	0.98	1.01	-0.20	-0.43	0.23	1.98
y5n10	1.01	1.02	1.01	-1.00	-0.78	-0.22	-1.53
y5n11	0.99	1.03	0.96	-0.99	-1.15	0.16	1.05
y5n12	1.15	1.17	1.13	0.67	0.74	-0.07	-0.85
y5n13	1.04	1.04	1.04	-0.17	0.10	-0.27	-2.52
y5n14	0.99	0.98	1.01	0.65	0.43	0.22	2.45
y5n15	1.00	1.00	0.99	-3.26	-3.41	0.16	0.35
y5n16	1.05	1.06	1.04	1.89	1.93	-0.03	-0.45
y5n17	1.04	1.03	1.06	0.35	0.34	0.02	0.17
y5n18	1.01	1.03	0.99	-1.00	-1.43	0.43	4.63
y5n19	1.09	1.10	1.08	2.39	2.69	-0.30	-4.02
y5n20	1.03	1.05	1.01	0.67	0.03	0.64 [a]	6.75 [b]
y5n21	0.97	0.97	0.97	1.08	1.33	-0.25	-3.20
y5n22	0.95	0.97	0.94	-0.77	-0.79	0.02	0.13
y5n23	1.01	1.01	1.02	2.38	2.98	-0.60 [a]	-8.03 [b]
y5n24	0.96	0.94	0.98	0.67	1.08	-0.41	-4.87
y5n25	1.03	1.03	1.03	0.20	0.45	-0.25	-2.66
y5n26	1.00	0.99	1.01	1.29	1.32	-0.04	-0.45
y5n27	1.02	1.00	1.03	0.27	0.10	0.17	1.71
y5n28	0.93	0.94	0.92	-0.37	-0.73	0.35	2.78
y5n29	0.95	0.94	0.96	-0.50	-0.69	0.20	1.53
y5n30	0.93	0.92	0.94	3.16	3.31	-0.15	-1.82
y5n31	0.95	0.97	0.92	0.75	0.89	-0.14	-1.63
y5n32	1.00	1.00	1.00	-2.19	-2.28	0.09	0.34
y5n33	0.97	0.97	0.96	0.51	0.50	0.01	0.12
y5n34	0.96	0.96	0.96	-1.45	-1.17	-0.29	-1.65
y5n35	0.97	0.95	0.98	-1.20	-1.23	0.03	0.17
y5n36	0.91	0.91	0.90	-0.98	-1.32	0.34	2.12
y5n37	0.98	0.97	0.99	-0.42	-0.28	-0.14	-1.14
y5n38	0.95	0.95	0.96	1.11	0.93	0.18	2.26
y5n39	0.98	0.99	0.97	-2.34	-1.94	-0.40	-1.55
y5n40	1.00	0.96	1.03	-0.90	-0.79	-0.11	-0.76
y5n41	0.97	1.00	0.94	-0.43	-0.55	0.12	0.98
y5n42	1.07	1.07	1.07	-0.72	-0.49	-0.23	-1.76
y5n43	0.88	0.87	0.89	1.52	1.52	-0.01	-0.07
y5n44	0.95	0.95	0.94	-0.45	-0.34	-0.11	-0.92
y5n45	0.95	0.95	0.95	1.02	0.90	0.11	1.41
y5n46	1.04	1.05	1.03	-0.08	0.03	-0.11	-1.06
y5n47	1.00	0.99	1.00	0.76	1.11	-0.35	-4.27
y5n48	1.05	1.05	1.05	-1.05	-1.15	0.10	0.65
				$\rho=$	0.97		

Notes:

All items had INFT MNSQ value within the range 0.77–1.30

a difference in item difficulty outside the range ±0.50

b adjusted st(d1–d2) outside the range ±6.16

ρ rank order correlation

All All students who answered all items (N = 3601)

Boys (N = 1685)

Girls (N = 1916)

```
------------------------------------------------------------------------------------------
Comparison of item estimates for groups boys and girls on the numeracy scale
L = 32      order = input
------------------------------------------------------------------------------------------
                                  Plot of standardised differences
                            Easier for boys                  Easier for girls

      -10  -9   -8   -7  -6   -5   -4   -3   -2   -1    0    1    2    3    4    5    6
-------+----+----+----+----+----+----+----+----+----+----+----+----+----+----+----+----+---
 y3n01                     |                    .     *   |         .                    |
 y3n02                     |                    .  *      |         .                    |
§y3n03        *            |                    .         |         .                    |
 y3n04                     |                    .     *   |         .                    |
 y3n05                     |                    .         |        *.                    |
 y3n06                     |                    .         *         .                    |
 y3n07                     |               *    .         |         .                    |
 y3n08                     |                *   .         |         .                    |
 y3n09                     |                    .         |         .    *               |
 y3n10                     |                    .         |         .  *                 |
 y3n11                     |                    .         |      *  .                    |
 y3n12                     |                    .         |*        .                    |
 y3n13                     |                    .         |  *      .                    |
 y3n14                     |            *       .         |         .                    |
 y3n15                     |                   *.         |         .                    |
 y3n16                     |                    .         |     *   .                    |
 y3n17                     |      *             .         |         .                    |
 y3n18                     |                    .         |  *      .                    |
 y3n19                     |                    .       * |         .                    |
 y3n20                     |                    .         |     *   .                    |
 y3n21                     |                    .     *   |         .                    |
 y3n22                     |                    .         |        *.                    |
 y3n23                     |                   *.         |         .                    |
 y3n24                     |                    .     *   |         .                    |
 y3n25                     |                    .         |         . *                  |
 y3n26                     |                    .         |         *                    |
 y3n27                     |                    .         |         .   *                |
 y3n28                     |                    .         |         .*                   |
 y3n29                     |                    .         |         .*                   |
 y3n30                     |                    .         |      *  .                    |
 y3n31                     |                    .         |     *   .                    |
 y3n32                     |                    .         |   *     .                    |
==========================================================================================
```

Notes:

All items had INFT MNSQ value within the range 0.83–1.20

§ item threshold adjusted standardised difference outside the range ± 6.16

Inner boundary range ± 2.0

Outer boundary range ± 6.16

Figure 8-1. Year 3 numeracy item analysis (gender comparison)

From Tables 8-3 and 8-4, it is evident that all the items in the numeracy tests recorded INFT MNSQ values within the predetermined range (0.77 to 1.30) in boys as well as in girls. Similarly, all the items in the reading and language tests recorded INFT MNSQ values within the desired range. Thus, based on item INFT MNSQ criterion, it is evident that gender bias was not a problem in the 1995 BSTP.

A negative value of difference in item threshold (or difference in standardised item threshold) in Tables 8-3 and 8-4 indicate that the item was relatively easier for the boys than for the girls, while a positive value implies the opposite. Using this criterion, it is obvious that a vast majority of the year 3 as well as the year 5 test items were apparently in favour of one gender or the other. However, it is important to remember that a mere difference between threshold values of an item for boys and girls may not be sufficient evidence to imply bias for or against a particular gender.

Nevertheless, a difference in item threshold outside the ±0.50 range is large enough to cause concern. Likewise, differences in adjusted standardised difference in item thresholds outside the ±6.16 ranges (for year 3 data) and ±6.00 range (for year 5 data) should raise concern.

```
--------------------------------------------------------------------------------
Comparison of item estimates for groups boys and girls on the numeracy scale
L = 48     order = input
--------------------------------------------------------------------------------
                     Plot of standardised differences
               Easier for boys                   Easier for girls

       -9  -8   -7  -6  -5  -4   -3  -2  -1   0    1   2   3   4    5   6   7
 -------+---+----+---+---+---+----+---+---+---+----+---+---+---+----+---+---+
 y5n01               |                .       |*       .                |
 y5n02               |                .       |     *  .                |
 y5n03               |                .       |        .            *   |
 y5n04               |       *        .       |        .                |
 y5n05               |                .       |*       .                |
 y5n06               |                .       |        . *              |
 y5n07               |                .       |     *  .                |
 y5n08               |              * .       |        .                |
 y5n09               |                .       |        *                |
 y5n10               |                . *     |        .                |
 y5n11               |                .       |    *   .                |
 y5n12               |                .    *  |        .                |
 y5n13               |              * .       |        .                |
 y5n14               |                .       |        . *              |
 y5n15               |                .       | *      .                |
 y5n16               |                .     * |        .                |
 y5n17               |                .       |*       .                |
 y5n18               |                .       |        .          *     |
 y5n19               |       *        .       |        .                |
§y5n20               |                .       |        .                |  *
 y5n21               |           *    .       |        .                |
 y5n22               |                .       |*       .                |
§y5n23      *        |                .       |        .                |
 y5n24               |    *           .       |        .                |
 y5n25               |             *  .       |        .                |
 y5n26               |                .     * |        .                |
 y5n27               |                .       |       *.                |
 y5n28               |                .       |        .  *             |
 y5n29               |                .       |      * .                |
 y5n30               |                *       |        .                |
 y5n31               |                .*      |        .                |
 y5n32               |                .       | *      .                |
 y5n33               |                .       |*       .                |
 y5n34               |                .*      |        .                |
 y5n35               |                .       |*       .                |
 y5n36               |                .       |        *                |
 y5n37               |                .  *    |        .                |
 y5n38               |                .       |        .*               |
 y5n39               |                . *     |        .                |
 y5n40               |                .    *  |        .                |
 y5n41               |                .       |   *    .                |
 y5n42               |                .*      |        .                |
 y5n43               |                .       *        .                |
 y5n44               |                .   *   |        .                |
 y5n45               |                .       |     *  .                |
 y5n46               |                .   *   |        .                |
 y5n47               |      *         .       |        .                |
 y5n48               |                .       |  *     .                |
================================================================================
```

Notes:
All items had INFT MNSQ value within the range 0.77–1.30
§ item threshold adjusted standardised difference outside the range ± 6.00,
Inner boundary range ± 2.0
Outer boundary range ± 6.16

Figure 8-2. Year 5 numeracy item analysis (gender comparison)

From the use of the above criteria, Item y3n03 (that is, Item 3 in the year 3 numeracy test), and Item y5n23 (that is, Item 23 in the year 5 numeracy test) were markedly easier for the boys compared to the girls (see Tables 8-3 and 8-4, and Figures 8-1 and 8-2). On the other hand, Item y5n20 (that is, Item 20 in the year 5 numeracy test) was markedly easier for the girls compared to the boys. There were no items in the years 3 and 5 reading and language tests that recorded differences in threshold values outside the desired range.

Figures 8-3 to 8-5 show the item characteristic curves of the numeracy items identified as suspects in the preceding paragraphs (that is, Items y3n03, y5n23 and y5n20 respectively) while Figure 8-6 is an example of an ICC of an non-suspect item (in this case y3n18). The ICCs in Figures 8-3 to 8-6 were obtained using RUMM (Andrich, Lyne, Sheridan & Luo, 2000) software because the current versions of QUEST do not provide these curves.

It can be seen from Figure 8-3 (Item y3n03) and Figure 8-4 (Item y5n23) that the ICCs for boys are clearly higher than those of girls, which means that boys stand greater chances than girls of getting these items correct at the same ability level. On the contrary, the ICC for girls for Item y5n20 (Figure 8-5) is mostly higher than that of boys for the low-achieving students, meaning that, for low achievers, this item is biased in favour of girls. However, it can further be seen from Figure 8-5 that Item y5n20 is non-uniformly biased along the ability continuum because, for high achievers, the ICC for boys is higher than that of girls. Nevertheless, considering the area under the curves, this item (y5n20) is mostly in favour of girls.

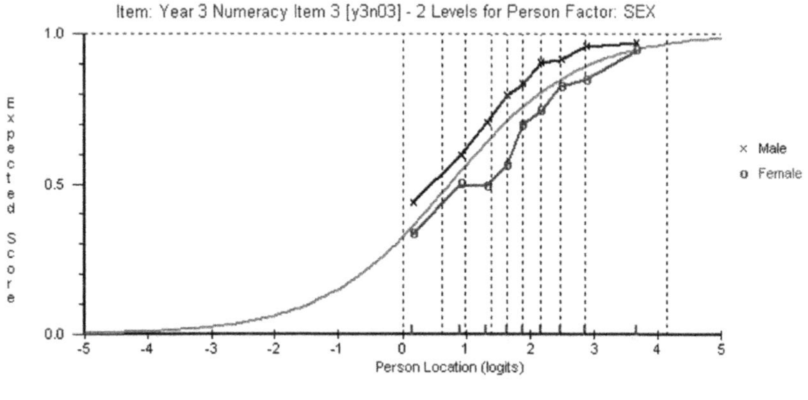

Figure 8-3. ICC for Item y3n03 (biased in favour of boys, $d_1 - d_2 = -0.78$)

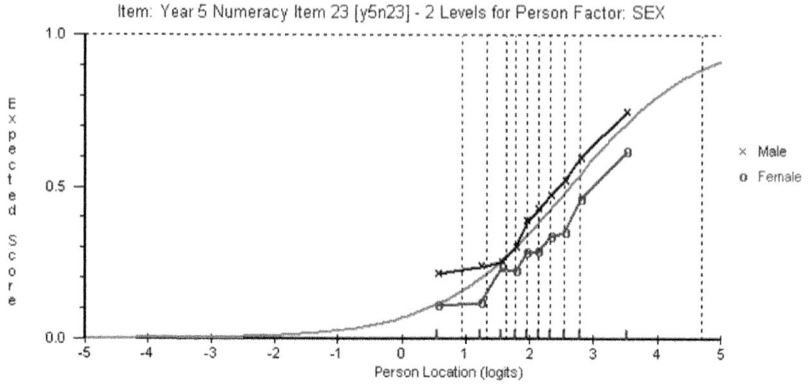

Figure 8-4. ICC for Item y5n23 (biased in favour of boys, d_1 - d_2 = -0.60)

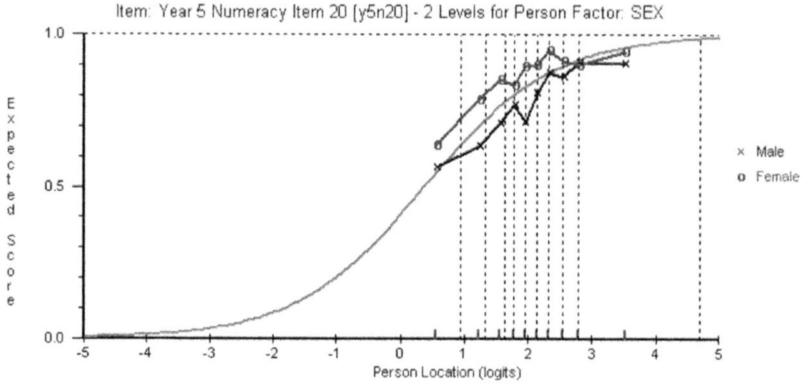

Figure 8-5. ICC for Item y5n20 (mostly biased in favour of girls, d_1 - d_2= 0.64)

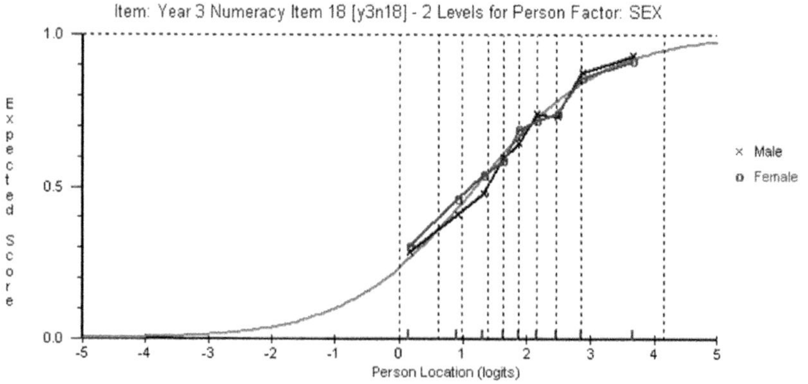

Figure 8-6. ICC for Item y3n18 (non-biased, d_1 - d_2= 0.05)

7. PLAUSIBLE EXPLANATION FOR GENDER BIAS

Another way of saying that an item is gender-biased is to say that there is some significant interaction between the item and the sex of the students (Scheuneman, 1979). Since bias is a characteristic of the item, then it is logical to ask whether there is something in the item that makes it favourable to one group and unfavourable to the other. It is common to examine the item's format and content in the investigation of item bias (Cole & Moss, 1989). Hence, to scrutinise why an item exhibits bias, there is a need to provide answers to the following questions:

1. Is the item format favourable or unfavourable to a given group?
2. Is the content of the item offensive to a given group to the extent of affecting the performance of the group on the test?
3. Does the content of the item require some experiences that are unique to a particular group and that gives its members an advantage in answering the item?

For the three items (y3n03, y5n20 and y5n23) identified as exhibiting gender bias in this study, it was difficult to establish from either their format or content as to why they showed bias (see Hungi, 1997, pp. 167–170). It is likely that these items were identified as bias just by mere chance and gender bias may not have been an issue in the 1995 Basic Skills Tests. Cole and Moss (1989) argue that it would be necessary to carry out replication studies before definite decisions could be made to eliminate items identified as biased in future tests.

8. CONCLUSION

In this study, data from the 1995 Basic Skills Testing Program are used to demonstrate two simple techniques for detecting gender-biased items based on Rasch measurement procedures. One technique involves an examination of differences in threshold values of items among gender groups and the other technique involves an examination of fit of item in different gender groups.

The analyses and discussion presented in this study are interesting for at least two reasons. Firstly, the procedures described in this chapter could be employed to identify biased items for different groups of students, divided by such characteristics as socioeconomic status, age, race, migrant status and school location (rural/urban). However, sizeable numbers of students are required within the subgroups for the two procedures described to provide a sound test for item bias.

Secondly, this study has demonstrated that magnitude of bias could be more relevant if expressed in terms of years of learning that a student spends at school. Obviously, expressing the extent of bias in terms of learning time lost or gained for the student could makes the information more useful to test developers, students and other users of test results.

9. REFERENCES

Ackerman, T. A., & Evans, J. A. (1994). The Influence of Conditioning Scores in Performing DIF Analyses. *Applied Psychological Measurement, 18*(4), 329-342.

Adams, R. J. (1992). Item Bias. In J. P. Keeves (Ed.), *The IEA Technical Handbook* (pp. 177-187). The Hague: IEA.

Adams, R. J., & Khoo, S. T. (1993). QUEST: The Interactive Test Analysis System. Hawthorn, Victoria: Australian Council for Education Research.

Adams, R. J., & Rowe, K. J. (1988). Item Bias. In J. P. Keeves (Ed.), *Educational Research, Methodology, and Measurement: An International Handbook* (pp. 398-403). Oxford: Pergamon Press.

Allen, N. L., & Donoghue, J. R. (1995). *Application of the Mantel-Haenszel Procedure to Complex Samples of Items*. Princeton, N. J.: Educational Testing Service.

Andrich, D., Lyne, A., Sheridan, B., & Luo, G. (2000). RUMM 2010: Rasch Unidimensional Measurement Models (Version 3). Perth: RUMM Laboratory.

Chang, H. H. (1995). *Detecting DIF for Polytomously Scored Items: An Adaptation of the SIBTEST Procedure*. Princeton, N. J.: Educational Testing Service.

Cole, N. S., & Moss, P. A. (1989). Bias in Test Use. In R. L. Linn (Ed.), *Education Measurement* (3rd ed., pp. 201-219). New York: Macmillan Publishers.

Dorans, N. J., & Kingston, N. M. (1985). The Effects of Violations of Unidimensionality on the Estimation of Item and Ability Parameters and on Item Response Theory Equating of the GRE Verbal Scale. *Journal of Educational Measurement, 22*(4), 249-262.

Hambleton, R. K. (1989). Principles and Selected Applications of Item Response Theory. In R. L. Linn (Ed.), *Education Measurement* (3rd ed., pp. 147-200). New York: Macmillan Publishers.

Hambleton, R. K., & J, R. H. (1989). Detecting Potentially Biased Test Items: Comparison of IRT Area and Mantel-Haenszel Methods. *Applied Measurement in Education, 2*(4), 313-334.

Hambleton, R. K., & Swaminathan, H. (1985). *Item Response Theory: Principles & Application*. Boston, MA: Kluwer Academic Publishers.

Hungi, N. (1997). *Measuring Basic Skills across Primary School Years*. Unpublished Master of Arts, Flinders University, Adelaide.

Hungi, N. (2003). *Measuring School Effects across Grades*. Adelaide: Shannon Research Press.

Kelderman, H. (1989). Item Bias Detection Using Loglinear IRT. *Psychometrika, 54*(4), 681-697.

Kino, M. M. (1995). *Differential Objective Function*. Paper presented at the Annual Meeting of the National Council on Measurement in Education, San Francisco, CA.

Klieme, E., & Stumpf, H. (1991). DIF: A Computer Program for the Analysis of Differential Item Performance. *Educational and Psychological Measurement, 51*(3), 669-671.

Lautenschlager, G. J. (1994). IRT Differential Item Functioning: An Examination of Ability Scale Purifications. *Educational and Psychological Measurement, 54*(1), 21-31.

Lord, F. M., & Stocking, M. L. (1988). Item Response Theory. In J. P. Keeves (Ed.), *Educational Research, Methodology, and Measurement: An International Handbook* (pp. 269-272). Oxford: Pergamon Press.

Mazor, K. M. (1995). Using Logistic Regression and the Mantel-Haenszel with Multiple Ability Estimates to Detect Differential Item Functioning. *Journal of Educational Measurement, 32*(2), 131-144.

McNamara, T. F. (1996). *Measuring Second Language Performance*. New York: Addison Wesley Longman.

Narayanan, P., & Swaminathan, H. (1994). Performance of the Mantel-Haenszel and Simultaneous Item Bias Procedures for Detecting Differential Item Functioning. *Applied Psychological Measurement, 18*(4), 315-328.

Osterlind, S. J. (1983). *Test Item Bias*. Beverly Hills: Sage Publishers.

Parshall, C. G., & Miller, T. R. (1995). Exact versus Asymptotic Mantel-Haenszel DIF Statistics: A Comparison of Performance under Small-Sample Conditions. *Journal of Educational Measurement, 32*(3), 302-316.

Pashley, P. J. (1992). *Graphical IRT-Based DIF Analyses*. Princeton, N. J: Educational Testing Service.

Potenza, M. T., & Dorans, N. J. (1995). DIF Assessment for Polytomously Scored Items: A Framework for Classification and Evaluation. *Applied Psychological Measurement, 19*(1), 23-37.

Rogers, H. J., & Swaminathan, H. (1993). A Comparison of the Logistic Regression and Mantel-Haenszel Procedures for Detecting Differential Item Functioning. *Applied Psychological Measurement, 17*(2), 105-116.

Scheuneman, J. (1979). A method of Assessing Bias in Test Items. *Journal of Educational Measurement, 16*(3), 143-152.

Scheuneman, J., & Bleistein. (1994). Item Bias. In T. Husén & T. N. Postlethwaite (Eds.), *The International Encyclopedia of Education* (2rd ed., pp. 3043-3051). Oxford: Pergamon Press.

Spray, J., & Miller, T. (1994). *Identifying Nonuniform DIF in Polytomously Scored Test Items*. Iowa: American College Testing Program.

Stocking, M. L. (1997). Item Response Theory. In J. P. Keeves (Ed.), *Educational Research, Methodology, and Measurement: An International Handbook* (2nd ed., pp. 836-840). Oxford: Pergamon Press.

Tang, H. (1994, January 27-29, 1994). *A New IRT-Based Small Sample DIF Method.* Paper presented at the Annual Meeting of the Southwest Educational Research Association, San Antonio, TX.

Thorndike, R. L. (1982). *Applied Psychometrics*. Boston, MA: Houghton-Mifflin.

Tittle, C. K. (1988). Test Bias. In J. P. Keeves (Ed.), *Educational Research, Methodology, and Measurement: An International Handbook* (pp. 392-398). Oxford: Pergamon Press.

Tittle, C. K. (1994). Test Bias. In T. Husén & T. N. Postlethwaite (Eds.), *The International Encyclopedia of Education* (2rd ed., pp. 6315-6321). Oxford: Pergamon Press.

Vijver, F. R., & Poortinga, Y. H. (1991). Testing Across Cultures. In H. R. K & J. N. Zaal (Eds.), *Advances in Education and Psychological Testing* (pp. 277-308). Boston, MA: Kluwer Academic Publishers.

Zwick, R. (1994). A Simulation Study of Methods for Assessing Differential Item Functioning in Computerized Adaptive Tests. *Applied Psychological Measurement, 18*(2), 121-140.

Chapter 9

RATERS AND EXAMINATIONS

Steven Barrett
University of South Australia

Abstract: Focus groups conduced with undergraduate students revealed general concerns about marker variability and the possible impact on examination results. This study has two aims: firstly, to analyse the relationships between student performance on an essay style examination, the questions answered and the markers; and, secondly, to identify and determine the nature and the extent of the marking errors on the examination. These relationships were analysed using two commercially available software packages, RUMM and ConQuest to develop the Rasch test model. The analyses revealed minor differences in item difficulty, but considerable inter-rater variability. Furthermore, intra-rater variability was even more pronounced. Four of the five common marking errors were also identified.

Key words: Rasch Test Model, RUMM, ConQuest, rater errors, inter-rater variability, intra-rater variability

1. INTRODUCTION

Many Australian universities are addressing the problems associated with increasingly scarce teaching resources by further increasing the casualisation of teaching. The Division of Business and Enterprise at the University of South Australia is no exception. The division has also responded to increased resource constraints through the introduction of the faculty core, a set of eight introductory subjects that all undergraduate students must complete. The faculty core provides the division with a vehicle through which it can realise economies of scale in teaching. These subjects have enrolments of up to 1200 students in each semester and are commonly taught by a lecturer, supported by a large team of sessional tutors.

S. Alagumalai et al. (eds.), Applied Rasch Measurement: A Book of Exemplars, 159–177.

The increased use of casual teaching staff and the introduction of the faculty core may allow the division to address some of the problems associated with its resource constraints, but they also introduce a set of other problems. Focus groups that were conducted with students of the division in the late1990s consistently raised a number of issues. Three of the more important issues identified at these meetings were:

- consistency between examination markers (inter-rater variability);
- consistency within examination markers (intra-rater variability); and
- differences in the difficulty of examination questions (inter-item variability).

The students who participated in these focus groups argued that, if there is significant inter-rater variability, intra-rater variability and inter-item variability, then student examination performance becomes a function of the marker and questions, rather than the teaching and learning experiences of the previous semester.

The aim of this paper is to assess the validity of these concerns. The paper will use the Rasch test model to analyse the performance of a team of raters involved in marking the final examination of one of the faculty core subjects. The paper is divided into six further sections. Section 2 provides a brief review of the five key rater errors and the ways that the Rasch test model can be used to detect them. Section 3 outlines the study design. Section 4 provides an unsophisticated analysis of the performance of these raters. Sections 5 and 6 analyse these performances using the Rasch test model. Section 7 concludes that these rater errors are present and that there is considerable inter-rater variability. However, intra-rater variability is an even greater concern.

2. FIVE RATINGS ERRORS

Previous research into performance appraisal has identified five major categories of rating errors, severity or leniency, the halo effect, the central tendency effect, restriction of range and inter-rater reliability or agreement (Saal, Downey & Lahey, 1980). Engelhard and Stone (1998) have demonstrated that the statistics obtained from the Rasch test model can be used to measure these five types of error. This section briefly outlines these rating errors and identifies the underlying questions that motivate concern about each type of error. The discussion describes how each type of rating error can be detected by analysing the statistics obtained after employing the Rasch test model. The critical values reported here, and in Table 9.1, relate to the rater and item estimates obtained from ConQuest. Other software packages may have different critical values. The present study extends this procedure by demonstrating how Item Characteristic Curves and Person Characteristic Curves can also be used to identify these rating errors.

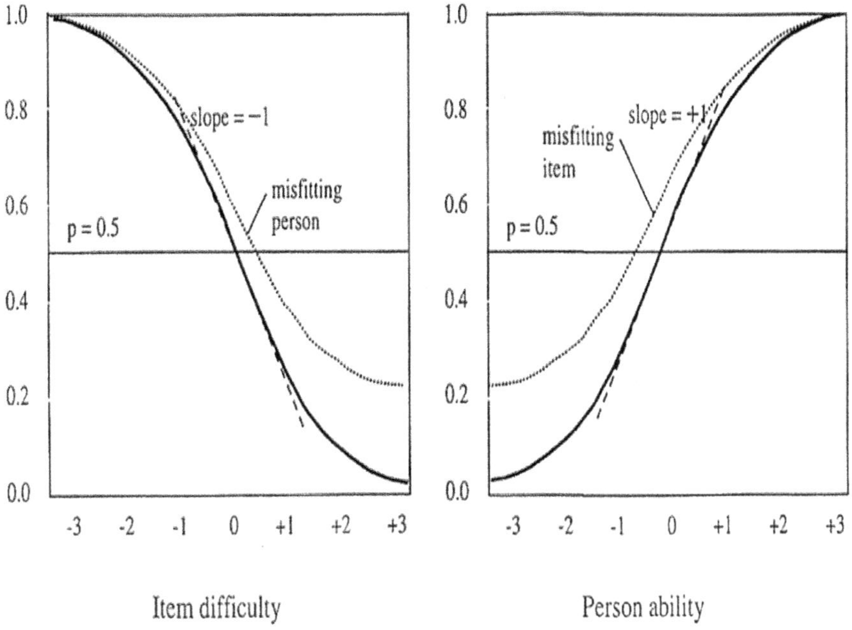

Source: Keeves and Alagumalai 1999, 30.

Figure 9-1. Item and Person Characteristic Curves

2.1 Rater severity or leniency

Rater severity or leniency refers to the general tendency on the part of raters to consistently rate students higher or lower than is warranted on the basis of their responses (Saal et al. 1980). The underlying questions that are addressed by indices of rater severity focus on whether there are statistically significant differences in rater judgments.

The statistical significance of rater variability can be analysed by examining the rater estimates that are produced by ConQuest (Tables 9.3 and 9.5 provide examples of these statistics). The estimates for each rater should be compared with the expert in the field: that is, the subject convener in this instance. If the leniency estimate is higher than the expert, then the rater is a harder marker and if the estimate is lower then the rater is an easier marker. Hence, the leniency estimates produced by ConQuest are reverse scored.

Evidence of rater severity of leniency can also be seen in the Person Characteristics Curves of the raters that are produced by software packages such as RUMM. If the Person Characteristic Curve for a particular rater lies to the right of that of the expert then that rater is more severe. On the other hand, a Person Characteristic Curve lying to the left implies that the rater is more lenient that the expert (Figure 9.1). Conversely, the differences in the difficulty of items can be determined from the estimates of discrimination produced by ConQuest. Tables 9.4 and 9.6 provide examples of these estimates.

2.2 The halo effect

The halo effect appears when a rater fails to distinguish between conceptually distinct and independent aspects of student answers (Thorndike 1920). For example, a rater may be rating items based on an overall impression of each answer. Hence, the rater may be failing to distinguish between conceptually essential or non-essential material. The rater may also be unable to assess competence in the different domains or criteria that the items have been constructed to measure (Engelhard 1994). Such a holistic approach to rating may also artificially create dependency between items. Hence, items may not be rated independently of each other. The lack of independence of rating between items can be determined from the Rasch test model.

Evidence of a halo effect can be obtained from the Rasch test model by examining the rater estimates: in particular, the mean square error statistics, or weighted fit MNSQ. See Tables 9.3 and 9.5 for examples. If these statistics are very low, that is less than 0.6, then raters may not be rating items independently of each other.

The shape of the Person Characteristic Curve for the raters can also be used to demonstrate the presence or absence of the halo effect. A flat curve, with a vertical intercept significantly greater than zero or which is tending towards a value significantly less than one as item difficulty rises, is an indication of the halo effect (Figure 9.1).

2.3 The central tendency effect

The central tendency effect describes situations in which the ratings are clustered around the mid-point of the rating scale and reflects reluctance by raters to use the extreme ends of the rating scale. This is particularly problematic when using a polycotomous rating scale, such as the one used in this study. The central tendency effect is often associated with inexperienced and less well-qualified raters.

This error can simply be detected by examining the marks of each rater using descriptive measures of central tendency, such as the mean, median, range and standard deviation, but as illustrated in Section 4, this can lead to errors. Evidence of the central tendency effect can also be obtained from the Rasch test model by examining the item estimates: in particular, the mean square error statistics, or unweighted fit MNSQ and the unweighted fit t. If these statistics are high (that is, the unweighted fit MNSQ is greater than 1.5 and the unweighted fit t is greater than 1), then the central tendency effect is present. Central tendency can also be seen in the Item Characteristic Curves, especially if the highest ability students consistently fail to attain a score of one on the vertical axis and the vertical intercept is significantly greater than zero.

2.4 Restriction of range

The restriction of range error is related to central tendency, but it is also a measure of the extent to which the obtained ratings discriminate between different students in respect to their different performance levels (Engelhard 1994; Engelhard & Stone, 1998). The underlying question that is addressed by restriction of range indexes focus on whether there is a statistical significance in item difficulty as shown by the rater estimates. Significant differences in these indices demonstrate that raters are discriminating between the items. The amount of spread also provides evidence relating to how the underlying trait has been defined. Again, this error is associated with inexperienced and less well-qualified raters.

Evidence of the restriction of range effect can be obtained from the Rasch test model by examining the item estimates: in particular, the mean square error statistics, or weighted fit MNSQ. This rating error is present if the weighted fit MNSQ statistic for the item is greater than 1.30 or less than 0.77.

These relationships are also reflected in the shape of the Item Characteristic Curve. If the weighted fit MNSQ statistic is less than 0.77, then the Item Characteristic Curve will have a very steep upward sloping section, demonstrating that the item discriminates between students in a very narrow ability range. On the other hand, if the MNSQ statistic is greater than 1.30, then the Item Characteristic Curve will be very flat with little or no steep middle section to give it the characteristic 'S' shape. Such an item fails to discriminate effectively between students of differing ability.

2.5 Inter-rater reliability or agreement

Inter-rater reliability or agreement is based on the concept that ratings are of a higher quality if two or more independent raters arrive at the same rating. In essence, this rating error reflects a concern with consensual or convergent validity. The model fit statistics obtained from the Rasch test model provides evidence of this type of error (Engelhard & Stone, 1998). It is unrealistic to expect perfect agreement between a group of raters. Nevertheless, it is not unrealistic to seek to obtain broadly consistent ratings from raters.

Indications of this type of error can be obtained by examining the mean square errors for both raters and items. Lower values reflect more consistency or agreement or a higher quality of ratings. Higher values reflect less consistency or agreement or a lower quality of ratings. Ideally these values should be 1.00 for the weighted fit MNSQ and 0.00 for the weighted fit t statistic. Weighted fit MNSQ greater than 1.5 suggest that raters are not rating items in the same order.

The unweighted fit MNSQ statistic is the slope at the point of inflection of the Person Characteristic Curve. Ideally, this slope should be negative 1.00. Increased deviation of the slope from this value implies less consistent and less reliable ratings.

Table 9.1: Summary table of rater errors and Rasch test model statistics

Rater error	Features of the curves if rater error present	Features of the statistics if rater error present
Leniency	Need to compare Person Characteristic Curve with that of the experts	Rater estimates Comparing estimate of leniency with the expert Lower error term implying more consistency
Halo effect	Person Characteristic Curve Maximum values do not approach 1 as student ability rises Vertical intercept does not tend to 0 as item difficulty rises	Rater estimates Weighted fit MNSQ < 1
Central tendency	Item Characteristic Curve Vertical intercept much greater than 0 Maximum values does not approach 1 as student ability rises	Item estimates: Unweighted fit MNSQ >> 1 Unweighted fit t >> 0
Restriction of range	Item Characteristic Curve Steep section of curve occurs over a narrow range of student ability or Curve is very flat with no distinct 'S' shape	Item estimates Weighted fit 0.77 <MNSQ < 1.30.
Relaibility	Person Characteristic Curve	Rater estimates:

Slope at point of inflection significantly greater than or less than 1.00	Weighted fit MNSQ >> 1 Weighted fit t >> 0

3. DESIGN OF THE STUDY

The aim of this study is to use the Rasch test model to determine whether student performance in essay examinations is a function of the person who marks the examination papers and the questions students attempt, rather than an outcome of the teaching and learning experiences of the previous semester. The study investigates the following four questions:

- To what extent does the difficulty of items in an essay examination differ?
- What is the extent of inter-rater variability?
- What is the extent of intra-rater variability? and
- To what extent are the five rating errors present?

The project analyses the results of the Semester 1, 1997 final examinations results in communication and the media, which is one of the faculty core subjects. The 833 students who sat this examination were asked to answer any four questions from a choice of twelve. The answers were arranged in tutor order and the eight tutors, who included the subject convener, marked all of the papers written by their students. The unrestricted choice in the paper and the decision to allow tutors to mark all questions answered by their students maximised the crossover between items. However, the raters did not mark answers written by students from other tutorial groups. Hence, the relationship between the rater and the students cannot be separated. It was therefore decided to have all of the tutors double-blind mark a random sample of all of the other tutorial groups in order to facilitate the separation of raters, students and items. In all, 19.4 per cent of the papers were double-marked. The 164 double-marked papers were then analysed separately in order to provide some insights into the effects of student performance by fully separating raters, items and students.

4. PHASE ONE OF THE STUDY: INITIAL QUESTIONS

At present, the analysis of examination results and student performance at most Australian universities tends to be not very sophisticated. An analysis of rater performance is usually confined to an examination of a range of measures of central tendency, such as the mean, median range and standard deviation of marks for each rater. If these measures vary too much, then the subject convener may be required to take remedial action, such as moderation, staff development or termination of employment of the sessional staff member. Such remedial action can have severe implications for both the subject convener: it is time-consuming, and the sessional staff members involved may lose their jobs for no good reason. Therefore, an analysis of rater performance needs to be done properly.

Table 9.2 presents the average marks for each item for every rater and the average total marks for every rater on the examination that is the focus of this study. An analysis of rater performance would usually involve a rather cursory analysis of data similar to the data present in Table 9.2. Such an analysis constitutes Phase One of this study. The data in Table 9.2 reveal some interesting differences that should raise some interesting questions for the subject convener to consider as part of her curriculum development process. Table 9.2 shows considerable differences in question difficulty and the leniency of markers. Rater 5 is the hardest and rater 6 the easiest, while Item 6 appears to be the easiest and Items 2 and 3 the hardest. But are these the correct conclusions to be drawn from these results?

Table 9.2: Average raw scores for each question for all raters

	Rater								
Item	**1**	**2**	**3**	**4**	**5**	**6**	**7**	**8**	**All**
1	7.1	6.6	7.2	7.2	5.4	7.1	6.5	6.6	6.8
2	7.0	6.2	6.7	7.1	6.4	7.1	6.8	6.4	6.5
3	6.8	6.5	6.4	6.9	6.0	6.8	6.5	6.5	6.5
4	7.0	6.8	7.3	7.3	5.5	6.8	6.7	6.5	6.7
5	7.2	6.7	7.0	7.6	6.0	7.7	7.4	7.2	7.1
6	7.4	7.2	8.0	7.3	6.5	7.7	6.5	7.0	7.2
7	7.0	6.7	6.1	7.2	5.8	7.3	6.6	6.8	6.8
8	7.2	6.9	6.5	7.0	5.8	7.6	8.0	7.0	6.9
9	7.0	6.8	7.2	7.0	6.7	7.3	7.9	7.2	7.0
10	7.3	6.8	6.1	6.9	5.6	7.2	7.4	6.9	6.8
11	7.5	6.5	6.0	7.0	5.7	6.8	6.9	6.6	6.6
12	7.1	6.8	5.9	7.2	5.9	7.6	7.3	6.9	6.8
mean*	28.4	26.8	26.5	28.6	23.8	29.1	28.5	27.8	27.4
n#	26	225	71	129	72	161	70	79	833

Note *: average total score for each rater out of 40; each item marked out of 10

Note #: n signifies the number of papers marked by each tutor; N = 833

5. PHASE TWO OF THE STUDY

An analysis of the results presented in Table 9.2 using the Rasch test model tells a very different story. This phase of the study involved an analysis of all 833 examination scripts. However, as the raters marked the papers belonging to the students in their tutorial groups, there was no crossover between raters and students.

An analysis of the raters (Table 9.3) and the items (Table 9.4), conducted using ConQuest, provides a totally different set of insights into the performance of both raters and items. Table 9.3 reveals that rater 1 is the most lenient marker, not rater 6, with the minimum estimate value. He is also the most variable, with the maximum error value. Indeed, he is so inconsistent that he does not fit the Rasch test model, as indicated by the rater estimates. His unweighted fit MNSQ is significantly different from 1.00 and his unweighted fit t statistic is greater than 2.00. Nor does he discriminate well between students, as shown by the maximum value for the weighted fit MNSQ statistic, which is significantly greater than 1.30. The subject convener is rater 2 and this table clearly shows that she is an expert in her field who sets the appropriate standard. Her estimate is the second highest, so she is setting a high standard. She has the lowest error statistic, which is very close to zero, so she is the most consistent. Her unweighted fit MNSQ is very close to 1.00 while her unweighted fit t statistic is closest to 0.00. She is also the best rater when it comes to discriminating between students of different ability as shown by her weighted fit MNSQ statistic which is not only one of the few in the range 0.77 to 1.30, but it is also very close to 1.00. Furthermore, her weighted fit t is very close to zero.

Table 9.3: Raters, summary statistics

			Weighted fit		**Unweighted fit**	
Rater	**Leniency**	**Error**	**MNSQ**	**t**	**MNSQ**	**t**
1	-0.553	0.034	1.85	2.1	1.64	3.8
2	0.159	0.015	0.96	-0.1	0.90	-1.3
3	0.136	0.024	1.36	1.2	1.30	2.5
4	-0.220	0.028	1.21	0.9	1.37	3.2
5	0.209	0.022	1.64	2.0	1.62	4.8
6	0.113	0.016	1.29	1.3	1.23	2.3
7	0.031	0.024	1.62	1.9	1.60	4.6
8	0.124					

N= 833

Table 9.4: Items, summary statistics

			Weighted fit		Unweighted fit	
Item	**Discrimination**	**Error**	**MNSQ**	**t**	**MNSQ**	**t**
1	0.051	0.029	0.62	-1.4	0.52	-8.1
2	-0.014	0.038	0.88	-.02	0.73	-3.7
3	0.118	0.025	0.64	-1.5	0.61	-6.7
4	-0.071	0.034	0.75	-0.8	0.68	-4.9
5	-0.128	0.023	0.67	-1.5	0.54	-8.8
6	-0.035	0.034	0.84	-0.5	0.76	-3.7
7	0.148	0.025	0.58	-1.9	0.53	-10.5
8	0.091	0.030	0.72	-1.0	0.65	-5.9
9	0.115	0.019	0.39	-3.8	0.34	-17.5
10	0.011	0.032	0.74	-0.9	0.63	-5.4
11	-0.144	0.034	0.77	-0.8	0.66	-4.6
12	-0.142					

N=833

Table 9.4 summarises the item statistics that were obtained from ConQuest. The results of this table also do not correspond well to the results presented in Table 9.2. 7, not Items 2 and 3, now appears to be the hardest item on the paper, while Item 11 is the easiest. Unlike the tutors, only items 2 and 3 fit the Rasch test model well. Of more interest is the lack of discrimination power of these items. Ten of the weighted fit MNSQ figures are less than the critical value of 0.77. This means that these items only discriminate between students in a very narrow range of ability. Figure 9.3, below, shows that these items generally only discriminate between students in a very narrow range in the very low student ability range. Of particular concern is Item 9. It does not fit the Rasch test model (unweighted fit t value of -3.80). This value suggests that the item is testing abilities or competencies that are markedly different to those that are being tested by the other 11 items. The same may also be said for Item 7, even though it does not exceed the critical value of –2.00 for this measure. Table 9.4 also shows that there is little difference in the difficulty of the items. The range of the item estimates is only 0.292 logits.

On the basis of this evidence there does not appear to be a significant difference in the difficulty of the items. Hence, the evidence in this regard does not tend to support student concerns about inter-item variability. Nevertheless, the specification if Items 7 and 9 needs to be improved.

```
          rater             item                  rater by item
+1   |               |                  |                        |
     |               |                  |                        |
     |               |                  |                        |
     |               |                  |                        |
     |               |                  |8.5 4.8 4.9             |
     |               |                  |1.2 6.4 5.5 7.5         |
     |2 3 5          |7                 |2.1 2.2 6.2 1.3         |
     |6 7 8          |1 3 8 9 10        |1.1 3.1 5.1 4.2         |
 0   |               |2 4 5 6           |4.1 6.1 7.1 8.1         |
     |4              |11 12             |3.2 5.2 8.2 2.3         |
     |               |                  |7.2 6.3 1.4 8.4         |
     |               |                  |1.10                    |
     |1              |                  |                        |
     |               |                  |4.5                     |
     |               |                  |                        |
-1   |               |                  |                        |
```

N = 833, vertical scale is in logits, some parameters could not be fitted on the display

Figure 9-2. Map of Latent Distributions and Response Model Parameter Estimates

Figure 9.2 demonstrates some other interesting points that tend to support the concerns of the students who participated in the focus groups. First, the closeness of the leniency of the majority of raters and the closeness in the difficulty of the item demonstrate that there is not much variation in rater severity or item difficulty. However, raters 1 and 4 stand out as particularly lenient raters. The range in item difficulty is only 0.292 logits. However, the most interesting feature of this figure is the maximum intra-rater variability. The intra-rater variability of rater 4 is approximately 50 per cent greater than the inter-rater variability of all eight raters as a whole: that is, the range of the inter-rater variability is 0.762 logits. Yet the intra-rater variability of rater 4 is much greater (1.173 logits), as shown by the difference in the standard set for Item 5 (4.5 in Figure 9.2) and Items 8 and 9 (4.8 and 4.9 in Figure 9.2). Rater 4 appears to find it difficult to judge the difficulty of the items he has been asked to mark. For example, Items 8 and 5 are about the same level of difficulty. Yet, he marked Item 8 as if it were the most difficult item on the paper and then marked Item 5 as if it were the easiest. It is interesting to note that the easiest rater, rater 1, is almost as inconsistent as rater 4, with an intra-rater variability of 0.848. With two notable exceptions, the intra-rater variation is less than the inter-rater variation. Nevertheless, intra-rater differences do appear to be significant. On the basis of this limited evidence it may be concluded that intra-rater variability is as much a concern as inter-rater variability. It also appears that intra-rater variability is directly related to the extent of the variation from the standard set by the subject convener. In particular, more lenient raters are also more likely to higher intra-rater variability.

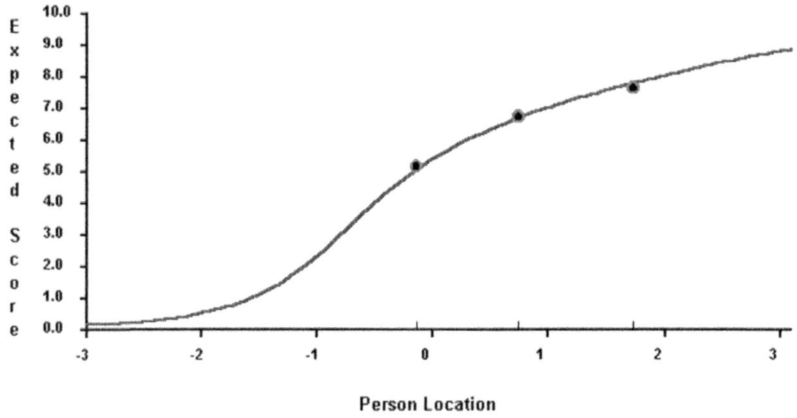

Figure 9-3. Item Characteristic Curve, Item 2

The Item Characteristics Curves that were obtained from RUMM confirm the item analyses that were obtained from ConQuest. Figure 9.3 shows the Item Characteristic Curve for Item 2, which is representative of 11 of the 12 items in this examination. These items discriminate between students in a narrow range at the lower end of the student ability scale, as shown by the weighted fit MNSQ value being less than 0.77 for most items. However, none of these 11 curves has an expected value much greater than 0.9: that is, the best students are not consistently getting full marks for their answers. This reflects the widely held view that inexperienced markers are unwilling to award full marks for essay questions. On the other hand, Item 4 (Figure 9.4) discriminates poorly between students regardless of their ability. The weakest students are able to obtain quite a few marks; yet the best students are even less likely to get full marks than they are on the other 11 items. Either the item or its marking guide needs to be modified, or the item should be dropped from the paper. Moreover, all of the items, or their marking guides, need to be modified in order to improve their discrimination power.

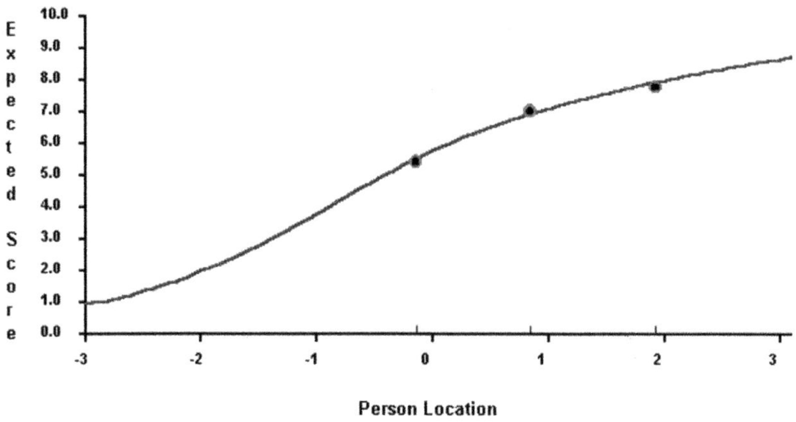

Figure 9-4. Item Characteristic Curve, Item 4

In short, there is little correspondence between the results obtained by examining the data presented in Table 9.2, using descriptive statistics, and the results obtained from the Rasch test model. Consequently, any actions taken to improve either the item or test specification based on an analysis of the descriptive statistics, could have rather severe unintended consequences. However, the analysis needs to be repeated with some crossover between tutorial groups in order to separate any effects of the relationships between students and raters. For example, rater 6 may only appear to be the toughest marker as his tutorials have an over-representation of weaker students, while rater 1 may appear to be the easiest marker as her after hours class may contain an over representation of more highly motivated mature-aged students. These interactions between the raters, the students and the items need to be separated from each other so that they can be investigated. This occurs in Section 6.

6. PHASE THREE OF THE STUDY

The second phase of this study was designed to maximise the crossover between raters and items, but there was no crossover between raters and students. The results obtained in relation to rater leniency and item difficulty may be influenced by the composition of tutorial groups as students had not been randomly allocated to tutorials. Hence, a 20 per cent sample of papers were double-marked in order to achieve the required crossover and to provide some insights into the effects of fully separating, raters, items and students. Results of this analysis are summarised in Tables 9.5 and 9.6 Figure 9.5.

The first point that emerges from Table 9.5 is that the separation of raters, items and students leads to a reduction in inter-rater variability from 0.762 logits to 0.393 logits. Nevertheless, rater 1 is still the most lenient. More interestingly, rater 2, the subject convener, has become the hardest marker, reinforcing her status as the expert. This separation has also increased the error for all tutors, yet at the same time reducing the variability between all eight raters. More importantly all eight raters now fit the Rasch test model as shown by the unweighted fit statistics. In addition, all raters are now in the critical range for the weighted fit statistics, so they are discriminating between students of differing ability.

Table 9-5: Raters, Summary Statistics

			Weighted Fit		Unweighted Fit	
Rater	**Leniency**	**Error**	**MNSQ**	**t**	**MNSQ**	**t**
1	-0.123	0.038	0.92	-0.1	0.84	-0.8
2	0.270	0.035	0.87	-0.2	0.83	-1.0
3	-0.082	0.031	0.86	-0.3	0.82	-1.1
4	0.070	0.038	1.02	0.2	0.91	-0.4
5	-0.105	0.030	1.07	0.3	1.09	0.6
6	0.050	0.034	0.97	0.1	0.95	-0.2
7	0.005	0.032	1.06	0.3	1.04	0.3
8	-0.085					

N= 164

Table 9-6: Items, Summary Statistics

			Weighted Fit		Unweighted Fit	
Item	**Discrimination**	**Error**	**MNSQ**	**t**	**MNSQ**	**t**
1	0.054	0.064	1.11	0.4	1.29	1.4
2	0.068	0.074	1.34	0.7	1.62	2.4
3	-0.369	0.042	0.91	-0.1	0.95	-0.3
4	0.974	0.072	1.33	0.7	1.68	2.8
5	-0.043	0.048	1.01	0.2	1.11	0.8
6	-0.089	0.062	1.10	0.4	1.23	1.2
7	-0.036	0.050	0.92	-0.1	1.02	0.2
8	-0.082	0.050	0.99	0.1	1.07	0.5
9	-0.146	0.037	0.75	-0.7	0.80	-1.5
10	0.037	0.059	1.01	0.2	1.13	0.7
11	-0.214	0.057	1.14	0.4	1.41	2.1

12	-0.154

N = 164

However, unlike the rater estimates, the variation in item difficulty has increased from 0.292 to 1.343 logits (Table 9.6). Clearly now decisions about which questions to answer may be important determinants of student performance. For example, the decision to answer Item 4 in preference to Items 3, 9, 11 or 12 could see a student drop from the top to the bottom quartile, such is the observed differences in item difficulties. Again the separation of raters, items and students has increased the error term: that is, it has reduced the degree of consistency between the marks that were awarded and student ability. All items now fit the Rasch test model. The unweighted fit statistics, MNSQ and t, are now very close to one and zero respectively. Finally, ten of the weighted fit statistics now lie in the critical range for the weighted MNSQ statistics. Hence, there has been an increase in the discrimination power of these items. They are now discriminating between students over a much wider range of ability.

Finally, Figure 9.5 shows that the increased inter-item variability is associated with an increase in the intra-rater by item variability, despite the reduction in the inter-rater variability. The range of rater by item variability has risen to about 5 logits. More disturbingly, the variability for individual raters has risen to over two logits. The double-marking of these papers and the resultant crossover between raters and students has allowed the raters to be separated from each other by student interactions. Figure 9.5 now shows that raters 1 and 4 are as severe as the other raters and are not the easiest raters, in stark contrast to what is shown in Figure 9.2. It can therefore be concluded that these two raters appeared to be easy markers because their tutorial classes contained a higher proportion of higher ability students. Hence, accounting for the student-rater interactions has markedly reduced the observed inter-rater variability.

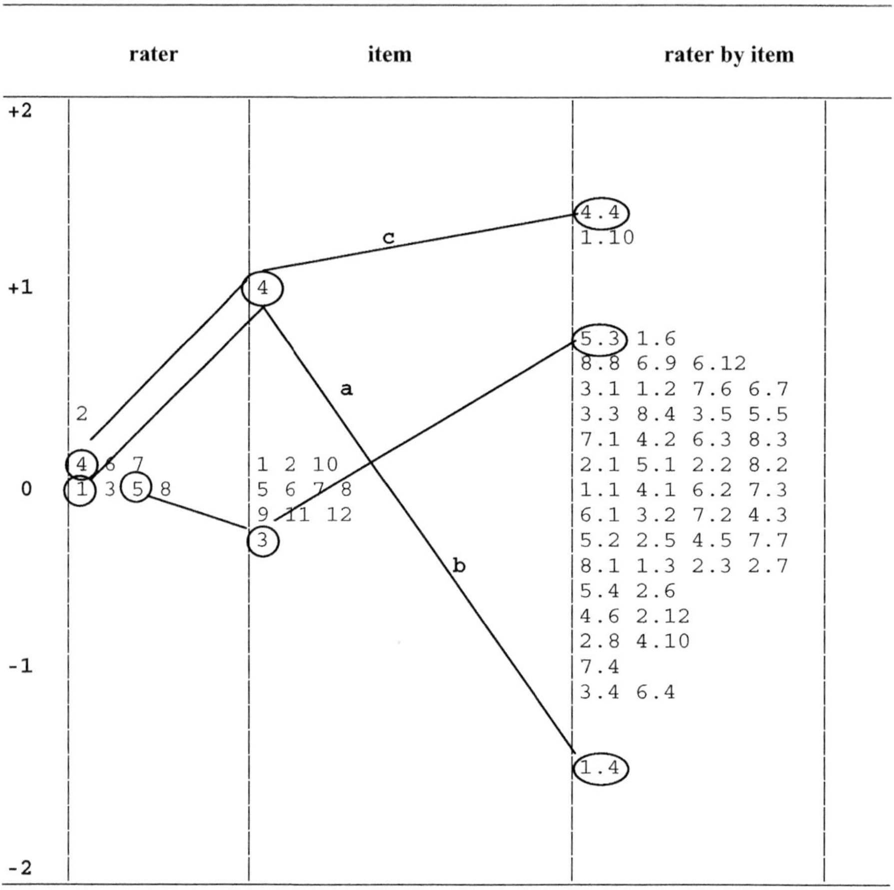

Notes:
Some outliers in the rater by item column have been delted from this figure.
N = 164

Figure 9-5. Map of Latent Distributions and Response Model Parameter Estimates

However, separating the rater by student interactions appears to have increased the levels of intra-rater variability. For example, Figure 9.5 shows that raters 1 and 4 are setting markedly different standards for items that are of the same difficulty level. This intra-rater variability is illustrated by the three sets of lines on Figure 9.5. Line (a) shows the performance of rater 4 marking Item 4. This rater has not only correctly identified that Item 4 is the hardest item in the test, but he is also marking it at the appropriate level, as indicated by the circle 4.4 in the rater by item column. On the other hand, line (b) shows that rater 1 has not only failed to recognise that Item 4 is the hardest item, but he has also identified it as the easiest item in the examination paper and has marked it as such, as indicated by the circle 1.4 in the rater by item column. Interestingly, as shown by line (c), rater 5 has not identified Item 3 as the easiest item in the examination paper and has marked it as if it were almost as difficult as the hardest item, as shown by the circle 5.3 in the rater by item column. Errors such as these can significantly affect the examination performance of students.

The results obtained in this phase of the study differ markedly from the results obtained during the preceding phase of the study. In general, raters and items seem to fit the Rasch test model better as a result of the separation of the interactions between raters, items and students. On the other hand, the intra-rater variability has increased enormously. However, the MNSQ and t statistics are a function of the number of students involved in the study. Hence, the reduction in the number of papers analysed in this phase of the study may account for much of the change in the fit of the Rasch test model in respect to the raters and items.

It may be concluded from this analysis that, when students are not randomly assigned to tutorial groups, then the clustering of students with similar characteristics in certain tutorial groups is reflected in the performance of the rater. However, in this case, a 20 per cent sample of double-marked papers was too small to determine the exact nature of the interaction between raters, items and students. More papers needed to be double-marked in this phase of the study to improve the accuracy of both the rater and item estimates. In hindsight, at least 400 papers needed to be analysed during this phase of the study in order to more accurately determine the item and rater estimates and hence more accurately determine the parameters of the model.

7. CONCLUSION

The literature on performance appraisal identifies five main types of rater errors, severity or leniency, the halo effect, the central tendency effect, restriction of range and inter-rater reliability or agreement. Phase 2 of this study identified four of these types of errors applying to a greater or lesser extent to all raters, with the exception of the subject convener. Firstly, rater 1, and to a lesser extent rater 4, mark far more leniently than either the subject convener or the other raters. Secondly, there was, however, no clear evidence of the halo effect being present in the second phase of the study (Table 9.3). Thirdly, there is some evidence, in Table 9.3 and Figures 9.2 and 3, of the presence of the central tendency effect. Fourthly, the weighted fit MNSQ statistics for the items (Table 9.4) show that the items discriminate between students over a very narrow range of ability. This is also strong evidence for the presence of restriction of range error. Finally, Table 9.2 provides evidence of unacceptably low levels of inter-rater reliability. Three of the eight raters exceed the critical value of 1.5, while a fourth is getting quite close. However, of more concern is the extent of the intra-rater variability.

In conclusion, this study provided evidence to support most of the concerns reported by students in the focus groups. This is because the Rasch test model was able to separate the complex interactions between student ability, item difficulty and rater performance from each other. Hence, each component of this complex relationship can be analysed independently. This in turn allows much more informed decisions to be made about issues such as mark moderation, item specification and staff development and training.

There is no evidence to suggest that the items in this examination differed significantly in respect to difficulty. The study did, however, find evidence of significant inter-rater variability, significant intra-rater variability and the presence of four of the five common rating errors present. However, the key finding of this study is that intra-rater variability is possibly more likely to lead to erroneous ratings that inter-rater variability.

8. REFERENCES

Adams, R.J. & Khoo S-T. (1993) *Conquest: The Interactive Test Analysis System,* ACER Press, Canberra.

Andrich, D. (1978) A Rating Formulation for Ordered Response Categories, *Psychometrica*, 43, pp. 561-573.

Andrich, D. (1985) An Elaboration of Guttman Scaling with Rasch Models for Measurement, in N. Brandon-Tuma (ed.) *Sociological Methodology*, Jossey-Bass, San Francisco.

Andrich, D. (1988) *Rasch Models for Measurement*, Sage, Beverly Hills.

Barrett, S.R.F. (2001) The Impact of Training in Rater Variability, *International Education Journal* 2(1), pp. 49-58.

Barrett, S.R.F. (2001) Differential Item Functioning: A Case Study from First Year Economics, *International Education Journal* 2(3), pp. 1-10.

Chase, C.L. (1978) *Measurement for Educational Evaluation*, Addison-Wesley, Reading.

Choppin, B. (1983) *A Fully Conditional Estimation Procedure for Rasch Model Parameters*, Centre for the Study of Evaluation, Graduate School of Education, University of California, Los Angeles.

Engelhard, G.Jr (1994) Examining Rater Error in the Assessment of Written Composition With a Many-Faceted Rasch Model, *Journal of Educational Measurement*, 31(2), pp 179-196.

Engelhard, G.Jr & Stone, G.E. (1998) Evaluating the Quality of Ratings Obtained From Standard-Setting Judges, *Educational and Psychological Measurement*, 58(2), pp 179-196.

Hambleton, R.K. (1989) Principles of Selected Applications of Item Response Theory, in R. Linn, (ed.) *Educational Measurement, 3rd ed.*, MacMillan, New York, pp. 147-200.

Keeves, J.P. & Alagumalai, S. (1999) New Approaches to Research, in G.N. Masters and J.P. Keeves, *Advances in Educational Measurement, Research and Assessment*, pp. 23-42, Pergamon, Amsterdam.

Rasch, G. (1968) A Mathematical Theory of Objectivity and its Consequence for Model Construction, European Meeting on Statistics, Econometrics and Management Science, Amsterdam.

Rasch, G. (1980) *Probabilistic Models for Some Intelligence and Attainment Tests*, University of Chicago Press, Chicago.

Saal, F.E., Downey, R.G. & Lahey, M.A (1980) Rating the Ratings: Assessing the Psychometric Quality of Rating Data, *Psychological Bulletin*, 88(2), 413-428.

van der Linden, W.J. & Eggen, T.J.H.M. (1986) An Empirical Bayesian approach to Item Banking, *Applied Psychological Measurement*, 10, pp. 345-354.

Sheridan, B., Andrich, D. & Luo, G. (1997) *RUMM User's Guide*, RUMM Laboratory, Perth.

Snyder, S. and Sheehan, R. (1992) The Rasch Measurement Model: An Introduction, *Journal of Early Intervention*, 16(1), pp. 87-95.

Weiss, D. (ed.) (1983) *New Horizons in Testing*, Academic Press, New York.

Weiss, D.J. & Yoes, M.E. (1991) Item Response Theory, in R.K. Hambleton and J.N. Zaal (eds) *Advances in Educational and Psychological Testing and Applications*, Kluwer, Boston, pp 69-96.

Wright, B.D. & Masters, G.N. (1982) *Rating Scale Analysis*, MESA Press, Chicago.

Wright, B.D. & Stone M.H. (1979) *Best Test Design*, MESA Press, Chicago.

Chapter 10

COMPARING CLASSICAL AND CONTEMPORARY ANALYSES AND RASCH MEASUREMENT

David D. Curtis
Flinders University

Abstract: Four sets of analyses were conducted on the 1996 Course Experience Questionnaire data. Conventional item analysis, exploratory factor analysis and confirmatory factor analysis were used. Finally, the Rasch measurement model was applied to this data set. This study was undertaken in order to compare conventional analytic techniques with techniques that explicitly set out to implement genuine measurement of perceived course quality. Although conventional analytic techniques are informative, both confirmatory factor analysis and in particular the Rasch measurement model reveal much more about the data set, and about the construct being measured. Meaningful estimates of individual students' perceptions of course quality are available through the use of the Rasch measurement model. The study indicates that the perceived course quality construct is measured by a subset of the items included in the CEQ and that seven of the items of the original instrument do not contribute to the measurement of that construct. The analyses of this data set indicate that a range of analytical approaches provide different levels of information about the construct. In practice, the analysis of data arising from the administration of instruments like the CEQ would be better undertaken using the Rasch measurement model.

Key words: classical item analysis, exploratory factor analysis, confirmatory factor analysis, Rasch scaling, partial credit model

1. INTRODUCTION

The constructs of interest in the social sciences are often complex and are observed indirectly through the use of a range of indicators. For constructs

S. Alagumalai et al. (eds.), Applied Rasch Measurement: A Book of Exemplars, 179–195.

that are quantified, each indicator is scored on a scale which may be dichotomous, but quite frequently a Likert scale is employed. Two issues are of particular interest to researchers in analyses of data arising from the application of instruments. The first is to provide support for claims of validity and reliability of the instrument and the second is the use of scores assigned to respondents to the instrument. These purposes are not achieved in separate analyses, but it is helpful to categorise different analytical methods. Two approaches to addressing these issues are presented, namely classical and contemporary, and they are shown in Table 10-1.

Table 10-1. Classical and contemporary approaches to instrument structure and scoring

	Instrument structure	Item coherence and case scores
Classical analyses	Exploratory factor analysis (EFA)	Classical test theory (CTT)
Contemporary analyses	Confirmatory factor analysis (CFA)	Objective measurement using the Rasch measurement model

In this paper, four analyses of a data set derived from the Course Experience Questionnaire (CEQ) are presented in order to compare the merits of both classical and contemporary approaches to instrument structure and to compare the bases of claims of construct measurement. Indeed, before examining the CEQ instrument, it is pertinent to review the issue of measurement.

2. MEASUREMENT

In the past, Steven's (1946) definition of measurement, that "...measurement is the assignment of numerals to objects or events according to a rule" (quoted in Michell, 1997, p.360) has been judged to be a sufficient basis for the measurement of constructs in the social sciences. Michell showed that Steven's requirement was a necessary, but not sufficient, basis for true measurement. Michell argued that it is necessary to demonstrate that constructs being investigated are indeed quantitative and this demonstration requires that assigned scores comply with a set of quantification axioms (p.357). It is clear that the raw 'scores' that are used to represents respondents' choices of response options are not 'numerical quantities' in the sense required by Michell, but reflect the order of response categories. They are not additive quantities and therefore cannot be used to generate 'scale scores' even though this has been a common practice in the social sciences.

Rasch (1960) showed that dichotomously scored items could be converted to an interval metric using a simple logistic transformation. His insight was subsequently extended to polytomous items (Wright & Masters, 1982). These transformations produce an interval scale that complies with the requirements of true measurement.

However, the Rasch model has been criticised for a lack of discrimination of noisy from precise measures (Bond & Fox, 2001, p.183-4; Kline, 1993, p.71 citing Wood (1978)). Wood's claim is quite unsound, but in order to deflect such criticism, it seems wise to employ a complementary method for ensuring that the instrument structure complies with the requirements of true measurement. Robust methods for examining instrument structure in support of validity claims may also support claims for a structure compatible with true measurement.

The application of both classical and contemporary methods for the analysis of instrument structure, for the refinement of instruments, and for generating individual scores are illustrated an analysis of data from the 1996 administration of the Course Experience Questionnaire (CEQ).

3. THE COURSE EXPERIENCE QUESTIONNAIRE

The Course Experience Questionnaire (CEQ) is a survey instrument distributed by mail to recent graduates of Australian universities shortly after graduation. It comprises 25 statements that relate to perceptions of course quality and responses to each item are made on a five-point Likert scale from 'Strongly disagree' to 'Strongly agree'. The administration, analysis and reporting of the CEQ is managed by the Graduate Careers Council of Australia.

Ramsden (1991), outlined the development of the CEQ. It is based on work that was done at Lancaster University in the 1980s and was developed as a measure of the quality of students' learning, which was correlated with measures of students' approaches to learning, rather than from an *a priori* analysis of teaching quality or institutional support and was intended to be used for formative evaluation of courses (Wilson, Lizzio, & Ramsden, 1996, pp.3-5). However, Ramsden pointed out that quality teaching creates the conditions under which students are encouraged to develop and employ effective learning strategies and that these lead to greater levels of satisfaction (Ramsden, 1998). Thus a logical consistency was established for using student measures of satisfaction and perceptions of teaching quality as indications of the quality of higher education programs. Hence, the CEQ can be seen as an instrument to measure graduates' perceptions of course quality.

In his justification for the dimensions of the CEQ, Ramsden (1991) referred to work done on subject evaluation. Five factors that were identified as components of perceived quality, namely, providing good teaching (GTS); establishing clear goals and standards (CGS); setting appropriate assessments (AAS); developing generic skills (GSS); and requiring appropriate workload (AWS). The items that were used in the CEQ and the sub-scales with which they were associated are shown in Table 10-2.

Table 10-2. The items of the 1996 Course Experience Questionnaire

Item	Scale	Item text
1	CGS	It was always easy to know the standard of work expected
2	GSS	The course developed my problem solving skills
3	GTS	The teaching staff of this course motivated me to do my best work
4*	AWS	The workload was too heavy
5	GSS	The course sharpened my analytic skills
6	CGS	I usually had a clear idea of where I was going and what was expected of me in this course
7	GTS	The staff put a lot of time into commenting on my work
8*	AAS	To do well in this course all you really needed was a good memory
9	GSS	The course helped me develop my ability to work as a team member
10	GSS	As a result of my course, I feel confident about tackling unfamiliar problems
11	GSS	The course improved my skills in written communication
12*	AAS	The staff seemed more interested in testing what I had memorised than what I had understood
13	CGS	It was often hard to discover what was expected of me in this course
14	AWS	I was generally given enough time to understand the things I had to learn
15	GTS	The staff made a real effort to understand difficulties I might be having with my work
16*	AAS	Feedback on my work was usually provided only as marks or grades
17	GTS	The teaching staff normally gave me helpful feedback on how I was going
18	GTS	My lecturers were extremely good at explaining things
19	AAS	Too many staff asked me questions just about facts
20	GTS	The teaching staff worked hard to make their subjects interesting
21*	AWS	There was a lot of pressure on me to do well in this course
22	GSS	My course helped me to develop the ability to plan my own work
23*	AWS	The sheer volume of work to be got through in this course meant it couldn't all be thoroughly comprehended
24	CGS	The staff made it clear right from the start what they expected from students
25	OAL	Overall, I was satisfied with the quality of this course

* Denotes a reverse scored item

4. PREVIOUS ANALYTIC PRACTICES

For the purposes of reporting graduates' perceptions of course quality, the proportion of graduates endorsing particular response options to the various propositions of the CEQ are often cited. For example, it might be said that

64.8 per cent of graduates either agree or strongly agree that they were "satisfied with the quality of their course" (item 25).

In the analysis of CEQ data undertaken for the Graduate Careers Council (Johnson, 1997), item responses were coded -100, -50, 0, 50 and 100, corresponding to the categories 'strongly disagree' 'disagree', 'neutral', 'agree', and 'strongly agree'. From these values, means and standard deviations were computed. Although the response data are ordinal rather than interval there is some justification for reporting means given the large numbers of respondents.

There is concern that past analytic practices have not been adequate to validate the hypothesised structure of the instrument and have not been suitable for deriving true measures of graduate perceptions of course quality. There had been attempts to validate the hypothesised structure. Wilson, Lizzio and Ramsden (1996) referred to two studies, one by Richardson (1994) and one by Trigwell and Prosser (1991) that used confirmatory factor analysis. However, these studies were based on samples of 89 and 35 cases respectively, far too few to provide support for the claimed instrument structure.

5. ANALYSIS OF INSTRUMENT STRUCTURE

The data set being analysed in this study was derived from the 1996 administration of the CEQ. The instrument had been circulated to all recent graduates (approximately 130,000) via their universities. Responses were received from 90,391. Only the responses from 62,887 graduates of bachelor degree programs were examined in the present study, as there are concerns about the appropriateness of this instrument for post-bachelor degree courses. In recent years a separate instrument has been administered to post-graduates. Examination of the data set revealed that 11,256 returns contained missing data and it was found that the vast majority of these had substantial numbers of missing items. That is, most respondents who had missed one item had also omitted many others. For this reason, the decision was taken to use only data from the 51,631 complete responses.

5.1 Using exploratory factor analysis to investigate instrument structure

Exploratory factor analyses have been conducted in order to show that patterns of responses to the items of the instrument reflect the constructs that were used in framing the instrument. In this study, exploratory factor analyses were undertaken using principal components extraction followed by varimax rotation using SPSS (SPSS Inc., 1995). The final rotated factor solution is represented in Table 10-3. Note that items that were reverse scored have been re-coded so that factor loadings are of the same sign. The five factors in this solution all had Eigen values >1 and together, they account for 56.9 per cent of the total variance.

Table 10-3. Rotated factor solution for an exploratory factor analysis of the 1996 CEQ data

Item no.	Sub-scale	Factor 1	Factor 2	Factor 3	Factor 4	Factor 5
8	AAS					0.7656
12	AAS					0.7493
16	AAS	0.5931				0.3513
19	AAS					0.7042
2	GSS		0.7302			
5	GSS		0.7101			
9	GSS		0.4891			
10	GSS		0.7455			
11	GSS		0.5940			
22	GSS		0.6670			
1	CGS			0.7606		
6	CGS			0.7196		
13	CGS			0.6879		
24	CGS	0.3818		0.6327		
3	GTS	0.6268	0.3012	0.3210		
7	GTS	0.7649				
15	GTS	0.7342				
17	GTS	0.7828				
18	GTS	0.6243				
20	GTS	0.6183				
4	AWS				0.7637	
14	AWS				0.5683	
21	AWS				0.7674	
23	AWS				0.7374	
25	Over all	0.4266	0.4544	0.4306		

Note: Factor loadings <0.3 have been omitted from the table. R^2 =0.57

From Table 10-3 it can be seen that items generally load at least moderately on the factors that correspond with the sub-scales that they were intended to reflect. There are some interesting exceptions. Item 16 was designed as an assessment probe, but loads more strongly onto the factor associated with the good teaching scale. This item referred to feedback on assignments, and the patterns of responses indicate that graduates associate this issue more closely with teaching than with other aspects of assessment raised in this instrument. Item 3, which made reference to motivation, was intended as a good teaching item but also had modest loadings onto factors associated with clear goals and generic skills. Item 25, an overall course satisfaction statement, has modest loadings on the good teaching, clear goals, and generic skills scales. However, its loadings onto the factors associated with appropriate workload and appropriate assessment items were quite low at .07 and 0.11 respectively. Despite these departures from what might have been hoped by its developers, this analysis shows a satisfactory pattern of loadings, suggesting that most items reflect the constructs that were argued by Ramsden (1991) to form the perceived course quality entity.

Messick (1989) argued that lack of adequate content coverage was a serious threat to validity. The exploratory factor analysis shows that most items reflect the constructs that they were intended to represent and that the instrument does show coverage of the factors that were implicated in effective learning. What exploratory factor analysis does not show is that the constructs that are theorised represent a quality of learning construct cohere to form that concept. In the varimax factor solution, each extracted factor is orthogonal to the others and therefore exploratory factor analysis does not provide a basis for arguing that the identified constructs form a unidimensional construct that is a basis for true measurement. Indeed, this factor analysis provides *prima facie* evidence that the construct is multi-dimensional. For this reason, a more flexible tool for examining the structure of the target construct is required, and confirmatory factor analysis provides this.

5.2 Using confirmatory factor analysis to investigate instrument structure

Although exploratory factor analysis has proven to be a useful tool in the examination of the structure of constructs in the social sciences, confirmatory factor analysis has come to play a more prominent role, as it is possible to hypothesise structures and to test those structures against observed data. In this study, the AMOS program (Arbuckle, 1999) was used to test the a series of plausible models.

Keeves and Masters (1999) have pointed out that constructs of interest in the social sciences are often complex, multivariate, multi-factored and multi-level. Tools such as exploratory factor analysis are limited in the extent to which they are able to probe these structures. Further, for a construct to be compatible with simple measurement – that is, to be able to report a single quantitative score that truly reflects a level of a particular construct – the structure of the construct must reflect ultimately a single underlying factor. The simplest case occurs when all the observed variables load onto a single factor. Acceptable alternatives include hierarchical structures in which several distinct factors are shown to reflect a higher order factor, or nested models in which variables reflect both a set of discrete and uncorrelated factors, but also reflect a single common factor. In these cases, it is expected that the loadings on the single common factor are greater than their loadings onto the discrete factors. As an alternative, if a model with discrete and uncorrelated factors was shown to provide a superior fit to the data, then this structure would indicate that a single measure could not reflect the complexity of construct.

Byrne (1998) has argued that confirmatory factor analysis should normally be used in an hypothesis testing mode. That is, a structure is proposed and tested against real data, then either rejected as not fitting or not rejected on the basis that an adequate degree of fit is found. However, she also pointed out that the same tool could be used to compare several alternatives. In this study, the purpose is to discover whether one of several alternative structures that are compatible with a single measurement is supported or whether an alternative model of discrete factors, that is not compatible with measurement, is more consistent with the data.

Four basic models were compared. It was argued in the development of the CEQ that course quality could be represented by five factors: good teaching, clear goals, generic skills development, appropriate assessment, and appropriate workload. It is feasible that these factors are undifferentiated in the data set and that all load directly onto an underlying perceived course quality factor. Thus the first model tested was a single factor model. A hierarchical model was tested in which the proposed five component constructs were first order factors and that they loaded onto a single second order perceived course quality factor. The third variant was a nested model in which the observed variables loaded onto the five component constructs and that they also loaded separately onto a single perceived course quality factor. Finally, an alternative, that is not compatible with a singular measure, has the five component constructs as uncorrelated factors. The structures corresponding to these models are shown in Figure 10-1.

Each of these models was constructed and then subject to a refinement process. Item 25, the 'overall course quality judgement', was removed from the models, as it was not meant to reflect any one of the contributing constructs, but rather was an amalgam of them all. In the refinement, variables were removed from the model if their standardised loading onto their postulated factor was below 0.4. Second, modification indices were examined, and some of the error terms were permitted to correlate. This was restricted to items that were designed to reflect a common construct. For example, the error terms of items that were all part of the good teaching scale were allowed to be correlated, but correlations were not permitted among error terms of items from different sub-scales. Finally, one of the items, Item 16, which was intended as an appropriate assessment item, was shown to be related also to the good teaching sub-scale. Where the modification index suggested that a loading onto the good teaching might provide a better model fit, this was tried.

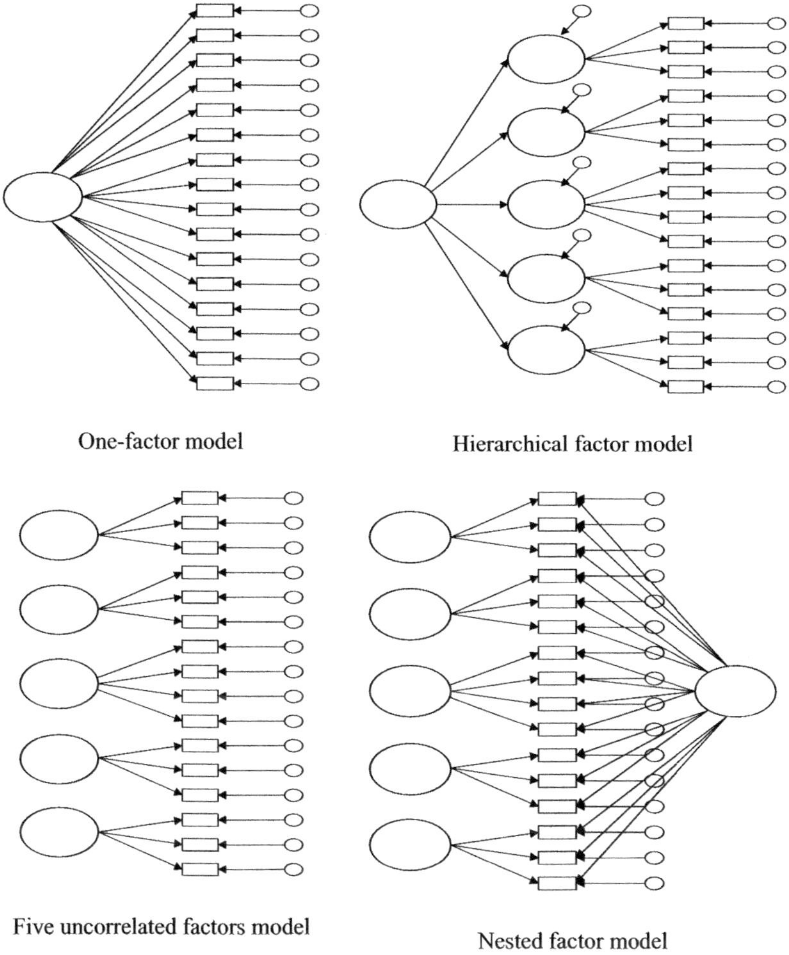

Figure 10-1. Structures of single factor, hierarchical factors, nested factors, and uncorrelated factor models for the CEQ

Summary results of the series of confirmatory factor analyses on these models are shown in Table 10-4. (More comprehensive tables of the results of these analyses are available in an Appendix to this chapter). The first conclusion to be drawn from these analyses is that the five discrete factor model, the one inconsistent with measurement, is inferior to the other three models, each of which is consistent with measurement of a unitary construct.

It is worth noting Bejar (1983, p.31) who wrote:

Unidimensionality does not imply the performance on items is due to a single psychological process. In fact, a variety of psychological processes are involved in responding to a set of items. However, as long as they function in unison - that is, performance on each item is affected by the same process in the same form - unidimensionality will hold.

Thus, the identification of discrete factors does not imply necessarily a departure from unidimensionality. However, it remains to be shown that the identified factors do operate in unison.

Second, the nested and single factor models are superior to the hierarchical structure. However, in the single factor model, only 18 of the original 25 items are retained. The set of fit indices favours slightly the nested model, but in this structure, factor loadings of some variables were greater on the discrete factor than on the common factor. Such items appear not to contribute well to a unidimensional construct that could be used as a basis for measurement. The removal of these items, as has been done in the single factor model, yields both an acceptable structural model and does provide a basis for claims that the construct in question does provide a basis for true measurement.

Table 10-4. Summary of model comparisons suing confirmatory factor analysis

Model	Variables retained	GFI	AGFI	PGFI	RMR	RMSEA
Single factor	18	0.962	0.943	0.674	0.041	0.054
Hierarchical five factor	23	0.942	0.926	0.741	0.069	0.055
Nested five factor	23	0.963	0.949	0.708	0.042	0.046
Discrete five factor	23	0.908	0.887	0.743	0.116	0.069

For detailed model fit statistics see Curtis (1999)

The confirmatory factor analyses suggest that the scale formed by 18 CEQ items does provide a reasonable basis for measurement of a latent 'perceived course quality' variable.

5.3 Conclusions about instrument structure

Conducting the exploratory factor analysis has been a useful exercise as it has provided evidence that the factors postulated to contribute to course quality are represented in the instrument and, largely, by the items that were intended to reflect those concepts. This analysis has provided important evidence to support validity claims as the construct of interest is represented by the full range of constituent concepts. Second, there are no extraneous factors in the model that would introduce ‘construct irrelevant variance’ (Messick, 1994). What exploratory factor analysis has not been able to provide is an indication that the component factors cohere sufficiently to provide a foundation for the measurement of a unitary construct. The exploratory analysis, in providing evidence for five discrete factors, seems to suggest that this may not be so.

Confirmatory factor analysis has enabled several alternative structures to be compared. Several structures that are potentially consistent with a unitary underlying construct have been compared among themselves and with an alternative structure of uncorrelated factors. The uncorrelated model is the one suggested by exploratory factor analysis. The confirmatory approach has shown that the uncorrelated model does not satisfactorily account for the response patterns of the observed data. Of the models that reveal reasonably good fit, the nested model, which retained 23 items, shows a pattern of loadings that suggests multidimensionality. The one-factor model, in which only 18 items were retained, shows acceptable fit to the data and supports a unidimensional underlying construct. A similar finding emerges from the application of the Rasch measurement model.

6. MEASUREMENT PROPERTIES OF THE COURSE EXPERIENCE QUESTIONNAIRE

In traditional analyses of survey instruments, it is common to compute the Cronbach alpha statistic as an indicator of scale reliability.

In order to test whether the complete instrument functions as a coherent indicator of perceived course quality and whether each of the sub-scales is internally consistent, the SPSS scale reliabilities procedure was employed. The results of these analyses are shown in Table 10-5. These scales all have quite satisfactory Cronbach alpha values.

Table 10-5. Scale coherence for the complete CEQ scale and its component sub-scales

Scale	Items	Cronbach alpha
GTS	6	0.8648
CGS	4	0.7768
AAS	4	0.6943
AWS	4	0.7154
GSS	6	0.7645
CEQ	25	0.8819

Note that all 51631 responses were used for all scales

In order to investigate the extent to which individual items contributed to the overall CEQ scale and to the sub-scale in which they were placed, the scale Cronbach alpha value, if the item were omitted from it, was calculated. Items for which the 'scale alpha if item deleted' value is higher than the scale alpha value with all items were judged not to have contributed to that scale and could be removed from it. The value of alpha for the CEQ scale with all

25 items is 0.8819 and the removal of each of items 4, 9 and 21 would improve the scale coherence. The point biserial correlations of other items were over 0.40 and ranged to 0.71, but for these items, the point biserial correlations were 0.32, 0.31 and 0.19 respectively. Together, the increase in scale alpha if these items are removed and the low correlations of scores on these items with scale scores indicate that these items do not contribute to the coherence of the scale.

Table 10-6. Contribution of items to the 25 item CEQ scale

Item number	Sub-scale	Corrected item-total correlation	Squared multiple correlation	Scale alpha if item deleted
1	CGS	.4779	.3821	.8769
2	GSS	.4202	.4572	.8783
3	GTS	.6333	.5142	.8727
4	AWS	.2496	.3116	.8825
5	GSS	.4178	.4451	.8784
6	CGS	.5483	.4448	.8750
7	GTS	.5976	.5167	.8735
8	AAS	.3076	.3001	.8819
9	GSS	.2277	.1892	.8841
10	GSS	.4454	.4048	.8777
11	GSS	.4164	.2686	.8784
12	AAS	.4617	.4147	.8773
13	CGS	.5086	.3741	.8761
14	AWS	.4572	.3116	.8774
15	GTS	.5890	.4680	.8738
16	AAS	.4056	.3218	.8791
17	GTS	.6167	.5346	.8731
18	GTS	.5974	.4673	.8740
19	AAS	.4124	.3206	.8785
20	GTS	.5824	.4547	.8742
21	AWS	.1044	.3149	.8869
22	GSS	.4022	.3308	.8787
23	AWS	.3242	.3541	.8812
24	CGS	.5268	.3967	.8756
25	Over all	.6690	.5216	.8721

7. RASCH ANALYSES

Rasch analyses of the 1996 CEQ data were undertaken using Quest (Adams & Khoo, 1999). Because all items used the same five response categories both the rating scale and the partial credit models were available. A comparison of the two models was undertaken using Conquest (Wu, Adams, & Wilson, 1998) and the deviances were 11,242.831, (24 parameters) for the rating scale model and 10,988.359 for the partial credit model (78 parameters) The reduction in deviance was 254.472 for 54 additional parameters, and on this basis the partial credit model was chosen for subsequent analyses. The 51,631 cases with complete data were used and all 25 items were included in analyses.

7.1 Refinement

The refinement process involved examining item fit statistics and item thresholds and removing those items that revealed poor fit to the Rasch measurement model. Given that the instruments is a low stakes survey for individual respondents but important for institutions, critical values chosen for the Infit Mean Square (IMS) fit statistics were 0.72 and 1.30, corresponding to "run of the mill" assessment (Linacre, Wright, Gustafsson, & Martin-Lof, 1994). More lenient critical values, of say 0.6 to 1.4, could have been used.

Item thresholds estimates (Andrich or tau thresholds in Quest) were examined for reversals. None were found. Reversals of item thresholds would indicate that response options for some items, and therefore the items, are not working as intended and would require revision of the items. On each iteration, the worst fitting item whose Infit Mean Square was outside the accepted range was deleted and the analysis re-run. In succession, items 21 (AWS), 9 (GSS), 4 (AWS), 23 (AWS), 8 (AAS) and 16 (AAS) were removed as underfitting a unitary construct. Item 25, the overall judgment item, was removed as it overfitted the scale and therefore added little unique information. This left a scale with 18 items, although the retained items were not identical to those that remained following the CFA refinement. The CFA refinement retained Item 16 but rejected Item 19, while in the Rasch refinement, Item 16 was omitted and Item 19 was preserved.

Summary item and case statistics for the 18-item scale following refinement are shown in Table 10-7. The item mean is constrained to 0. The item estimate reliability (reliability of item separation, Wright & Masters, 1982, p.92) of 1.00 indicates that the items are well separated relative to the errors of their locations on the scale and thus define a clear scale. The high values for this index may be influenced by the relatively large number of cases used in the analysis. The mean person location of 0.49 indicates that the instrument is reasonably well-targeted for this population. Instrument targeting is displayed graphically in the default Quest output in a map showing the distribution of item thresholds adjacent to a histogram of person locations. The reliability of case estimates is 0.89 and this indicates that responses to items are consistent and result in the reliable estimation of person locations on the scale. Andrich (1982) has shown that this index is numerically equivalent to Cronbach alpha, which, under classical item analysis, was 0.88 for all 25 items.

Estimated items locations, Masters thresholds (absolute estimate of threshold location) and the Infit Mean Square fit statistic for each of the 18 fitting items are shown in Table 10-8. Item locations range from -0.55 to +0.64 and thresholds from -2.06 (item 19) to +2.70 (item 14). It is useful to examine these ranges, and in particular the threshold range. If a person is located at a greater distance than about two logits from a threshold the probability of the expected response is about 0.9 and little information can be gleaned from the response. The threshold range of the CEQ at approximately 5 logits gives the instrument a useful effective measurement range, sufficient for the intended population.

Table 10-7. Summary item and case statistics from Rasch analysis

	N	Mean	Std Dev	Reliability
Items	18	0.00	0.35	1.00
Cases	51631	0.49	0.89	0.89

Items were retained in the refinement process on the basis of their Infit Mean Square values. These statistics, which range from 0.77 for item 3 to 1.23 for item 12 and have a mean of 1.00 and a standard deviation of 0.13, are shown in Table 10-8.

Table 10-8. Estimated item thresholds and Infit Mean Square fit indices for 18 fitting CEQ items

Item	Locat'n	Std err	T'hold 1	T'hold 2	T'hold 3	T'hold 4	IMS
1	0.05	0.01	-1.82	-0.73	0.41	2.36	1.03
2	-0.47	0.01	-1.95	-1.13	-0.35	1.56	1.05
3	0.12	0.00	-1.51	-0.65	0.64	2.00	0.77
5	-0.55	0.01	-1.96	-1.33	-0.39	1.50	1.03
6	-0.04	0.00	-1.65	-0.67	0.07	2.08	0.92
7	0.64	0.00	-0.99	-0.01	1.13	2.43	0.88
10	-0.13	0.01	-1.68	-1.06	0.15	2.06	1.05
11	-0.39	0.00	-1.46	-0.86	-0.39	1.16	1.14
12	-0.09	0.00	-1.20	-0.73	0.23	1.33	1.23
13	0.06	0.00	-1.70	-0.67	0.44	2.17	1.04
14	0.16	0.01	-1.79	-0.67	0.41	2.70	1.15
15	0.37	0.00	-1.22	-0.40	0.86	2.23	0.90
17	0.41	0.00	-1.36	-0.30	0.82	2.48	0.86
18	0.32	0.01	-1.45	-0.79	0.98	2.54	0.83
19	-0.43	0.01	-2.06	-1.73	0.41	1.67	1.18
20	0.16	0.01	-1.43	-0.83	0.55	2.35	0.86
22	-0.43	0.01	-1.67	-1.22	-0.34	1.50	1.09
24	0.23	0.01	-1.60	-0.62	0.77	2.39	0.95

(For clarity, standard errors of threshold estimates have not been shown but range from 0.01 to 0.04)

7.2 Person estimates

Having refined the instrument and established item parameters, it was possible to generate estimates on the scale formed by the 18 retained items. Person estimates and their standard errors were generated and were available for use in other analyses. For example, the Rasch scaled data were used in a three-level model designed to explore the influences of individual characteristics such as gender and age, of course type, and of institution on individual perceptions of course quality (Curtis & Keeves, 2000). In cases where data are missing or where different groups have responded to different sub-sets of items, the Rasch scaled score is able to provide a reliable metric that is not possible if only raw scores are used.

The standard errors that are estimated for scaled scores also provide important evidence for validity claims. Since validity "refers to the degree to which evidence and theory support the interpretation of test scores" (American Educational Research Association, American Psychological Association, & National Council on Measurement in Education, 1999, p.9) it is necessary to know the precision of the score in order to claim that particular interpretations are warranted.

8. SUMMARY

The purpose of this chapter was to compare classical and contemporary approaches to both the examination of instrument structures and the exploration of the measurement properties of instruments.

It has been shown that exploratory factor analysis helps to provide evidence for validity claims. However, it is not able to provide evidence to support a claim for a structure that is compatible with the measurement of a unitary construct. Confirmatory factor analysis has enabled the comparison of alternative structures, and is able both to provide evidence for validity and to compare structures for conformity with the requirements of measurement. It is also possible to use confirmatory factor analysis to refine instruments by removing items that do not cohere within a measurement-compatible structure.

Classical item analysis was able to provide some information about instrument coherence, but appears not to be sensitive to items that fail to conform to the demands of measurement. Elsewhere (Wright, 1993) classical item analysis has been criticised for not being able to deal with missing data nor for situations in which different groups of respondents have different item subsets. By using the Rasch measurement model, the measurement properties of the CEQ instrument have been investigated, it has been shown that an instrument can be refined by the removal of misfitting items, and item independent estimates of person locations have been made. Such measures, with known precision, are available as inputs to other forms of analysis and also contribute to claims of test validity.

9. REFERENCES

Adams, R. J., & Khoo, S. T. (1999). Quest: the interactive test analysis system (Version PISA) [Statistical analysis software]. Melbourne: Australian Council for Educational Research.

American Educational Research Association, American Psychological Association, & National Council on Measurement in Education. (1999). *Standards for educational and psychological testing*. Washington, DC: American Educational Research Association.

Andrich, D. (1982). An index of person separation in latent trait theory, the traditional KR-20 index, and the Guttman scale response pattern. *Educational Research and Perspectives, 9*(1), 95-104.

Arbuckle, J. L. (1999). AMOS (Version 4.01) [CFA and SEM analysis program]. Chicago, IL: Smallwaters Corporation.

Bejar, I. I. (1983). *Achievement testing. Recent advances*. Beverly Hills: Sage Publications.

Bond, T. G., & Fox, C. M. (2001). *Applying the Rasch model. Fundamental measurement in the human sciences*. Mahwah, NJ: Lawrence Erlbaum and Associates.

Byrne, B. M. (1998). *A primer of LISREL: basic applications and programming for confirmatory factor analytic models*. New York: Springer-Verlag.

Curtis, D. D. (1999). *The 1996 Course Experience Questionnaire: A Re-Analysis*. Unpublished Ed. D. dissertation, The Flinders University of South Australia, Adelaide.

Curtis, D. D., & Keeves, J. P. (2000). The Course Experience Questionnaire as an Institutional Performance Indicator. *International Education Journal, 1*(2), 73-82.

Johnson, T. (1997). *The 1996 Course Experience Questionnaire: a report prepared for the Graduate Careers Council of Australia*. Parkville: Graduate Careers Council of Australia.

Keeves, J. P., & Masters, G. N. (1999). Issues in educational measurement. In G. N. Masters & J. P. Keeves (Eds.), *Advances in measurement in educational research and assessment* (pp. 268-281). Amsterdam: Pergamon.

Kline, P. (1993). *The handbook of psychological testing*. London: Routledge.

Linacre, J. M., Wright, B. D., Gustafsson, J.-E., & Martin-Lof, P. (1994). Reasonable mean-square fit values. *Rasch Measurement Transactions, 8*(2), 370.

Messick, S. (1989). Validity. In R. L. Linn (Ed.), *Educational measurement* (pp. 13-103). New York: American Council on Education, Macmillan.

Messick, S. (1994). The interplay of evidence and consequences in the validation of performance assessments. *Educational Researcher, 23*(2), 13-23.

Michell, J. (1997). Quantitative science and the definition of measurement in psychology. *British Journal of Psychology, 88*, 355-383.

Ramsden, P. (1991). Report on the Course Experience Questionnaire trial. In R. Linke (Ed.), *Performance indicators in higher education* (Vol. 2). Canberra: Commonwealth Department of Employment, Education and Training.

SPSS Inc. (1995). SPSS for Windows (Version 6.1.3) [Statistical analysis program]. Chicago: SPSS Inc.

Wilson, K. L., Lizzio, A., & Ramsden, P. (1996). *The use and validation of the Course Experience Questionnaire* (Occasional Papers 6). Brisbane: Griffith University.

Wright, B. D. (1993). Thinking with raw scores. *Rasch Measurement Transactions, 7*(2), 299-300.

Wright, B. D., & Masters, G. N. (1982). *Rating scale analysis.* Chicago: MESA Press.

Wu, M. L., Adams, R. J., & Wilson, M. R. (1998). ConQuest generalised item response modelling software (Version 1.0) [Statistical analysis software]. Melbourne: Australian Council for Educational Research.

Chapter 11

COMBINING RASCH SCALING AND MULTI-LEVEL ANALYSIS

Does the playing of chess lead to improved scholastic achievment?

Murray Thompson
Flinders University

Abstract: The effect of playing chess on problem solving was explored using Rasch scaling and hierarchical linear modelling. It is suggested that this combination of Rasch scaling and multilevel analysis is a powerful tool for exploring such areas where the research design has proven difficult in the past.

Key words: Rasch scaling, multi-level analysis, chess

1. INTRODUCTION

Combining the tools of educational measurement can be used to overcome complex educational issues. This chapter presents an example of a solution to one such problem, illustrating how these measurement tools can be used to answer complex questions.

One of the most difficult problems in educational research has been that of dealing with structured data that arise from school-based research. Students are taught in groups and very often the researcher has to be content with these intact groups and so random assignment to groups is simply not possible and the assumption of simple random samples does not hold.

The problem explored in this chapter relates to the question of whether the playing of chess leads to improved scholastic achievement. Those involved with chess often make the claim that the playing of chess leads to improved grades in school. Ferguson (n.d.) and Dauvargne (2000) summarise the research which supports this view.

S. Alagumalai et al. (eds.), Applied Rasch Measurement: A Book of Exemplars, 197–206.

Essentially, there are a number of problems associated with any research in this area. Perhaps this reflects the quasi-experimental nature of much of the research or perhaps it is a result of the view that those who play chess are the smart students who would have performed equally well without chess. The learning and the playing of chess take a considerable period of time and practice and any improvement in scores in cognitive tests may be confused with the normal development of the students. The usual experimental design for investigating the effects of an instructional intervention has an experimental group and a control group and utilizes a pre-test and post-test arrangement, which compares one group with the other. In school situations, such designs can be very difficult to maintain effectively, with so many other intervening factors. For example, the two groups will often be two intact class groups and so the "random assignment of students" to the groups is in reality a random assignment of treatments to the groups. Moreover, as intact groups, it is likely that their treatments may differ in a number of other ways. For example, they may have different teachers, or the very grouping of the students themselves may have an effect. In addition, there is the risk that any positive finding may be a Hawthorne effect rather than a consequence of the treatment itself.

It seems that the traditional pre-test and post-test experimental designs have led to results which, while encouraging, have not been conclusive and need further support.

An alternative approach that is discussed in this chapter makes use of statistical control to take into account the effect of the confounding variables.

2. A STUDY INTO THE SCHOLASTIC EFFECTS OF CHESS

If, as has been argued, the playing of chess confers an academic advantage to students then those students who play chess should, when controlling for other variables, such as intelligence and grade level, perform better than those who do not. In this study, the performance of 508 students from grades 6-12 in the Australian Schools Science Competition was analyzed. The Australian Schools Science Competition is an Australia wide competition that is held every year. Students in Grades 3 - 12 compete in this multiple-choice test. The competition is administered by the Educational Testing Centre of the University of New South Wales. Faulkner (1991) outlined the aims of the competition and gave a list of items from previous competitions. Among its aims are the promotion of interest in science and awareness of the relevance of science and related areas to the lives of the students, and the recognition and encouragement of excellence in science. An emphasis is placed on the ability of students to apply the processes and skills of science. Since science syllabi throughout Australia vary, the questions that are asked are essentially independent of any particular syllabus and are designed to test scientific thinking. Thus, the questions are not designed to test knowledge, but rather to test the ability of the candidates to interpret and examine information in scientific and related areas. Students may be required to analyze, to measure, to read tables, to interpret graphs, to draw conclusions, to predict, to calculate and to make inferences from the data given in each of the questions. It seems logical then that involvement in chess should confer an advantage to individual students in the Australian Schools Science Competition.

It is hypothesised then that students who play chess regularly should perform better in the Australian Schools Science Competition than those who do not, when controlling for the other variables that may be involved.

2.1 Rasch Scaling and the Australian Schools Science Competition

The Australian Schools Science Competition is a multiple choice test in which students select from four alternatives. There is a separate test for each year level, but many of the items are common to more than one year level. Rasch scaling allows the conversion of the performance scores into an interval scale which allows for the equating of the scores across the grade levels, to provide an input for the multi-level analysis. Thompson (1998, 1999) showed that the data from the Science Competition fit the Rasch model well and that concurrent equating could be used to equate the results from the different year levels to allow the results from all of participating students to be put on the one scale.

The process of undertaking the Rasch analysis can be a complex one. In particular, it is necessary to manipulate a great deal of data to get these into an appropriate form. In this case the data consist of the individual letter responses of each of the students for each of the questions. To the original four responses, A, B, C and D, has been added the response N, indicating that the student did not attempt this item. The first task is to identify the items common to more than one grade level and arrange the responses into columns. In all, in this study, there were 249 different items across the seven grade levels from 6-12. The responses from all the students for each separate item had to be arranged into separate columns in a spread-sheet and then converted into a text file for input into the Rasch analysis program. This sorting process is very time consuming and requires a great deal of patience. It can be readily seen that for 508 students and 249 items, the spread sheet to be manipulated had 249 columns and 508 rows, plus the header rows. This spread-sheet file, 99ScienceComp.xls, can be requested from the author through email. This spread-sheet file is then converted into a text file for input into the QUEST program for Rasch analysis (Adams and Khoo, 1993). The QUEST program has been used to analyse these data and to estimate the difficulty parameters for all items and the ability parameters for all students. The submit file used to initiate the QUEST program, 99con2.txt can be requested from the author. The initial analysis indicated that a few of the items needed to be deleted. A quick reference to the item fit map in this file indicates that of the 249 items 8 items, 24, 63, 150, 180, 239, 241, 243 and 249 failed to fit the Rasch model and did not meet the infit mean square criteria. This is seen with each of these items lying outside the accepted limits indicated by the vertical lines drawn on the item fit map. QUEST suggests that the item fit statistics for each item should lie between 0.76 and 1.30. These values are within with the generally accepted range for a normal multiple-choice test, as suggested by Bond and Fox (2001, p. 179). Consequently, the 8 items were deleted from the analysis. The data were run once again using the QUEST program and the output files (9DINTAN.txt, 9DSHO2.txt, 9DSHOCA2.txt, 9DSHOIT2.txt) can be requested from the author through email. These files have been converted to text files for easy reference. They include the show file, the item analysis file, the show items files and the show case file. Of particular interest is the show case file because it gives the estimates of the performance ability of each student in the science competition, and since these scores are Rasch scaled using concurrent equating, all of the scores from grade 6-12 have been placed on a single scale. It is these Rasch scaled scores for each of the students that we now wish to explain in terms of the hypothesised variables.

2.2 Preparing the data for multi-level analysis

The performance ability score for each student was then transferred to another spread-sheet file and the IQ data for each student was added. This file, Chesssort.xls, which can be requested from the author on the CD ROM, includes information on the the student group numbers that corresponds to the 22 separate groups who undertook the test, the individual student ID codes, their IQ scores, their performance scores and a dichotomous variable to indicate whether or not the student played chess. Those individual students for whom no IQ score was available have been deleted from the sample. This leaves a group of 508 students, of whom 64 were regular chess players.

Multi-level analysis using hierarchical linear modelling and the HLM program is then used to analyse the Rasch scaled data. (Bryk & Raudenbush,

1992, Bryk, Raudenbush, & Congdon,1996, Raudenbush, & Bryk, 1996, 1997)

This HLM requires that the data be arranged into at least two levels. In this case, the Level 1 data has the individual students arranged into class groupings with the performance data, the IQ data and the chess data for each individual. Level 1 data includes information specific to each individual and includes chess playing (Y/N), IQ and group membership. At level 2, data on groups are represented and include Grade level. These data sets both need to be converted to files suitable for input into the HLM program for the multi-level analysis phase of the study. In this case, the data format was specified using FORTRAN style conventions and this necessitates arranging the text file into appropriate columns. It is vital that when this is done, careful checks are made that the data line up in their appropriate columns. This can be difficult when the data has a varying number of digits, such as IQ, which could be, for example, 95.5 or 112.0. Often spaces need to be inserted into the final text file. It is critical that when this check is done, a type-writer style font with every character taking the same space is used. Fonts such as "Courier" or "Courier New" are ideal. The final data set is given in two files. The level 1 data is shown as Chess.txt, that can be requested from the author through email.

The Level 2 data appear as Chesslev2.dat. Once the data has been input into HLM, it is necessary to construct a sufficient statistics matrix.

2.3 Summary of the Research methods

This study uses data from an independent boys school with a strong tradition of chess playing. The school fields teams in competitions at both the primary and secondary levels and so a significant and identifiable group of the students plays competitive chess in the organized inter-school competition and practises chess regularly. Each of these students played a regular fortnightly competition and was expected to attend weekly practice, where they received chess tuition from experienced chess coaches. The students had also taken part in the Australian Schools Science Competition as part of intact groups and data from 1999 for Grades 6 - 12 were available for analysis. IQ data were readily available for the students in Grades 6 -12. Subjects, then, were all boys (n= 508) in Grades 6 –12, for whom IQ data were available. Of these 508 students 64 were competitive chess players. Rasch scaling, with concurrent equating, was used to put all of the scores on a single scale. These scores were then used as the outcome variable to be explained using a hierarchical linear model, and the variables of IQ, chess playing, other class level factors, grouping and grade to see if the playing of chess made a significant contribution to Science Competition achievement.

A dichotomous variable was used to indicate the playing of chess, with chess players being given 1 and non-players 0. Chess players were defined as those who represented the school in competitions on a regular basis.

The HLM program is then used to build up a model to explain the data and this final model is compared with the null model to determine the variance explained by variable included in the model.

3. RESULTS

The performance ability scores were used as the outcome variable in a hierarchical linear model to be explained by the various parameters involved.

The final model was as follows in equations (1), (2), (3) and (4).

In this Level 1 model, the outcome variable Y, the Rasch scaled performance scores measured by the Science Competition test is modelled using an intercept or base level B_0, plus a term that expresses the effect of IQ, with its associated slope, B_1, and a term which expresses the effect of playing chess and its associated slope B_2. There is also an error term R. Thus the outcome variable Y is explained in terms of IQ and involvement in chess at Level 1.

$$Y = B_0 + B_1*(IQ) + B_2*(CHESS) + R \tag{1}$$

In the Level 2 model, the effect of the Level 2 variables on each of the B terms in the Level 1 model is given in equations (2), (3) and (4).

$$B_0 = G_{00} + G_{01}*(GRADE) + U_0 \tag{2}$$

$$B_1 = G_{10} + U_1 \tag{3}$$

$$B_2 = G_{20} + U_2 \tag{4}$$

Thus in equation (2), the constant term B_0 is expressed as a function of Grade, with an associated slope G_{01}. Values of each of these terms are estimated and the level of statistical significance evaluated to assess the effect of each of the terms.

Initially, the HLM program makes estimates of the various values of the slopes and intercepts, using a least squares regression procedure and then in an iterative process, improves the estimation using a maximum likelihood estimation and the empirical Bayes procedure. The final output from the

HLM program. Table 11-1 shows the reliability estimates of the Level 1 data.

Table 11-1. Relaibility estimates of the Level 1 data.

Random Level-1 coefficient	Reliability estimate
INTRCPT1, B_0	0.664
IQ, B_1	0.324
CHESS, B_2	0.019

Table 11-2 shows the least -squares regression estimates of the fixed effects.

Table 11-2. The least-squares regression estimates of the fixed effects.

Fixed Effect	Coefficient	Std Error	T-ratio	Approx degrees of freedom	P-value
For **INCPT1**, B_0					
INCPT2, G_{00}	-1.65	0.18	-9.22	504	0.00
GRADE, G_{01}	0.22	0.200	11.03	504	0.00
For **IQ** slope, B_1					
INCPT2, G_{10}	0.04	0.002	19.09	504	0.00
For **CHESS** slope, B_2					
INTRCPT2, G_{20}	0.12	0.09	1.32	504	0.17

The final estimations of the fixed effects are shown in Table 11-3.

Table 11-3. The final estimations of the fixed effects.

Fixed Effect	Coefficient	Standard Error	T-ratio	Approx degrees of freedom	P-value
For **INTRCPT1**, B_0					
INTRCPT2, G_{00}	-1.57	0.33	-4.81	20	0.00
GRADE, G_{01}	0.21	0.06	5.69	20	0.00
For **IQ** slope, B_1					
INTRCPT2, G_{10}	0.04	0.00	13.67	21	0.00
For **CHESS** slope, B_2					
INTRCPT2, G_{20}	0.06	0.09	0.62	21	0.54

The final estimations of the variance components are shown in Table 11-4.

In order to calculate the amount of variance explained by the model, a null model, with no predictor variables was formulated. The estimates of the variance components for the null model are shown in Table 11-5.

Table 11-4. The final estimations of the variance components.

Random Effect	Std Dev.	Variance Component	df	Chi-square	P-value
INTRCPT1, U_0	0.222	0.049	15	52.09	0.000
IQ slope, U_1	0.007	0.000	16	29.84	0.019
CHESS slope, U_2	0.053	0.003	16	20.48	0.199
Level-1, R	0.606	0.367			

Table 11-5. Estimated variance components for the null model.

Random Effect	Std Dev.	Variance Component	df	Chi-square	P-value
INCPT1, U0	0.602	0.362	21	357.7	0.000
Level-1, R	0.749	0.561			

Using the data from Tables 11-4 and 11-5, the amount of variance explained is calculated as follows:

$$\text{Variance explained at Level 2} = \frac{0.362 - 0.049}{0.362} = 0.865$$

$$\text{Variance explained at Level 1} = \frac{0.561 - 0.367}{0.561} = 0.346$$

In addition, the intraclass correlation can be calculated.

$$\rho = \frac{\tau_{00}}{\tau_{00} + \sigma^2} = \frac{0.362}{0.362 + 0.561}$$

$$= 0.392$$

This intraclass correlation represents the variance within groups compared to the total variance between and within groups. Thus the model is explaining 33.9 per cent (0.392 x 0.865) of the variance in terms of grade levels. The remaining 21.0 per cent ((1 - 0.392) x 0.346) is explained as the variation brought about by IQ and the playing of chess. In all 54.9 per cent of the variance in scores is explained by the model and 45.1 per cent is unexplained.

4. DISCUSSION AND INTERPRETATION OF THE RESULTS

In order to interpret the results, Table 11-2 is examined. The term G_{00} represents the baseline level, to which is added the effect of the grade level to determine the value of the intercept B_0. The value G_{00} represents the effect of the grade level and since this is statistically significant, it can be concluded from this that the students improve by 0.21 of a logit over one grade level, taking into account the effect of IQ and playing chess. The next important value is the term G_{10}, which indicates the effect of IQ on the performance in the Science Competition. Clearly this has a significant effect and even though the value seems very small, being 0.036, it must be remembered that it involves a metric coefficient for a variable whose mean value is in excess of 100 and has a range of over 50 units.

Of particular interest in this study is the value G_{20}. This represents the effect of playing competitive chess on the Science Competition achievement. It suggests that, taking into account the effects of IQ and grade level, students who play chess competitively, are performing at a level of 0.056 of a logit better than others, when controlling for the other variables of grade and IQ. This is approximately equivalent to one quarter of a year's work. However this result was not found to be significant.

This study has examined a connection between the playing of chess and the cognitive skills involved in science problem solving. The results have not shown a significant effect of the playing of chess on the Science Competition achievement of the students, when controlling for IQ and grade level.

5. CONCLUSION

The purpose of this study is to explore the relationship between the playing of chess and improved scholastic achievement and to illustrate the value of combining Rasch measurement with multi-level modelling. The difficulty in the research design associated with the intact groups of students has been overcome using the combination of Rasch scaling to place scores on a single scale and statistical control using a hierarchical linear model to obtain an estimate of the effect of playing chess and its statistical significance. The results of this study do not provide support for the hypothesis that the playing of chess leads to improved scholastic achievement. It is possible that the methodology of controlling for both grade level and IQ has removed the effect that has traditionally been attributed to chess, suggesting that those students who have been interested in chess have tended to be the more capable students. That is, the students who performed more ably at a particular grade level tended to have a higher IQ and there did not seem to be any significant effect of the playing of chess. This study provides a very useful application of both Rasch scaling and HLM and this method of analysis could be repeated easily in other situations.

6. REFERENCES

Adams, R. J. & Khoo, S-T. (1993). *QUEST the interactive test analysis system* Hawthorn Vic, Australia: ACER.

Bond, T. G. & Fox, C. M (2001). *Applying the Rach model: Fundamental measurement in the human sciences.* NJ: Lawrence Erlbum

Bryk, A. S. & Raudenbush, S.W. (1992). *Hierarchical linear models: applications and data analysis methods* Beverly Hills, Ca: Sage.

Bryk, A.S., Raudenbush, S. W., & Congdon, R.T. (1996). *HLM for Windows version 4.01.01* Chicago: Scientific Software.

Dauvergne, P. (2000). *The case for chess as a tool to develop our children's minds* Retrieved May 8, 2004, from http://www.auschess.org.au/articles/chessmind.htm

Faulkner, John (ed) (1991). *The Best of the Australian Schools Science Competition* Rozelle, NSW, Australia: Science Teachers' Association of New South Wales.

Ferguson, R. (n.d.). *Chess in education research summary*. Retrieved May 8, 2004, from http://www.easychess.com/chessandeducation.htm

Raudenbush, S.W. & Bryk, A. S. (1997) Hierarchical linear models. In J. P. Keeves, (ed) *Educational research, methodology and measurement* (2nd ed.), Oxford: Pergamon, pp. 2590-2596.

Raudenbush, S.W. & Bryk, A. S. (1996) *HLM Hierarchical linear and nonlinear modeling with HLM/2L and HLM/3L programs* Chicago: Scientific Software.

Thompson, M. J. (1998) The Australian Schools Science Competition - A Rasch analysis of recent data. Unpublished paper, The Flinders University of South Australia.

Thompson, M. (1999). An evaluation of the implementation of the Dimensions of Learning program in an Australian independent boys school. *International Education Journal,* 1 (1) 45-60. Retrieved May 9, 2004, from http://iej.cjb.net

Chapter 12

RASCH AND ATTITUDE SCALES: EXPLANATORY STYLE

Shirley M. Yates
Flinders University, Adelaide, Australia

Abstract: Explanatory style was measured with the Children's Attributional Style Questionnaire (CASQ) in 243 students from Grades 3 to 9 on two occasions separated by almost three years. The CASQ was analysed with the Rasch model, with separate analyses also being carried out for the Composite Positive (CP) and Composite Negative (CN) subscales. Each of the three scales met the requirements of the Rasch model, and although there was some slight evidence of gender bias, particularly in CN, no grade level differences were found.

Key words: Rasch, Explanatory Style, Gender Bias, Grade and Gender Differences

1. INTRODUCTION

Analyses of attitude scales with the Rasch model allows for the calibration of items and scales independently of the student sample and of the sample of items employed (Wright & Stone, 1979). Joint location of students and items on the same scale are important considerations in attitude measurement, particularly in relation to attitudinal change over time (Anderson, 1994). In this study, items in the *Children's Attributional Style Questionnaire* (CASQ) (Seligman, Peterson, Kaslow, Tanenbaum, Alloy & Abramson, 1984) and student scores were analysed together on the same scale, but independently of each other with *Quest* (Adams & Khoo, 1993) and the data compared over time. The one parameter item response Rasch model employed in the analyses of the CASQ assumes that the relationship between an item and the student taking the item is a conjoint function of student attitude and item difficulty level on the same latent trait dimension of

S. Alagumalai et al. (eds.), Applied Rasch Measurement: A Book of Exemplars, 207–225.

explanatory style (Snyder & Sheehan, 1992). In estimating item difficulty in the CASQ, QUEST takes into account the explanatory style of students in the calibration sample and then frees the item difficulty estimates from these attitudes. Likewise, student attitude is estimated by freeing it from estimates of item difficulty (Snyder & Sheehan, 1992). Response possibilities reflect the level of items on the underlying scale (Green, 1996). Differential item functioning was considered in relation to student gender. Grade and gender differences were also examined.

The Rasch model employs the notion of a single specified construct (Snyder & Sheehan, 1992) or inherent latent trait dimension (Weiss & Yoes, 1991; Hambleton, 1989), referred to as the requirement for unidimensionality (Wolf, 1994). While items and persons are multifaceted in any measurement situation, explanatory style measures need to be thought of and behave as if the different facets act in unison (Green, 1996). Scores on Rasch calibrated instruments represent the probabilistic estimation of the attitude level of the respondent based on the proportion of correct responses and the mean difficulty level of the items attempted. This has a distinct advantage over classical test theory procedures in which scores are created by summing responses, as the scale from which such scores have been obtained are built with items that satisfy unidimensionality. Use of Rasch scaling procedures also addresses shortcomings of classical test theory in which estimates of item difficulty, item discrimination, item quality and spread of subjects' ability or attitude levels associated with raw scores are confounded mathematically (Snyder & Sheehan, 1992).

1.1 Explanatory Style

Students differ in their characteristic manner of explaining cause and effect in their personal world, a trait referred to as explanatory style (Peterson & Seligman, 1984). This attribute habitually predisposes them to view everyday interactions and events from a predominantly positive (optimistic) or negative (pessimistic) framework (Eisner & Seligman, 1994). An optimistic explanatory style is characterised by explanations for good events as due to permanent, personal and pervasive causes while bad events are attributed to temporary, external and specific causes (Seligman, 1990; 1995). Conversely, a pessimistic explanatory style is characterised by explanations for causes of bad events as stable, internal or global and causes of good events as unstable, external and specific in nature (Seligman, 1990; 1995).

Explanatory style in school-aged students has been assessed principally with the CASQ (Seligman *et al.,* 1984). This forced choice pencil and paper instrument consists of 48 items of hypothetically good or bad events involving the child, followed by two possible explanations. For each event, one of the permanent, personal or pervasive explanatory dimensions is varied while the other two are held constant. Sixteen questions pertain to each of the three dimensions, with half referring to good events and half referring to bad events. The CASQ is scored by the assignment of 1 to each permanent, personal and pervasive response, and 0 to each unstable, external or specific response. Scales are formed by summing the three scores across the appropriate questions for the three dimensions, for composite positive (CP) and composite negative (CN) events separately (Peterson, Maier & Seligman, 1993) and by subtracting the CN score from CP for a composite total score (CT) (Nolen-Hoeksema, Girgus, & Seligman, 1986).

Psychometric properties of the CASQ have been investigated with classical test theory. Concurrent validity was established with a study in which CP and CN correlated significantly ($p < 0.001$) with the *Children's Depression Inventory* (Seligman *et al.*, 1984). Moderate internal consistency indices have been reported for CP and CN (Seligman *et al.*, 1984; Nolen-Hoeksema *et al.*, 1991, 1992), and CT (Panak & Garber 1992). The CASQ has been found to be relatively stable in the short term (Peterson, Semmel, von Baeyer, Abramson, Metalsky, & Seligman, 1982; Seligman *et al.*, 1984; Nolen-Hoeksema *et al.*, 1986), but in the longer term, test-retest correlations decreased, particularly for students as they entered adolescence. These lower reliabilities may be attributable to changes within students, but they could also be reflective of unreliability in the CASQ measure (Nolen-Hoeksema & Girgus, 1995).

Estimations of the CASQ's validity and reliability through classical test theory have been hampered by their dependence upon the samples of children who took the questionnaire (Osterlind, 1983; Hambleton & Swaminathan, 1985; Wright, 1988; Hambleton, 1989; Weiss & Yoes, 1991). Similarly, information on items within the CASQ has not been sample free, with composite scores calculated solely from the number of correct items answered by subjects. CASQ scores have been combined in different ways in different studies (for example, Curry & Craighead, 1990; Kaslow, Rehm, Pollack & Siegel, 1988; McCauley, Mitchell, Burke, & Moss, 1988) and although a few studies have reported the six dimensions separately, the majority have variously considered CP, CN and CT (Nolen-Hoeksema *et al.*, 1992). While CP and CN scores tend to be negatively correlated with each other, Nolen-Hoeksema *et al.* (1992) have asserted that the difference between these two scores constitutes the best measure of explanatory style. However, this suggestion has not been substantiated by any detailed analysis of the scale. Items have not been examined to determine the extent to which they each contribute to the various scales, or indeed whether they can be aggregated meaningfully into the respective positive, negative and composite scales. Furthermore, the extent to which the CASQ measured the psychological construct of explanatory style and indeed whether it is meaningful to examine the construct in terms of a style has not been established. No clear guidelines or cutoff scores for the determination of optimism and pessimism have been reported.

The CASQ's psychometric properties were investigated more rigorously with the one-parameter logistic or Rasch model of item response theory, to determine the relative contributions of each of the 48 items, as well as the most consistent and meaningful estimations of student scores. The CASQ met conceptually the Rasch model requirement of unidimensionality as it had been designed to measure the construct of a single trait of explanatory style (Seligman *et al.,* 1984). Use of the Rasch procedure was highly appropriate for analysis of the CASQ, as the model postulates that estimates of task difficulties are independent of the particular persons whose performances are used to estimate them, and that estimates of the performance of persons are independent of the particular tasks that they attempt (Wright & Stone, 1979; Wright, 1988; Hambleton, 1989; Kline 1993). Thus, different persons can attempt different sets of items, yet their performances can be estimated on the same scale. Questions as to whether the construct of explanatory style is measured adequately by the 48 questionnaire items, whether those items can be assigned meaningfully to positive or negative dimensions and the determination of the most appropriate delineation of student scores could all be addressed from these analyses. Not only was Rasch analysis used to examine whether the scale is best formed by a single latent construct of explanatory style, but it also provided a means by which the feasibility of the most meaningful and robust scores could be determined. Gender and item biases and gender and grade level differences were also examined.

2. RASCH SCALING OF THE CASQ

The CASQ was administered to students on two occasions (Time 1, Time 2) separated by three years. Two hundred and ninety three students in Grades 3-7 in two metropolitan primary schools in South Australia participated at Time 1 (T1) and 335 students from Grades 5-9 at Time 2 (T2), with 243 students participating on both occasions. Rasch analyses were carried out with all students who took part in the study at T1, as item characteristic curves, used in the examination of the relationship between a student's observed performance on an item and the underlying unobserved trait or ability being measured by the item, are dependent upon a large number of subjects taking the item. These analyses were then checked with the T2 sample. Initial inspection of the T1 data indicated some students had omitted some items. In order to determine if these missing data affected the overall results, the data were analysed with the missing data included and then with it excluded. Since differences with the missing items included or excluded were trivial, the analysis proceeded without the missing data being included.

The 24 CP items, 24 CN items and composite measure (CT) in which CN item scores were reversed, were analysed separately with the Rasch procedure using the *Quest* program (Adams & Khoo, 1993) to determine whether the items and scales fitted the Rasch model. With *Quest*, the fit of a scale to the Rasch model is determined principally through item infit and outfit statistics which are weighted residual-based statistics (Wright & Masters 1982; Wright, 1988). In common with most confirmatory model fitting, the tests of fit provided by *Quest* are sensitive to sample size, so use of mean square fit statistics as effect measures in considerations of model and data compatibility is recommended (Adams & Khoo, 1993). The infit statistic, which indicates item or case discrimination at the level where $p = 0.5$, is more robust as outfit statistics are sensitive to outlying observations and can sometimes be distorted by a small number of unusual observations, (Adams & Khoo, 1993). Accordingly, infit statistics only, with infit mean square (IMS) ranges set from 0.83 to 1.20 were considered. In all analyses the probability level for student responses to an item was set at 0.50 (Adams & Khoo, 1993). Thus, the threshold or difficulty level of any item reflected the relationship between student attitude and difficulty level of the item, such that any student had a 50 per cent chance of attaining that item. Results for the CP and CN scales are presented first, followed by those for CT.

2.1 CP and CN Scales

Rasch analyses of CP and CN at T1 and T2 indicated the scales could be considered independently as the items on both scales fitted the Rasch model. The IMS statistics, which measured consistency across performance levels and the discriminating power of an item, indicated that the fit of items to CP and CN, independently of sample size, lay within the range of 0.84-1.12, establishing a high degree of fit of all items to the two separate scales. These IMS statistics for both scales for T1 and T2 are shown in Table 12-1, with the data for CP presented in the left hand columns and the data for CN in the right hand columns. For each item on the two occasions there is very little difference if any in the IMS values.

Estimates of item difficulty are represented by thresholds in the *Quest* program (Adams & Khoo, 1993). The threshold value for each item is the ability or attitude level required for a student to have a 50 per cent probability of passing that step. As there is very little difference in the thresholds for the items at T1 and T2, the results of the latter only are presented in Figure 12-1. Respective item estimate thresholds, together with the map of case estimates for CP at T2 are combined with those for the T2 CN results in this figure. Case estimates (student scores) were calculated concurrently, using the 243 students for whom complete data were available for T1 and T2. The concurrent equating method, which involves pooling of the data, has been found to yield stronger case estimates than equating based on anchor item equating methods (Morrison & Fitzpatrick, 1992; Mahondas, 1996).

Table 12-1. Infit mean squares for CP and CN for Time 1 and Time 2

CP Item number	**T1 IMS** (N = 293)	**T2 IMS** (N = 335)	**CN** Item number	**T1 IMS** (N = 293)	**T2 IMS** (N = 335)
1 Item 1	0.95	0.96	Item 6	1.01	1.00
2 Item 2	0.99	1.01	Item 7	0.96	0.99
3 Item 3	1.09	1.04	Item 10	0.96	1.05
4 Item 4	1.08	1.12	Item 11	1.06	1.07
5 Item 5	0.90	1.00	Item 12	0.98	0.98
6 Item 8	1.04	0.94	Item 13	1.02	0.99
7 Item 9	1.03	0.98	Item 14	1.00	1.07
8 Item 16	0.99	0.97	Item 15	0.98	0.96
9 Item 17	1.01	1.09	Item 18	0.91	0.98
10 Item 19	0.97	1.02	Item 20	0.99	0.94
11 Item 22	0.91	0.96	Item 21	0.93	1.02
12 Item 23	0.89	0.88	Item 24	1.03	1.07
13 Item 25	1.02	1.04	Item 26	1.10	1.08
14 Item 30	1.06	1.03	Item 27	1.03	0.97
15 Item 32	1.06	1.09	Item 28	1.01	1.00
16 Item 34	1.00	0.95	Item 29	1.03	1.03
17 Item 37	0.98	1.00	Item 31	1.06	1.00
18 Item 39	1.02	1.02	Item 33	0.95	0.95
19 Item 40	1.05	1.04	Item 35	0.99	0.94
20 Item 41	1.01	0.97	Item 36	0.93	0.93
21 Item 42	0.98	0.98	Item 38	1.02	0.98
22 Item 43	0.89	0.84	Item 46	1.03	1.01
23 Item 44	1.06	0.99	Item 47	1.04	1.00
24 Item 45	1.00	1.05	Item 48	0.93	1.00
Mean	1.00	1.00		1.00	1.00
SD	0.06	0.06		0.05	0.04

```
CP T2                                         CN T2
--------------------------------------------------------------------------------
Item Estimates (Thresholds)    (N = 335 L = 24 Probability Level=0.50)
--------------------------------------------------------------------------------
All on CP T2                                  || All on CN T2
--------------------------------------------------------------------------------
 3.0                      |                   ||                          |
                          |                   ||                          |
                          |                   ||                          |
                          |                   ||                          |
                          |                   ||                          |
                          |                   ||                          |
                          |                   ||                          |
                          |   1               ||                          |
                          |                   ||                          |
 2.0                      |                   ||                          |
                        X |                   ||                          |
                          |                   ||                          |
                          |                   ||                          |
                       XX |                   ||                          |
                          |                   ||                          |
                          |                   ||                          |
                      XXX |                   ||                          |
                          |                   ||                          |
 1.0              XXXXXXX |                   ||                          |  21  36
                          |  16  34  39       ||                          |  18
                 XXXXXXXX |  44               ||                          |  15  48
                          |                   ||                          |  12
                  XXXXXXX |   4               ||                        X |
                        X |  23               ||                          |  13  20
         XXXXXXXXXXXXXXXX |   5  41  42  45   ||                       XX |
         XXXXXXXXXXXXXXXX |  22               ||                          |  33
                          |  40  43           ||                       XX |
 0.0       XXXXXXXXXXXXXX |                   ||                    XXXXX |  27
                        X |   17              ||                          |  38  46
      XXXXXXXXXXXXXXXXXXX |                   ||                  XXXXXXX |
                   XXXXXX |                   ||                        X |   6  24
             XXXXXXXXXXXX |  32               ||                 XXXXXXXX |
                        X |   9  30  37       ||                       XX |  35
               XXXXXXXXXX |                   ||                XXXXXXXXX |   7
                        X |                   ||                       XX |  10  31  47
                XXXXXXXXX |  19  25           ||           XXXXXXXXXXXXXX |  29
                          |                   ||                       XX |  14
-1.0                 XXXX |                   ||        XXXXXXXXXXXXXXXXX |  11
                          |                   ||                        X |
                      XXX |                   ||    XXXXXXXXXXXXXXXXXXXXX |
                        X |    3              ||                       XX |
                          |                   ||           XXXXXXXXXXXXXX |
                        X |                   ||                        X |
                          |                   ||                          |  26
                          |   8               ||           XXXXXXXXXXXXXX |
                          |                   ||                          |
-2.0                      |                   ||                        X |
                          |                   ||                XXXXXXXXX |
                          |                   ||                        X |
                          |                   ||                        X |
                        X |   2               ||                        X |
                          |                   ||                          |
                          |                   ||                     XXXX |
                          |                   ||                        X |
                          |                   ||                          |
-3.0                      |                   ||                          |
                          |                   ||                          |
                          |                   ||                          |
                          |                   ||                          |
                          |                   ||                      XXX |
                          |                   ||                          |
                          |                   ||                          |
                          |                   ||                          |
                          |                   ||                          |
-4.0                      |                   ||                          |
--------------------------------------------------------------------------------
  Each X represents    2 students
================================================================================
```

Figure 12-1. Item threshold and case estimate maps for CP and CN at T2

Maps of item thresholds generated by *Quest* (Adams & Khoo, 1993) are useful as both the distribution of items and pattern of student responses can be discerned readily. With Rasch analysis both item and case estimates can be presented on the same scale, with each independent of the other. In Rasch scale maps, the mean of the item threshold values is set at zero, with more difficult items positioned above the item mean and easier items below the item mean. As items increase in difficulty level they are shown on the map relative to their positive logit value, while as they become easier they are positioned on the map relative to their negative logit value. In attitude scales, difficult items are those with which students are probably less likely to respond favourably, while easier items are those with which students have a greater probability of responding favourably.

In the CP scale in Figure 12-1, 14 of the 24 items were located above 0, the mean of the difficulty level of the items, with Item 1 being particularly difficult. Students' scores were distributed relatively symmetrically around the scale mean. Eighteen students had scores below -1.0 logits, indicating low levels of optimism. Two students had particularly low scores as evidenced by their placement below -2.0 logits. In the CN scale, nine items were above the mean of the difficulty level of the items, indicating that the probability of students agreeing with these statements was less likely. Students' scores, however, clustered predominantly below the scale zero, indicating their relatively optimistic style. Approximately 86 students were more pessimistic as evidenced by their scores above the scale mean of zero, and a further 20 students had scores above the logit of +1.0.

2.1.1 Gender bias in CP and CN

As Rasch analysis is based on notions of item and sample independence, and as the calibrated items in a test measure a single underlying trait, the procedure readily lends itself to the detection of item bias (Stocking, 1994). Differential item functioning (DIF) (Kelderman & Macready, 1990) or bias is evident when test items of the item scores of two comparable groups, matched with respect to the construct being measured by the questionnaire, are systematically different. In order to investigate gender differences in explanatory style, it was first necessary to establish whether the items in either CP or CN had any inherent biases, using the scores from the 293 students at T1.

Male and female bias estimates for the CP scale were examined in terms of the standardised differences in item difficulties. It was evident in Figure 12-2 that Item 1 [*Suppose you do very well on a test at school: (A - scored as 1) I am smart (B - scored as 0) I am good in the subject that the test was in*] and Item 44 [*You get a free ice-cream: (A -scored as 1) I was nice to the ice-cream man that day (B -scored as 0) The ice-cream man was feeling friendly that day*] were biased significantly in favour of males as their standardised difference was greater than +2.0 or less than -2.0. This bias indicated that when the CP scale was considered independently, boys were more likely than girls to respond positively [response (*A*)] to these two items. Estimates of optimism in boys may therefore have been slightly enhanced relative to that of girls because of bias in these items. There were no items biased significantly in favour of females.

```
                           Plot of Standardised Differences

                  Easier for male                Easier for female

      -3         -2         -1          0          1          2          3
-------+----------+----------+----------+----------+----------+----------+
item 1         *  .                     |                     .
item 2            .                     |                  *  .
item 3            .                     |        *            .
item 4            .                   * |                     .
item 5            .               *     |                     .
item 8            .                     *                     .
item 9            .                     *                     .
item 16           .                     |    *                .
item 17           .                     |                     .
item 19           .                     |     *               .
item 22           .                     |    *                .
item 23           .             *       |                     .
item 25           .                 *   |                     .
item 30           .                    *|                     .
item 32           .                   * |                     .
item 34           .                     |   *                 .
item 37           .                     |                  *  .
item 39           .  *                  |                     .
item 40           .             *       |                     .
item 41           .                  *  |                     .
item 42           .                     |               *     .
item 43           .                     |                 *   .
item 44 *         .                     |                     .
item 45           .                     |   *                 .
=====================================================================
```

Figure 12-2. Gender comparisons of standardised differences of CP item estimates

When differences in item difficulties between males and females in the CN scale were standardised, only Item 26 [*You get a bad mark on your school work: (A - scored as 1) I am not very clever (B -scored as 0) Teachers are unfair*], as shown in Figure 12-3, was significantly biased against girls, with pessimism (*A- scored as 1*) high and optimism (*B -scored as 0*) low on the scale axis. Thus, in the measurement of pessimism, girls were more likely than boys to respond unfavourably [response (*A*)] to Item 26, thus potentially increasing slightly the reported level of pessimism in girls. No items on this scale were significantly biased in favour of boys.

2.1.2 Gender differences in CP and CN

Item IMS values were examined separately in the T1 data for males and females, with the results presented in Table 12-2. The Rasch model requires the value of this statistic to be close to unity. Ranges for females (N = 130)

extended from 0.87 - 1.13 for CP and 0.88 - 1.13 for CN, and were clearly within the acceptable limits of 0.83 and 1.20. For males (N = 162) the IMS values of CP were generally acceptable, ranging from 0.88 - 1.38 with only Item 44 misfitting. However, CN values ranged from 0.78 - 1.78 with six items, presented in Table 12-3, beyond the acceptable range. Items 18 and 20 are underfitting and provide redundant information, while Items 27, 28, 31 and 33 are overfitting and may be tapping facets other than negative explanatory style. These latter findings are of significance, especially if results of the CN scale alone were to be reported as the index of explanatory style. While results for females would not be affected by the inclusion of these items, the overfitting items in particular would need to be deleted before the case estimates of males could be determined.

```
                    Plot of Standardised Differences

                 Easier for male                Easier for female

      -3        -2        -1          0         1         2         3
-------+---------+---------+---------+---------+---------+---------+
item 6           .          *        |                   .
item 7           .    *              |                   .
item 10          .       *           |                   .
item 11          .                   |             *     .
item 12          .                   |             *     .
item 13          .                 * |                   .
item 14          .                 * |                   .
item 15          .                   |   *               .
item 18          .   *               |                   .
item 20          .                   |           *       .
item 21          .               *   |                   .
item 24          .                   *                   .
item 26          .                   |                   .        *
item 27          .      *            |                   .
item 28          .               *   |                   .
item 29          .                   |               *   .
item 31          .     *             |                   .
item 33          .         *         |                   .
item 35          .                   |         *         .
item 36          .                   |       *           .
item 38          .                   |               *   .
item 46          .                   |*                  .
item 47          .                   |         *         .
item 48          .                   | *                 .
====================================================================
```

Figure 12-3. Gender comparisons of standardised differences of CN item estimates

2.1.3 Grade level differences in CP and CN

Infit mean squares for the T1 CP and CN data were also examined for possible differences between Grade levels, with the results presented in Tables 12-4 and 12-5 respectively. While there were very few differences between Grades 5, 6 and 7 in both CP and CN scales, some variability was evident for students in Grades 3 and 4. As the size of the student sample in

Grade 3 was too small, it was necessary to collapse the data for Grades 3 and 4 students. In Tables 12-4 and 12-5 data for both Grade 4 (N = 72) and the combined Grade 3/4 (N = 92) is given.

Table 12-2. Gender differences in infit statistics for CP and CN

CP	Male (N = 163)	Female (N = 130)	CN	Male (N = 163)	Female (N = 130)
1 Item 1	0.88	0.93	1 Item 6	0.86	1.13
2 Item 2	1.09	1.01	2 Item 7	0.91	0.97
3 Item 3	1.13	1.13	3 Item 10	0.90	1.00
4 Item 4	1.04	1.11	4 Item 11	1.05	1.03
5 Item 5	0.91	0.88	5 Item 12	0.94	0.88
6 Item 8	1.12	1.04	6 Item 13	0.94	1.03
7 Item 9	1.03	1.07	7 Item 14	0.95	1.02
8 Item 16	1.00	0.96	8 Item 15	0.88	0.99
9 Item 17	1.06	0.99	9 Item 18	0.78	0.93
10 Item 19	1.02	0.98	10 Item 20	0.95	0.94
11 Item 22	0.89	0.92	11 Item 21	0.81	0.95
12 Item 23	0.90	0.87	12 Item 24	1.07	1.04
13 Item 25	0.99	0.99	13 Item 26	1.12	1.03
14 Item 30	1.06	1.05	14 Item 27	1.78	0.97
15 Item 32	1.08	1.02	15 Item 28	1.61	1.05
16 Item 34	0.99	1.01	16 Item 29	1.04	1.01
17 Item 37	1.02	1.00	17 Item 31	1.27	1.08
18 Item 39	1.20	1.01	18 Item 33	1.44	0.95
19 Item 40	1.08	1.05	19 Item 35	1.04	0.98
20 Item 41	1.00	1.05	20 Item 36	0.96	0.92
21 Item 42	0.99	0.97	21 Item 38	0.93	1.05
22 Item 43	0.88	0.94	22 Item 46	1.14	1.04
23 Item 44	1.38	1.03	23 Item 47	1.07	1.01
24 Item 45	0.98	1.02	24 Item 48	1.12	0.97

In the combined Grade 3 and 4 levels in these analyses, 13 CP items presented in Table 12-4, and nine CN items presented in Table 12-5, yielded IMS values outside the acceptable range, but this was not the case when the data for Grade 4 children were examined separately. Thus some degree of instability, in terms of the coherent scalability of the items, was evident for the youngest children within the present sample. However, lack of fit of items to the scale may well be a consequence of the relatively few numbers of students involved in the estimation. Anderson (1994) has recommended a minimum of 100 cases for consistent estimates to be made.

Table 12-3. Misfitting CP and CN items for males at T1

Scale	IMS	Item numbers and item statements
CP	1.38	44. You get a free ice-cream
(N=24)		a) I was nice to the ice-cream man that day
		b) The ice-cream man was feeling friendly that day
CN	0.78	18. You almost drown when swimming in a river
(N=24)		a) I am not a very careful person
		b) Some days I am not very careful
	0.81	21. You do a project with a group of kids and it turns out badly
		a) I don't work well with the people in the group
		b) I never work well with the group
		27. You walk into a door, and hurt yourself
	1.78	a) I wasn't looking where I was going
		b) I can be rather careless
		28. You miss the ball, and your team loses the game
	1.61	a) I didn't try hard while playing ball that day
		b) I usually don't try hard when I am playing ball
		31. You catch a bus, but it arrives so late that you miss the start of
	1.27	the movie film
		a) Sometimes the bus gets held up
		b) Buses almost never run on time
		33. A team that you are on loses a game
	1.44	a) The team does not try well together
		b) That day the team members didn't try well

3. COMPOSITE TOTAL SCALE

An examination of the item fit statistics, presented in Table 12-6 for both T1 and T2, showed that all items, with IMS values lying in the range 0.94 - 1.17, clearly fitted a single (CT) scale of explanatory style. With respect to the item threshold and student response values for both occasions presented in Figure 12-4, the range of the students' responses indicated that the majority were optimistic as their scores were above the scale zero (0). Thirty-four students had scores which fell between zero and -1.0 logits.

3.1.1 Gender bias in the CT scale

The CT scale was examined for gender bias for the T1 sample, with the results shown in Figure 12-5. Standardised differences indicated that three items (Items 1, 26, 44) were biased significantly in favour of males, but there was no evidence for bias for females. The evidence of bias for Item 26 for females on the CN scale alone, noted earlier, became a male biased item on the CT scale, because of the reversal of the CN scale to obtain the total. The scale as a whole was thus slightly biased in favour of males, providing males with a score that might be more optimistic than would be observed with unbiased items.

Table 12-4. Infit mean squares for each Grade level for CP at T1

Item Number	Grade 4 (N = 72)	Grade 3/4 (N = 92)	Grade 5 (N = 52)	Grade 6 (N = 97)	Grade 7 (N = 72)
1 Item 1	0.85	0.82	0.96	1.02	1.01
2 Item 2	1.30	1.81	0.99	0.89	1.02
3 Item 3	1.26	1.47	1.09	0.99	1.06
4 Item 4	1.16	1.32	1.26	1.12	0.92
5 Item 5	1.09	1.36	0.85	0.86	0.92
6 Item 8	1.27	1.65	1.10	1.01	1.04
7 Item 9	1.04	1.39	1.20	1.00	1.02
8 Item 16	1.04	1.24	1.06	1.03	0.95
9 Item 17	1.28	1.55	0.94	0.99	1.09
10 Item 19	1.17	1.55	0.97	1.03	0.94
11 Item 22	0.77	0.91	0.89	1.01	0.85
12 Item 23	0.92	1.07	0.91	0.92	0.81
13 Item 25	1.07	1.06	1.06	1.08	1.02
14 Item 30	1.01	1.19	1.07	0.97	1.10
15 Item 32	1.02	1.09	1.02	1.12	1.24
16 Item 34	1.49	1.67	1.14	0.94	1.00
17 Item 37	1.09	1.26	0.92	1.14	1.06
18 Item 39	1.04	1.08	0.96	1.17	1.04
19 Item 40	1.00	1.06	1.02	1.04	1.09
20 Item 41	1.02	1.13	0.90	1.04	1.01
21 Item 42	1.05	1.23	0.92	1.00	0.96
22 Item 43	0.91	0.94	0.80	0.95	0.93
23 Item 44	1.13	1.12	1.12	0.98	1.02
24 Item 45	1.17	1.32	0.90	1.14	0.94

3.1.2 Gender and grade level differences in CT

Gender differences were not evident in the CT scale infit statistics estimates which ranged from 0.94 to 1.07 for females and 0.90 to 1.07 for males. Similarly, marked differences were not evident between grade levels, with the IMS value ranging from 0.90 to 1.12 for students in Grades 3 and 4, from 0.87 to 1.09 for Grade 5, from 0.90 to 1.12 for Grade 6, and from 0.87

to 1.09 for Grade 7. All of these values were clearly within the predetermined acceptable range of 0.83 to 1.20.

Table 12-5. Infit mean squares for each Grade level for CN at T1

Item Number	Grade 4 (N = 72)	Grade 3/4 (N = 92)	Grade 5 (N = 52)	Grade 6 (N = 97)	Grade 7 (N = 72)
1 Item 6	1.05	0.93	1.11	0.94	0.93
2 Item 7	0.89	0.82	0.90	0.94	1.03
3 Item 10	0.93	0.83	0.87	0.92	1.03
4 Item 11	0.99	1.00	1.08	1.10	1.10
5 Item 12	0.78	0.77	1.10	1.08	0.95
6 Item 13	1.07	1.00	1.02	0.94	1.01
7 Item 14	0.98	0.89	1.05	1.02	0.99
8 Item 15	0.86	0.89	0.99	0.95	1.01
9 Item 18	0.87	0.77	1.00	0.83	0.91
10 Item 20	0.99	0.92	1.03	0.98	1.01
11 Item 21	0.91	0.85	0.90	0.86	0.99
12 Item 24	0.91	1.11	1.05	1.15	0.88
13 Item 26	1.09	1.05	1.04	1.54	1.03
14 Item 27	0.75	0.70	1.05	1.03	1.01
15 Item 28	1.37	1.34	0.94	0.40	1.08
16 Item 29	1.10	1.09	1.07	1.11	0.99
17 Item 31	1.18	1.41	0.88	1.24	1.06
18 Item 33	1.10	1.21	1.01	1.44	1.00
19 Item 35	0.98	0.97	0.99	0.85	1.03
20 Item 36	1.34	1.39	0.92	1.06	0.93
21 Item 38	0.92	0.82	0.93	0.96	1.08
22 Item 46	1.04	1.26	1.09	1.34	1.01
23 Item 47	1.04	1.23	0.92	1.15	0.97
24 Item 48	3.01	2.96	1.02	1.67	0.98

4. SUMMARY OF RASCH ANALYSIS OF THE CASQ

The CP, CN and CT are all scalable as they each independently meet the requirements of the Rasch model. With reference to the question as to whether the CP, CN, or CT scales should be used either alone or in combination, the Rasch analyses clearly indicate that the CT scale could be used in preference to either the CP or CN alone, because all items in the CT scale have satisfactory item characteristics for both the total group and the sub groups of interest. Scores can be meaningfully aggregated to form a composite scale of explanatory style which is psychometrically robust. In this total scale there is some evidence of gender bias in three items, such that the pessimism of males may be slightly under-represented, but this bias is more evident if the CN scale only were to be reported. While some instability or the small number of cases may have affected the scalability of the items for students at the Grade 3 level in the CP and CN scales, there were otherwise no grade level differences in item properties in the scales.

Table 12-6. Infit mean squares for CT at T1 and T2

CT Item number	T1 IMS (N = 293)	T2 IMS (N = 335)	Item number	T1 IMS (N = 293)	T2 IMS (N = 335)
1 Item 1	0.96	0.98	25 Item 25	1.00	1.02
2 Item 2	1.00	0.99	26 Item 26	1.06	1.02
3 Item 3	1.08	1.00	27 Item 27	1.05	0.96
4 Item 4	1.03	1.10	28 Item 28	1.01	0.97
5 Item 5	0.96	1.00	29 Item 29	1.04	1.02
6 Item 6	1.00	1.04	30 Item 30	1.03	1.04
7 Item 7	1.01	0.99	31 Item 31	1.03	0.99
8 Item 8	1.02	0.96	32 Item 32	1.05	1.07
9 Item 9	1.02	0.97	33 Item 33	0.99	0.97
10 Item 10	0.99	1.04	34 Item 34	1.00	0.99
11 Item 11	1.00	1.02	35 Item 35	1.01	1.00
12 Item 12	0.99	0.98	36 Item 36	0.98	0.95
13 Item 13	1.01	1.00	37 Item 37	1.00	1.00
14 Item 14	1.01	1.00	38 Item 38	1.01	0.98
15 Item 15	0.99	0.98	39 Item 39	0.99	1.05
16 Item 16	0.98	1.00	40 Item 40	0.99	1.04
17 Item 17	1.03	1.09	41 Item 41	0.97	1.01
18 Item 18	0.95	1.01	42 Item 42	1.01	0.92
19 Item 19	1.00	1.03	43 Item 43	0.96	0.90
20 Item 20	1.02	0.98	44 Item 44	0.99	1.03
21 Item 21	0.96	0.99	45 Item 45	0.99	1.03
22 Item 22	0.94	0.96	46 Item 46	0.98	0.99
23 Item 23	0.96	0.91	47 Item 47	0.98	0.98
24 Item 24	0.96	1.05	48 Item 48	0.96	0.99
			Mean	1.00	1.00
			SD	0.03	0.04

```
CT T1                                       CT T2
------------------------------------------------------------------------------
Item Estimates (Thresholds)   (L = 48, Probability Level 0.50)
------------------------------------------------------------------------------
All on CT T1(N = 293)                            All on CT T2  (N = 335)
------------------------------------------------------------------------------
 3.0                   |                    ||                  |
                       |                    ||                  |
                       |                    ||                  |  1
                     X |                    ||                  |
                       |                    ||                  |
                       |                    ||                  |
                       |                    ||                X |
                     X |  1                 ||                  |
                       |                    ||                X |
 2.0                XX |                    ||                  |
                       |                    ||               XX |
                    XX |                    ||                X |
                     X |                    ||                  |
                       |                    ||                X |
                    XX |                    ||               XX |  16  34  39
                    XX |                    ||               XX |  44
                XXXXXX |  39                ||           XXXXXX |
            XXXXXXXXXX |                    ||               XX |   4
 1.0           XXXXXXX |   4  16  34  44    ||        XXXXXXXXX |  23  26
             XXXXXXXXX |  22                ||      XXXXXXXXXXX |   5  41  42 45
          XXXXXXXXXXXX |  26                ||        XXXXXXXXX |  22
            XXXXXXXXXX |  23  42            ||          XXXXXXX |  40  43
     XXXXXXXXXXXXXXXXX |   5  31  40 41 47  ||   XXXXXXXXXXXXXX |
        XXXXXXXXXXXXXX |  24  43  45        ||     XXXXXXXXXXXX |  11  17
        XXXXXXXXXXXXXX |                    ||    XXXXXXXXXXXXX |  14
            XXXXXXXXXX |   9 46             ||     XXXXXXXXXXXX |  29
                XXXXXX |  11  17  30  37    || XXXXXXXXXXXXXXXX |  10  31  32 37 47
 0.0 XXXXXXXXXXXXXXXXX |  32  35            ||           XXXXXX |   7   9  30
                  XXXX |  29                ||            XXXXX |  35
                    XX |  14  25            ||            XXXXX |
                     X |   3  19  20        ||              XXX |   6  19  24 25
                     X |                    ||                X |  38
                    XX |  10  13            ||                  |  46
                       |  12  15  33        ||               XX |  27
                       |   7                ||                  |
                       |   6   8  27  38    ||                X |   3  33
                       |  21  48            ||                  |
-1.0                   |                    ||                  |  13  20
                       |                    ||                  |
                       |                    ||                  |   8  12
                       |                    ||                  |  15  18  48
                       |   2  18  36        ||                  |  21
                       |                    ||                  |  36
                       |                    ||                  |
                       |                    ||                  |
                       |                    ||                  |   2
-2.0                   |  28                ||                  |
                       |                    ||                  |
                       |                    ||                  |
                       |                    ||                  |
                       |                    ||                  |
                       |                    ||                  |  28
                       |                    ||                  |
                       |                    ||                  |
                       |                    ||                  |
-3.0                   |                    ||                  |
------------------------------------------------------------------------------
  Each X represents   2 students

==============================================================================
```

Figure 12-4. Item threshold and case estimate maps for CT at T1and T3

As each of the three scales met the requirements of the Rasch model, the logit scale which is centred at the mean of the items and therefore not sample dependent was used to determine cutoff scores for optimism and pessimism. Students whose scores lay above a logit of +1.0 on the CP and CT scales are considered to be high on optimism, while those below a logit of -1.0 are considered to explain uncontrollable events from a negative or pessimistic framework.

Plot of Standardised Differences

```
                  Easier for male                Easier for female

                -3         -2         -1          0          1          2          3
-----------------+----------+----------+----------+----------+----------+----------+
item 1                  *   .                     |                     .
item 2                      .                     |                  *  .
item 3                      .                     |      *              .
item 4                      .                  *  |                     .
item 5                      .                *    |                     .
item 6                      .                     |           *         .
item 7                      .                     |                   * .
item 8                      .                   * |                     .
item 9                      .                  *  |                     .
item 10                     .                     |                 *   .
item 11                     .         *           |                     .
item 12                     .         *           |                     .
item 13                     .                     |  *                  .
item 14                     .                     |   *                 .
item 15                     .                   * |                     .
item 16                     .                     | *                   .
item 17                     .              *      |                     .
item 18                     .                     |                    *.
item 19                     .                     |  *                  .
item 20                     .           *         |                     .
item 21                     .                     |      *              .
item 22                     .                     |*                    .
item 23                     .          *          |                     .
item 24                     .                     | *                   .
item 25                     .              *      |                     .
item 26                *    .                     |                     .
item 27                     .                     |                 *   .
item 28                     .                     |     *               .
item 29                     .         *           |                     .
item 30                     .                 *   |                     .
item 31                     .                     |                   * .
item 32                     .                *    |                     .
item 33                     .                     |           *         .
item 34                     .                     |*                    .
item 35                     .             *       |                     .
item 36                     .              *      |                     .
item 37                     .                     |                 *   .
item 38                     .       *             |                     .
item 39                     .*                    |                     .
item 40                     .            *        |                     .
item 41                     .                     |                     .
item 42                     .                *    |                     .
item 43                     .                     |              *      .
item 44        *            .                     |                *    .
item 45                     .                     |*                    .
item 46                     .                     |*                    .
item 47                     .             *       |                     .
item 48                     .                     *                     .
================================================================================
```

Figure 12-5. Gender comparisons of standardised differences of CT item estimates

On the CN scale students who are above a logit +1.0 are considered to be high on pessimism, while those below -1.0 are low on that scale. Any students whose scores fell above or below a logit of -2.0 or -2.0 would hold even stronger causal explanations for uncontrollable events, such that those who scored below -2.0 logits on CP were considered to be highly pessimistic, while those in this range on CN were highly optimistic. The use of the logit as a cutoff score for each of the scales could also be used to facilitate an examination of trends in student scores from T1 to T2.

Use of the Rasch model in the CASQ analyses had clear advantages, overcoming many of the limitations of classical test theory that had been employed previously.

5. REFERENCES

Adams, R. J. & Khoo, S. K. (1993). *Quest: The interactive test analysis system.* Hawthorn, Victoria: Australian Council for Educational Research.

Anderson, L W. (1994). Attitude measures. In T. P. Husen and Postlethwaite, T, N. (Eds.), *The international encyclopedia of education.* (Vol. 1, pp. 380-390). Oxford: Pergamon.

Curry, J. F. & Craighead, W. E. (1990). Attributional style and self-reported depression among adolescent inpatients. *Child and Family Behaviour Therapy, 12,* 89-93.

Eisner, J. P. & Seligman, M. E. P. (1994). Self-related cognition, learned helplessness, learned optimism, and human development. In T. Husen, & T. N. Postlethwaite, (Eds.), *International encyclopedia of education.* (second edition), (Vol. 9, pp. 5403-5407). Oxford: Pergamon.

Green, K. E. (1996). Applications of the Rasch model to evaluation of survey data quality. *New Directions for Evaluation, 70,* 81-92.

Hambleton, R. K. (1989). Principles and selected applications of item response theory. *Education measurement.* (third edition), New York: Macmillan.

Hambleton, R. K. & Swaminathan, H. (1985). *Item response theory: Principles and application.* Boston: Kluwer.

Kaslow, N. J., Rehm, L. P., Pollack, S. L. & Siegel, A. W. (1988). Attributional style and self-control behavior in depressed and nondepressed children and their parents. *Journal of Abnormal Child Psychology, 16,* 163-175.

Kelderman, H., and Macready, G B. (1990). The use of uoglinear models for assessing differential item functioning across manifest and latent examinee groups. *Journal of Educational Measurement, 27,* (4), 307-327.

Kline, P. (1993). Rasch scaling and other scales. *The handbook of psychological testing.* London: Routledge.

Mahondas, R. (1996). *Test equating, problems and solutions: Equating English test forms for the Indonesian Junior Secondary final examination administered in 1994.* Unpublished Master of Education thesis. Flinders University of South Australia.

McCauley, E., Mitchell, J. R., Burke, P. M. & Moss, S. (1988). Cognitive attributes of depression in children and adolescents. *Journal of Consulting and Clinical Psychology, 56,* 903-908.

Morrison, C. A. & Fitzpatrick, S. J. (1992). *Direct and indirect equating: A comparison of four methods using the Rasch model.* Measurement and Evaluation Center: The University of Texas at Austin. ERIC Document Reproduction Services No. ED 375152.

Nolen-Hoeksema, S. & Girgus, J. S. (1995). Explanatory style and achievement, depression and gender differences in childhood and early adolescence. In G. McC. Buchanan, & M. E. P. Seligman, (Eds.), *Explanatory style* (pp. 57-70). Hillsdale, NJ: Lawrence Erlbaum Associates.

Nolen-Hoeksema, S., Girgus, J. S. & Seligman, M. E. P. (1986). Learned helplessness in children: A longitudinal study of depression, achievement, and explanatory style. *Journal of Personality and Social Psychology, 51,* 435-442.

Nolen-Hoeksema, S., Girgus, J. S. & Seligman, M. E. P. (1991). Sex differences in depression and explanatory style in children. *Journal of Youth and Adolescence, 20,* 233-245.

Nolen-Hoeksema, S., Girgus, J. S. & Seligman, M. E. P. (1992). Predictors and consequences of childhood depressive symptoms: A five year longitudinal study. *Journal of Abnormal Psychology, 101,* (3), 405-422.

Osterlind, S. J. (1983). Test item bias. *Sage University paper series on quantitative application in the social sciences*, 07-001. Beverly Hills: Sage Publications.

Panak, W. F. & Garber, J. (1992). Role of aggression, rejection, and attributions in the prediction of depression in children. *Development and Psychopathology, 4,* 145-165.

Peterson, C., Maier, S. F. & Seligman, M. E. P. (1993). *Learned helplessness: A theory for the age of personal control.* New York: Oxford University Press.

Peterson, C., Semmel, A., von Baeyer, C., Abramson, L. Y., Metalsky, G. I. & Seligman, M. E. P. (1982). The Attributional Style Questionnaire. *Cognitive Therapy and Research, 6,* 287-299.

Peterson, C. & Seligman, M. E. P. (1984). Causal explanation as a risk factor in depression: Theory and evidence. *Psychological Review, 91,* 347-374.

Seligman, M. E. P. (1990). *Learned optimism.* New York: Pocket Books.

Seligman, M. E. P. (1995). *The optimistic child.* Australia: Random House.

Seligman, M. E. P., Peterson, C, Kaslow, N. J., Tanenbaum, R. L., Alloy, L. B. & Abramson, L. Y. (1984). Attributional style and depressive symptoms among children. *Journal of Abnormal Psychology, 93,* 235-238.

Snyder, S. & Sheehan, R. (1992). Research methods The Rasch measurement model: An introduction. *Journal of Early Intervention, 16,* (1), 87-95.

Stocking, M. L. (1994). Item response theory. In T. Husén, & N. T. Postlethwaite, (Eds.), *The international encyclopaedia of education.* (pp. 3051-3055). Oxford: Pergamon Press.

Weiss, D. J. & Yoes, M. E. (1991). Item response theory. In R. K. Hambleton. and N. J. Zaal, (Eds.), *Advances in education and psychological testing.* Boston: Kluwer Academic Publishers.

Wolf, R. M. (1994). Rating scales. In T. Husen & T. N. Postlethwaite (Eds.), *The international encyclopedia of education.* (Second edition). (Vol. 8, pp. 4923-4930), Pergamon: Elsevier Science.

Wright, B. D. (1988). Rasch measurement models. In J. P. Keeves (Ed.), *Educational research, methodology and measurement: An international handbook.* Oxford: Pergamon Press.

Wright, B. D. & Masters, G. (1982). *Rating scales analysis.* Chicago: MESA Press.

Wright, B. D. & Stone, M. H. (1979). *Best test design.* Chicago: Mesa Press.

Chapter 13

SCIENCE TEACHERS' VIEWS ON SCIENCE, TECHNOLOGY AND SOCIETY ISSUES

Debra K. Tedman
Flinders University; St John's Grammar School

Abstract: This Australian study developed and used scales to measure the strength and coherence of students', teachers' and scientists' views, beliefs and attitudes in relation to science, technology and society (STS). The scales assessed views on: (a) science, (b) society and (c) scientists. The consistency of the views of students was established using Rasch scaling. In addition, structured group interviews with teachers provided information for the consideration of the problems encountered by teachers and students in the introduction of STS courses. The strength and coherence of teachers' views on STS were higher than the views of scientists, which were higher than those of students on all three scales. The range of STS views of scientists, as indicated by the standard deviation of the scores, was consistently greater than the range of teachers' views. The interviews indicated that a large number of teachers viewed the curriculum shift towards STS positively. These were mainly the younger teachers, who were enthusiastic about teaching the issues of STS. Some of the teachers focused predominantly upon covering the content of courses in their classes rather than discussing STS issues. Unfortunately, it was found in this study that a significant number of teachers had a limited understanding of both the nature of science and STS issues. Therefore, this study highlighted the need for the development of appropriate inservice courses that would enable all science teachers to teach STS to students in a manner that would provide them with different ways of thinking about future options. It might not be possible to predict with certainty the skills and knowledge that students would need in the future. However, it is important to focus on helping students to develop the ability to take an active role in debates on the uses of science and technology in society, so that they can look forward to the future with optimism.

Key words: Rasch scaling, views, STS, teachers, VOSTS, positivism

S. Alagumalai et al. (eds.), Applied Rasch Measurement: A Book of Exemplars, 227–249.

1. INTRODUCTION

The modern world is increasingly dependent upon science and technology. The growing impact of science and technology on the lives of citizens in contemporary societies is evidenced by an increased discussion of value-laden scientific and technological issues. These issues include: the use of nuclear power, the human impact of genetic engineering, acid rain, the greenhouse effect, desertification and advances in reproductive technology such as in-vitro fertilisation (IVF).

It is extremely important to recognise that science provides the knowledge base for technology and dominates the culture of developed nations, while technology provides the tools for science and shapes social networks (Lowe, 1995). Consequently, as many eminent researchers have argued (Bybee, 1987; Cross, 1990; Yager, 1990a, 1990b; Heath, 1992; Lowe, 1993), public debate and consideration of issues with respect to the relationships between science, technology and society (STS) are necessary in order to develop a national policy to guide the use of science and technology in contemporary and future societies. Public understanding of the relationships between science, technology and society is necessary for citizens to engage in informed debate on the use of science and technology in society. Scientific and technological knowledge has immense potential to improve the lives of humans in modern societies if the imparting of this knowledge is accompanied by open public debate about the risks, benefits and social cost of scientific and technological innovations (Gesche, 1995).

1.1 Worldwide recognition of the need to present a new model of science

The worldwide demand for changes in education and a shift in the emphasis of science courses in order to equip students for their lives in a technology-permeated society have led to a global revolution in science, technology and mathematics education. In many countries, there have been considerable efforts made to forge lasting educational reforms in the rapidly changing subject areas of science, technology and mathematics (*Education Week*, 10 April 1996, p. 1). The OECD case studies analysed efforts to change the focus of learning in science, technology and mathematics subjects from pure knowledge of subject matter to practical applications, with closer connections to students' everyday lives. These changes towards the presentation of a different model of science were driven both by serious concerns about the economic competitiveness of these countries, and distress about social and community-based issues, such as environmental deterioration.

In order for science education to have a positive influence on the limitation of environmental damage caused by acid rain, by way of example, an understanding of issues resulting from the interaction between science, technology and society is necessary. These factors include community reaction to the challenge of reducing pollution, the influence of different interest groups, and public policies about environmental protection. Awareness of the influence of the interaction between these social factors and the development and use of science and technology in society is furthered by science education which presents a different model of science from that presented in the past by including discussion of the issues of STS.

Science curricula, which include consideration of the issues of STS, seek to develop scientific literacy and prepare students to act as vital participants in a changing world in which science and technology are all pervasive. As Parker, Rennie and Harding (1995, p.186) have argued, the move toward the provision of 'science for all' reflects the aims of science education which include:

> (1) to educate students for careers in science and technology; and
> (2) to create a scientifically and technologically literate population, capable of looking critically at the development of science and technology, and of contributing to democratic decisions about this development.

The vital role of science teachers in the development of community attitudes towards science and technology was another issue identified in the report by the National Board of Employment, Education and Training in Australia (National Board of Employment, Education and Training, 1993).

1.2 Factors that determine the success of curriculum innovations

Since the Australian study, discussed in this chapter, focused upon the shift toward the inclusion of STS objectives in secondary science curricula, it was important in this study to consider the factors that determine the success of such curriculum innovations. In order for a curriculum shift to be successful, teachers should see the need for the proposed change, and both the personal and social benefits should be favourable at some point relatively early in its implementation. A major change which teachers consider to be complex, prescriptive and impractical, is likely to be difficult to implement. Fullan and Stiegelbauer (1991) suggested that factors such as characteristics of the change, need, clarity, complexity and practicality interact to determine the success or failure of an educational change. Analysis of these factors at the early stages of the introduction of a curriculum shift would enable teachers to assess students' performance and evaluate the changed curriculum.

An investigation of the adequacy of Australian teachers' training to enable them to teach the interrelations between science, technology and society effectively to their senior secondary students should aim to provide information for those involved in the provision of both undergraduate and in-service education of teachers. Teachers' assessment of the shift in the curriculum towards STS in this study would also be an important area for investigation.

In the early stages of the study it was considered important to ask how science teachers, who were traditionally trained and skilled in the linear, deductive model (Parker, 1992) would assist students to deal with value-laden STS issues. In order to deal adequately with discussions of value-laden STS issues in science classes, students must also receive some education involving attitudes and values, and the way in which to deal with such potentially volatile issues. If a curriculum change included a strong emphasis upon STS, then some secondary teachers might experience difficulty in enabling their students to meet the STS objectives. This could be because some teachers might not have strong and coherent views in regard to the nature of science and the interactions of science, technology and society.

It was clear that an investigation of teachers' views on STS was required. The findings of such an investigation might, as a consequence, form the basis for the development of appropriate and effective programs of professional development for teachers.

1.3 The influence of teachers' views on the success of the curriculum shift

In any investigation of the inclusion of STS in secondary science courses, it is important to investigate teachers' views and attitudes towards STS issues. Attitudes affect individuals' learning by influencing what they are prepared to learn. Bloom (1976, p. 74) wrote that:

> individuals vary in what they are emotionally prepared to learn as expressed in their interests, attitudes and self-views. Where students enter a task with enthusiasm and evident interest, the learning should be much easier, and all things being equal they should learn it more rapidly and to a higher level of attainment or achievement than will students who enter the learning task with lack of enthusiasm and evident disinterest.

The attitudes and views of teachers and students would, therefore, affect the chances of a successful implementation of the curriculum shift towards STS, since the predisposition of individuals from both of these groups to learn about the issues of STS would depend upon their views on STS.

More recently, the researchers Lumpe, Haney and Czerniak (1998, p. 3) supported the need to consider teacher beliefs in relation to STS when they argued that: 'Since teachers are social agents and possess beliefs regarding professional practice and since beliefs may impact actions, teachers' beliefs may be a crucial change agent in paving the way to reform'.

Evidence for the importance of examining the views, attitudes, beliefs, opinions and understandings of teachers in relation to the curriculum change was provided by an OECD study (*Education Week*, April 10, 1996). At the onset of a curriculum change, such as the shift towards the inclusion of STS objectives in secondary science courses in South Australia, teachers, who already have vastly changing roles in the classroom, are required to reassess their traditional classroom practices and teaching methods carefully. These teachers may then feel uneasy about their level of understanding of the new subject matter, and refuse to cooperate with a curriculum change which requires them to take on more demanding roles. While some teachers value the challenge and educational opportunities presented by the shift in objectives of the curricula, others object strongly to such a change. The success of such a curriculum change therefore requires the provision of opportunities for both in-service and preservice professional development and for regular collaboration with supportive colleagues (*Education Week*, 10 April 1996, p. 7).

In Australia, the Commission for the Future summarised the need for in-service and preservice education of teachers in relation to science and technology with the suggestion that, even with the best possible curriculum, students do not participate effectively unless it is delivered by teachers who instill enthusiasm by their interest in the subject. The further suggestion was advanced that, unless improved in-service and preservice education was provided for teachers, students would continue to move away from science and technology at both the secondary and tertiary levels (National Board od Employment, Education and Training, 1994).

2. IMPORTANCE OF THE STUDY

This investigation comprised a detailed examination of the inclusion of the study of science, technology and society (STS) issues into senior secondary science courses. The purpose of this study was to investigate a curriculum reform that had the potential to change science education markedly and to motivate students, while addressing gender imbalance in science education.

This investigation of the shift towards STS in secondary science courses sought information in order to guide: (a) the provision of effective in-service education, (b) the writing of courses, curricula and teaching resources, and (c) the planning of future curriculum developments. It was considered in the Australian study that this information would best be provided by the development and use of scales to measure teachers', students' and scientists' views on STS. The development of these scales and demonstration in this study that they were able to be used to provide a valid measure of respondents' views on STS represents a significant advance in the area of quantitative educational research in the field.

The study was located in 29 South Australian colleges and schools. South Australia was chosen as the location of the study, not merely for convenience, but also because the new courses introduced by the Senior Secondary Assessment Board of South Australia in 1993 and 1994 at Years 11 and 12 respectively included a substantial new STS orientation. Thus the study was undertaken in a school system at the initial stage of reform with the introduction of new science curricula.

2.1 The need to examine views on STS

Review of the published literature on previous studies further demonstrated the need to examine views on STS since there had been no previous Australian studies that had measured views on STS. Nevertheless, several previous overseas studies have investigated students' views towards STS, and one of these studies used scaled Views on Science, Technology and Society (VOSTS) items at a similar time to the scaling of the VOSTS items in this Australian study.

This study by Rubba and Harkness (1993) involved an empirically developed instrument to examine preservice and in-service secondary science teachers' STS beliefs. The authors began their article with the assertion that, in light of the increased STS emphasis in secondary science curricula, it was important to investigate the adequacy of teachers' understandings of the issues of STS. The authors proceeded with the suggestion that it was important for teachers to have adequate conceptions of STS, since science teachers were held accountable for the adequacy of student conceptions of the nature of science. They concluded that:

> the results showed that large percentages of the in-service and pre-service teachers in the samples held misconceptions about the

nature of science and technology and their interactions within society. (Rubba and Harkness, 1993, p. 425)

After Rubba and Harkness's (1993) findings of misconceptions in teachers' beliefs of STS, they recommended that these teachers study STS courses, since the college science courses that had been studied by these teachers did not appear to have developed accurate conceptions of STS issues in these teachers. Rubba and Harkness argued that it was important for teachers to have strong and coherent views in relation to the issues of STS.

Similarly, to this Australian study, the American researchers Rubba, Bradford and Harkness (1996) scaled a sample of VOSTS items to measure views on STS. Rubba, Bradford and Harkness asserted that it was incorrect to label item statements as right or wrong, and proceeded to charge a panel of judges to classify the responses so that a numerical scale value was allocated to each of the responses. This scaling was used to determine the adequacy of teachers' views on STS as assessed by a scale that gave partial credit to particular views, and greater or lesser credit to others.

In another previous study (Rennie & Punch, 1991), which involved the scaling of attitudes, like this Australian study, the authors decried the state of attitude research in many former studies. Rennie and Punch contended that one problem in this previous research was that the size of the effects of attitude on achievement was distorted by grouping similar scales together. The second problem was that there was no theoretical framework to direct the development of appropriate scales. In this study, the theoretical framework was considered carefully in order to inform the development of appropriate scales and questions, which would guide effective interviews with teachers to determine their views on the curriculum shift towards STS.

The discussion in this chapter shows why, at the time this Australian study was undertaken, the accurate measurement of views towards STS was extremely important. The results could provide information to teachers and curriculum developers, thus guiding curriculum modification in relation to STS, as well as informing administrators about the progress and significant issues relating to the shift towards the inclusion of STS objectives in senior secondary science courses in South Australia.

2.2 Specific questions addressed in this study

The following specific questions were addressed in regard to science teachers' views on STS issues:

1. Do teachers have strong and coherent views in relation to STS? What are the differences in the strength and coherence of the views towards

STS of secondary science teachers, their students at the upper secondary school level, and scientists?
2. What are teachers' views in relation to the recent shift in the emphasis of the South Australian secondary science curricula towards STS? How do the teachers translate the STS curriculum into what happens in the classroom? What are the gaps in their views and what are the implications of this for the training of teachers?

3. RESEARCH METHODS AND MATERIALS

In this investigation of the shift towards STS of the South Australian senior secondary science curricula, the views towards STS of students, teachers and scientists were measured. The overall aim, which guided this study, was to construct a 'master scale for STS' by scaling a selection of *Views on science, technology and society* (VOSTS, Aikenhead, Ryan & Fleming, 1989) items. It was necessary to develop and calibrate scales to measure and compare views on STS. The three scales, which were used in this Australian study, were developed from three of the VOSTS domains, which related to:

1. the effects of society on science and technology (Society);
2. the effects of science and technology on society (Science); and
3. the characteristics of scientists (Scientists).

Systematic methods of data collection and analysis were necessary. This method of data collection using scales also required careful consideration of the issues of strength and consistency. In order to address the research items in a valid manner, it was important to establish first the strength and consiste first the strength and consistech scaling. The validity of the scales used in this present study was confirmed on the basis of a review of the literature on the philosophy of science and the issues of STS, as well as by using the viewpoints of the experts (Tedman, 1998; Tedman & Keeves, 2000).

Twenty-nine metropolitan and non-metropolitan schools from the government, independent and Catholic sectors were visited, so that the scales could be administered to 1278 students and 110 science teachers. The scales were also administered to 31 scientists, to enable a comparison of the strength and coherence of the STS views of students, teachers and scientists. Details of the selection of schools are reported in Tedman (1998).

The study was conducted in South Australia during the early stages of a shift towards STS in the objectives of senior secondary science courses, and the teachers in the state were experiencing considerable pressures due to increased workloads and time demands. Consequently, the development of a master scale to measure views on STS required a sequence of carefully considered research methods and materials.

3.1 Scales used in the study

The main objective of the administration of the scaled instrument during this study was to measure the strength and coherence of students', teachers' and scientists' views in relation to STS. Information on students' and teachers' views was gathered by using an adaptation of the VOSTS instrument developed by Aikenhead, Fleming and Ryan in 1987. Following the review of the literature and careful consideration of issues relating to validity, it was argued that, since the VOSTS instrument was constructed using students' viewpoints, it was likely to produce valid results. This instrument to monitor students' views on STS was developed by Aikenhead and Ryan (1992) in an attempt to ensure valid student responses so that the meaning students read into the VOSTS choices was the same meaning as they would express if they were interviewed.

Aikenhead and Ryan (1992) suggested, however, that the items used in the VOSTS instrument could not be scaled as the previously used Likert-type responses did. Aikenhead had worked within a qualitative, interpretative approach, rather than the quantitative approach, and he had not realised that item response theory had advanced to the stage of encompassing a range of student responses for which partial credit could be given, rather than just two (Aikenhead, personal comment, 1997). The instrument used to gather viewpoints on STS in this Australian study was significantly strengthened by adding a measurement component to the work that had previously been done. This served to fortify the conclusions arrived at as a result of the study. Mathematical and statistical procedures to do this had been advanced (Masters, 1988) and acceptance of Aikenhead's challenge to develop a measurement framework for the VOSTS items was of considerable significance.

Details of the scales used in the present study, including calibration and validation, are presented in Tedman and Keeves (2000).

3.2 Statistical procedures employed in this study

In this Australian study, Rasch scaling was used. Rasch (1960) proposed a simplified model of the properties of test items, which, if upheld adequately, permitted the scaling of test items on a scale of the latent attribute that did not depend on the population from which the scaling data were obtained. This system used the logistic function to relate probability of success on each item to its position on the scale of the latent attribute (Thorndike, 1982, p. 96).

Thus, the Rasch scaling procedure employs a model of the properties of test items, which enables the placing of respondents and test items on a common scale. This scale of the latent attribute, which measures the strength and coherence of respondents' views towards STS, is independent of the sample from which the scaling data were obtained, as well as being independent of the items or statements employed. In order to provide partial credit for the different alternative responses to the VOSTS items, the Partial Credit Rasch model developed by Masters (1988) was used. Furthermore, Wright (1988) has argued that Rasch measurement models permitted a high degree of objectivity, as well as measurement on an interval scale.

A further advantage of the Rasch model was that, although the slope parameter was considered uniform for all of the items used to measure the strength and direction of students' views towards STS, the items differed in their location on the scale and could be tested for agreement to the slope parameter, and this aided the selection of items for the final scales.

Before proceeding with the main data collection phase of this study, it was necessary to calibrate the scales and establish the consistency of the scales. The consistency of the scaling of the instrument was established using the data obtained from the students in a pilot study on their views towards STS. The levels of strength and coherence of students' views were plotted so that the students who had strong and coherent views on STS when compared with the views of the experts were higher up on the scale. At the same time, the items to which they had responded were also located on the common scale. In this way, the consistency of the scales was established.

It was considered important to validate the scales used in this study to measure the respondents' views in relation to science, technology and society. In order to validate the scales, a sample of seven STS experts from the Association for the History, Philosophy and Social Studies of Science each provided an independent scaling of the instrument. Thus, the validation of the scales ensured that the calibration of the scales was strong enough to establish the coherence of respondents' views with those of the experts. Thus validation tested how well the views of respondents compared with the views of the experts. Furthermore, the initial scaling of the responses associated with each item was specified from a study of STS perspectives in relation to the STS issues addressed by the items in the questionnaire. The consistency and coherence of the scales and each item within a scale was tested using the established procedures for fit of the Rasch model to the items (Tedman & Keeves, 2001).

As a consequence of the use of Rasch scaling during the study, the scales that were developed were considered to be independent of the large sample of students who were used to calibrate the scales, and were independent of the particular items or statements included in the scales.

3.3 Analysis of data

The strength and coherence of teachers', students', and scientists' views on STS were measured using the scales constructed in the course of the study. Means and standard deviations were calculated using SPSS Version 6.1 (Norusis, 1990). The standard errors were calculated using a jackknife procedure with the WesVarPC program (Brick, Broene, James & Severynse, 1996). This type of calculation allowed for the fact that a cluster sample, rather than a simple random sample, was used in this Australian study of students' views.

3.4 Teacher interviews

In addition to the use of scales to gather data, group interviews and discussions enabled the collection of information on teachers' opinions and concerns in relation to this shift in the objectives of secondary science curricula in South Australia. Fairly open-ended questions formed the basis of the group interviews with teachers in each school with the aim of obtaining information on how the teachers saw the curriculum changes that were taking place. These interviews or group discussions in 23 schools with 101 teachers were important methods of gathering information for this study. Structured group interviews with teachers were considered preferable to individual interviews, since they provided the observations of the group of teachers discussing the planning of the curriculum within schools.

The discussion, in these interviews, of teachers' opinions and concerns in relation to the shift towards STS, was based upon the following six questions:

What is your understanding of STS? Why teach STS?
How do you incorporate STS into your subject teaching?
What resources do you use for teaching STS?
Do the students respond well to STS material? Do girls respond particularly well to STS material?
A strand of the Statements and Profiles for Australian Schools is ‘Working Scientifically’. What is science, and how do scientists work?
What do you see as the main issues in STS courses?

4. RESULTS AND DISCUSSION

The mean scale scores for students, scientists and teachers for the Science, Society, and Scientists Scales are presented in Table 13-1.

4.1 Mean scale scores

It can be seen from Table 13-1 that the mean scores for teachers on the Science, Society and Scientists Scales are substantially higher than the mean scores for scientists. The mean scores for scientists, in turn, are higher than the mean scores for students. The higher scores for teachers might indicate that teachers have had a greater opportunity to think about the issues of STS than scientists have. This has been particularly true in recent years, since there has been an increasing shift towards the inclusion of STS objectives in secondary science curricula in Australia, and, in fact around the world.

Reflection upon the sociology of science provides a further possible explanation for the discrepancy between the level of the STS views of scientists and teachers. Scientists interact and exchange ideas in unacknowledged collegial groups (Merton, 1973), the members of which are working to achieve common goals within the boundaries of a particular paradigm. Scientific work also receives validation through external review, and the reviewers have been promoted, in turn, through the recommendations of fellow members of the invisible collegial groups to which they belong.

Radical ideas and philosophies are, therefore, frequently discouraged or quenched. The underlying assumptions of STS are that science is an evolutionary body of knowledge that seeks to explain the world and that scientists as human beings are affected by their values, and cannot, therefore always be completely objective (Lowe, personal comment, 1995).

STS ideas might be regarded by many traditional scientists as radical or ill-founded. Thus, scientists in this study appear to have not thought about STS issues enough, since they might not have been exposed sufficiently to informed and open debate on these issues.

The suggestion that scientists construct their views from input they receive throughout their lives is also a possible explanation for the level of scientists' views being lower than that of teachers. The existing body of scientists in senior positions has received, in the greater part, a traditional science education. During these scientists' studies, science was probably depicted as an objective body of fact, and the ruling paradigms within which they, as students, received their scientific education defined the problems which were worthy of investigation (Kuhn, 1970). Educational establishments are therefore responsible for guarding or maintaining the existing positions or views on the philosophy, epistemology and pedagogy of science (Barnes, 1985).

Table 13-1. Mean scale scores, standard deviations, standard errors and 95 per cent confidence intervals for the mean - Science, Society and Scientists Scales

Group	Count	Mean	Standard Deviation	Standard Error[j]	95 Pct Conf. Int. for Mean
Science Students	1278	0.317	0.548	0.033	0.251 to 0.383
Scientists	31	0.516	0.506	0.091	0.334 to 0.698
Teachers	110	0.874	0.444	0.042	0.792 to 0.958
Society Students	1278	0.210	0.590	0.028	0.154 to 0.266
Scientists	31	0.339	0.499	0.090	0.159 to 0.519
Teachers	110	0.695	0.497	0.047	0.601 to .789
Scientists Students	1278	0.408	0.582	0.032	0.344 to 0.472
Scientists	31	0.596	0.932	0.167	0.262 to 0.930
Teachers	110	0.965	0.733	0.070	0.825 to 1.105

Notes:
[j] jackknife standard error of mean obtained using Wes Var PC

A comparison of the standard deviations of the teachers' and scientists' mean scores shows that the range of scientists' views on STS on the Science scale is greater than the range of teachers' views. The standard deviation of students' views demonstrates that students' views on STS range more widely than the views of either scientists or teachers.

In a way that is similar to the results for the Society scale, the standard deviation of scientists' views is higher than the standard deviation of teachers' views. In this case, however, the magnitude of the standard deviations for teachers' and scientists' views is quite similar.

For the Scientists Scale, unlike the Science and Society Scales, scientists have the largest standard deviation. The standard deviation of teachers exceeds that of the students involved in the study. Thus, of the three groups of respondents, scientists have the greatest range of views on the characteristics of professional scientists. This might be due to the suggestion made in the above discussions that peoples' life experiences determine their views, understandings and attitudes towards the issues of STS. Thus, the diverse experience scientists have of fellow scientists' characteristics could make it difficult for them to decide between the alternative responses to statements on the characteristics of scientists.

The fact that on the Scientists Scale, the standard deviation of scientists' views exceeds that of the teachers' views substantially, also indicates that some scientists had coherent views on STS issues, whereas others had just continued with their scientific work, or problem-solving within the boundaries of the prevailing paradigm, without considering the social relevance or context of science and technology. It would appear that some scientists were positivistic in their ideas, and others viewed the philosophy of science from more of an STS and post-positivistic perspective. This could be due to the different educational backgrounds of some of the younger scientists, since STS has been included as a component of some undergraduate science courses in Australian universities such as Griffith University from 1974. It is also possible that some scientists have considered STS issues, while some others have just continued with their fairly routine problem-solving and research.

Among this group of scientists there were some who could not answer several of the questions, perhaps because they had not thought about the issues sufficiently or, as discussed above, could not decide between alternative responses. This was recorded as a 0 response, and contributed to the wide spread of scores.

4.2 The strength and coherence of respondents' views on STS

The measurement of the strength and coherence of respondents' views on STS which was accomplished during this Australian study is important, since consideration of teachers' and students' views in relation to the objectives of new curricula is a crucial component of an investigation of a curriculum change. The standard deviations of the three groups of respondents also provide useful information.

Perhaps the consistency and coherence of teachers' views was due to the nature of the teacher groups who responded to the survey. The teachers who volunteered most willingly to complete the survey were quite young. Young teachers were often more likely to participate in an open debate than their older colleagues who held firmly entrenched views. In recent years, teacher education courses in science in Australia have moved to include a component on STS ideas. However, this 'add on' approach (Fensham, 1990) is a long way from a teaching focus that uses STS as a foundation upon which to build students' understandings of science, so further development is needed in these teacher education courses.

The standard deviation, and therefore range of teachers' views on STS, is not as great as the range of scientists' views. The smaller standard deviations for teachers' scores on the three scales, relative to the standard deviations for scientists' scores, might be due to young teachers' openness to debate, coupled with the recent changes in teacher education courses. This has resulted in the development of fairly coherent views on STS by the science teachers surveyed in this study. Teachers' higher mean scores on all three scales also support this suggestion.

The high level of the scores for teachers' views on STS is unexpected on the basis of the published findings of previous studies. Students' and teachers' views on the nature of science were assessed in the United States by Lederman (1986), for example, using the Nature of Scientific Knowledge scale (Rubba, 1976), and a Likert scale response format. Unlike the findings of the survey in this South Australian study, which used scales to measure students' and teachers' views and understandings in relation to STS, this American study found misconceptions in pre-service and in-service teachers' views and beliefs about STS issues (Lederman, 1986). However, the use in Lederman's study of comparison with the most commonly accepted attributes as a way of judging misconceptions was vague, and raised serious questions in regard to the validity of his instrument. The instrument used in this present study overcame this problem, since the validity was established first by testing the fit between the data and the Rasch scale as well as by an independent scaling of the instrument by seven experts.

The results of another survey (Duschl & Wright, 1989) in the United States of teachers' views on the nature of science and STS issues, led to the assertion that all of the teachers held the hypothetico-deductive philosophy of logical positivism. Thus, the authors concluded that commitment to this view of science explained the lack of effective consideration of the nature of science and STS in teachers' classroom science lessons. A reason suggested was that teachers of senior status probably received instruction and education in science that did not include any discussion of the nature of science. This explanation concurs with the explanation offered of the responses of teachers to questions on the nature of science in the structured interview component of the South Australian study.

A further possible explanation for the finding of gaps in studies over the past 20 years on teachers' understandings of the nature of science and STS is that some teachers might have relied on text books to provide them with ideas and understandings for their science lessons. It appears that these textbooks contained very little discussion of the nature of science or STS issues. This suggestion was supported by an examination by Duschl and Wright (1989), of textbooks used by teachers, since this 1989 study showed that the nature of science and the nature of scientific knowledge were not emphasised in these books. Although most of the text books began with an attempt to portray science as a process of acquiring knowledge about the world, the books failed to give any space to a discussion of the history of the development of scientific understanding, the methodology of science, or the relevance of science for students' daily lives. Gallagher (1991) suggested that these depictions of science were empirical and positivistic and that most teachers believed in the objectivity of science. In regard to the reasons for this belief, Gallagher (1991, p. 125) reached the cogent conclusion that:

> Science was portrayed as objective knowledge because it was grounded in observation and experiment, whereas the other school subjects were more subjective because they did not have the benefit of experiment, and personal judgments entered into the conclusions drawn. In the minds of these teachers, the objective quality of science made science somewhat 'better' than the other subjects.

It is possible that the finding, in the present study, of strong and coherent views, beliefs, and attitudes in regard to STS held by teachers is due, at least partially, to the shift towards the inclusion of STS issues and social relevance in science text books. Discussions with South Australian senior secondary science teachers indicated that the science textbooks used in secondary schools now included examples and discussions of the social relevance of science in many instances.

The traditional inaccurate and inappropriate image of science has been attributed (Gallagher, 1991) to science and teacher education courses, which placed great emphasis upon the rapid coverage of a large body of scientific knowledge, but gave prospective teachers little or no time to learn about the nature of science or to consider the history, philosophy and sociology of science. Fortunately, this situation has now changed, to an extent, in increasing numbers of tertiary courses in Australia. The coherent views of South Australian science teachers might be due in part to this change in the emphasis of tertiary courses.

4.3 Comparison with United States data (teachers and scientists)

The STS views of teachers in this South Australian study were stronger and of greater coherence than the views of scientists on all three scales. In a similar way to the South Australian study, Pomeroy's (1993) American study also used a well-validated survey instrument (Kimball, 1968) to explore the views, beliefs and attitudes of a sample of American research scientists and teachers. In the analysis of the results of this American study, the views which were identified in groups of statements included: (a) the traditional logico-positivist view of science, and (b) a non-traditional view of science characteristic of the philosophy of STS. Thus, consideration of the findings and an analysis of Pomeroy's study provide a useful basis for the discussion of the findings of the South Australian study.

The comparison of the responses of the American scientists and teachers showed that the scientists (mean = 3.14, SD = 0.52) had significantly ($p = 0.02$, $F = 0.41$) [sic] more traditional views of science than the teachers (mean = 2.92, SD = 0.63) (Pomeroy, 1993, p. 266). In the South Australian study, the scientists would also appear to have had more traditional views of science, or certainly views that were less favourable towards STS than teachers, as shown by their lower average mean scores on all three scales. Nevertheless, since the American group of teachers included elementary school teachers in addition to secondary science teachers, a direct comparison is not possible. However, this consideration of the findings of Pomeroy's study is still extremely useful, since the difference in the mean scores of the American secondary and elementary teachers was not significant ($p = 0.2$). For questions in Pomeroy's study, which gauged the non-traditional views of science characteristic of the philosophy of STS, the views of teachers were of higher average mean score than the views of scientists, although the difference was not significant (Pomeroy, 1993, p. 266).

Pomeroy's 1993 study showed that scientists, and to a lesser extent teachers, expressed the traditional view of science quite strongly. The suggestion was advanced that these findings 'add to the continuing dialogue in the literature as to the persistence of positivistic thought in scientists and educators today' (Pomeroy, 1993, p. 269). The findings of this Australian study appear to indicate that, by way of contrast, South Australian science teachers held views of science that were strong and coherent towards STS. Thus, many Australian teachers did not hold the traditional logico-positivist view of science. This may well have arisen from the courses studied by South Australian science teachers, in which there has been a strong emphasis upon STS issues in the biological sciences, in contrast to the emphasis in the physical sciences that existed in former years.

A further similarity between Pomeroy's scale and the scale used for this South Australian study is that they both found support for their statements from the writings of the philosophers of science. However, this South Australian study framed the statements of its scale in terms of language used by students and the views of the students. Thus, the difference lay in the fact that Pomeroy's items consisted of statements written exclusively in the language of the investigators after consideration of the works of the philosophers of science. Furthermore, Kimball's and Pomeroy's scales used a Likert-type response format.

It is important to consider whether any differences in the scientists' and teachers' understandings of the nature of science might be the result of training or experience. Kimball asserted that answers to this question were needed in order to guide the revision of science teacher education programs. Kimball's study compared the views on the nature of science of scientists and teachers of similar academic backgrounds and sought teachers' and scientists' views during the years after graduation. The fact that the educational backgrounds of respondents were taken into account for the comparisons made in the earlier American study highlights a limitation of the South Australian study, since, due to the restraints of confidentiality, it was not possible to gauge the academic backgrounds of respondents for a comparison of the views of Australian scientists and South Australian teachers of similar academic backgrounds.

4.4 Comparison with Canadian data (teachers and students)

The views of teachers were found to be stronger and more coherent on all three scales than the views of students in this South Australian study. Previous studies involving a comparison of students' and teachers' views, beliefs, and positions on STS issues also have a worthwhile place in this discussion to highlight the meaning of the results of this study. An assessment of the pre-post positions on STS issues of students exposed to an STS-oriented course, as compared with the positions of students and teachers not exposed to this course, was undertaken in Zoller, Donn, Wild and Beckett's (1991) study in British Colombia. This Canadian study used a scale consisting of six questions from the VOSTS inventory (Aikenhead, Ryan & Fleming, 1989) to compare the beliefs and positions of groups of STS-students with their teachers and with non-STS Grade 11 students. The study addressed whether the STS positions of students and their teachers were similar or different. Since the questions for the scale used in the South Australian study were also adapted from the VOSTS inventory, consideration of the results of Zoller et al.'s study is of particular interest in this discussion of the findings on South Australian teachers' and students' STS views. One of the six questions in the scale used in the Canadian study was concerned with whether scientists should be held responsible for the harm that might result from their discoveries, and this was also in the scales used in the Australian study. In the Canadian study, students' and teachers' views were compared by grouping the responses to each of the statements into clusters, which formed the basis for the analysis. The STS response profile of Grade 11 students was found to differ significantly from that of their teachers.

A critical problem in the report of Zoller's Canadian study in regard to the production of possible biases through non-random sampling and self-selection of teachers and scientists is also relevant for the South Australian study. The selection processes, which were used in both studies, might have produced some bias as a consequence of those who chose to respond being more interested in philosophical issues or more confident about their views on the philosophy, pedagogy and sociology of science. In the South Australian study, the younger teachers volunteered more readily, and this self-selection of teachers might have introduced some bias into the results.

4.5 Structured interviews with teachers

In order to gain richer and more interesting information on the views of teachers in relation to the curriculum shift towards the inclusion of STS, in each school, group interviews with teachers were conducted.

Teachers in 16 of the 23 interview groups viewed the curriculum shift towards STS positively, since they believed there was a need to include a discussion of STS issues in science classes in order to make the scientific concepts more socially relevant. These teachers suggested that, in order to prepare students to take their places as vital participants in the community, it was important to make the links between science, technology and society. Discussions with some teachers centred on the provision of motivation for students by discussing STS issues. Teachers believed that both male and female students responded well to STS.

The South Australian teachers discussed a diverse variety of teaching strategies with which they translated the STS curriculum into their classroom teaching. Many teachers used classroom discussions of STS issues as a strategy to help students to meet the STS objectives of their courses. Teachers were particularly enthusiastic about using discussions of knowledge of STS issues that students had gained from the media as a starting point for the construction of further understandings of scientific concepts.

Unfortunately, most of the teachers stressed that there was a lack of appropriate resources for teaching STS, and more than half suggested that it was difficult to find the time to research STS issues to include in their classes. The teachers suggested that access to a central resource pack and a variety of other resources would assist them in translating the STS curriculum into their classroom practice, because they believed that resources for teaching STS were very expensive and needed continual updating.

The findings of this Australian study supported Thomas' (1987) argument that teachers who had been trained in the traditional courses to teach pure science found it difficult to include discussions of STS issues in their science classes, because they believed this approach to science teaching questioned the absolute certainty of scientific knowledge. Moreover, some of the experienced science teachers in this Australian study had probably taught in basically the same way, and had used the same teaching techniques, for many years. These teachers believed that the STS emphasis of the curricula was an intrusion and expressed concern about not being able to cover all of the concepts, and having to vary their teaching methods. Hence, they were probably not comfortable about using innovative techniques such as role-plays, brainstorming and community-oriented projects in small groups.

The understanding of the nature of science and the nature of STS by some teachers in the interviews in this present study was found to be quite limited. It would appear that teachers with an educational background that provided them with an understanding of the interactions between science, technology and society, were quite well-informed and eloquent in regard to both the nature of science and STS issues. These were fairly young teachers, as well as teachers who had worked in other professions outside of teaching and more experienced teachers who had shown flexibility and openness to new teaching innovations consistently throughout their careers. They believed that there was a need to stop students believing in a simple technological fix for every problem.

The inclusion of consideration of STS in secondary science courses certainly does not water down the content, but it offers many benefits for both the teaching and learning of science. Researchers such as Ziman (1980) have argued that in science taught from the foundation of STS the science content was embedded in a social context, which provided meaning for students.

While the teachers in these interviews were concerned about demands on their time, they were generally quite supportive of the move to include STS objectives in secondary science courses. The vast majority of teachers believed that the inclusion of STS in secondary science curricula was a very positive move. However, the interviews with teachers pointed to the need for the provision of effective in-service courses for teachers in relation to the STS objectives in the senior secondary science curricula.

5. CONCLUSIONS

In this Australian study, a master scale was employed to measure the extent to which teachers, students and scientists held strong and coherent views of STS. The South Australian teachers' views of STS were shown in this study to be quite strong and coherent.

It appears from a consideration of the literature on teachers' views of the nature of science and the issues of STS, that many senior secondary science teachers throughout the years have depicted science incorrectly. However, it would also appear from the results of this present South Australian study that in recent years this situation might have begun to change. In this South Australian study, the STS views of secondary science teachers were shown to be quite strong and coherent. It is possible that the move towards the inclusion of STS objectives in senior secondary science courses in South Australia has precipitated the development of these more coherent views.

Another possible explanation for the finding of quite strong and coherent views on STS among South Australian secondary science teachers is that the media have provided Australian citizens with the opportunity to consider and reflect upon the issues of STS. However, the structured interviews in this Australian study showed that the results might have been due to the large proportion of younger teachers who volunteered to respond to the instrument that was used during this present study. The interviews included a greater range of both experienced and young teachers than the group that responded to the instrument in this study. It appears that the undergraduate training of science teachers has changed sufficiently to address the need created by the changed objectives, so that consideration of the issues of STS is included to a greater extent in secondary science courses in South Australia.

6. REFERENCES

Aikenhead, G.S., Fleming, R.W. & Ryan, A.G. (1987). High school graduates' beliefs about Science-Technology-Society. 1. Methods and issues in monitoring student views. *Science Education*, 71, 145-161.

Aikenhead, G.S. & Ryan, A.G. (1992). The development of a new instrument: 'Views on Science-Technology-Society' (VOSTS). *Science Education*, 76, 477-491.

Aikenhead, G.S., Ryan, A.G. & Fleming, R.W. (1989). *Views on Science-Technology-Society.* Department of Curriculum Studies, College of Education: Saskatchewan.

Barnes, B. (1985). *About Science.* Basil Blackwell: Oxford.

Bloom, B.S. (1976). *Human Characteristics and School Learning*. McGraw-Hill Book Company: New York.

Brick, J.M., Broene, P., James, P. & Severynse, J. (1996). *A User's Guide to WesVarPC program*. Westat Inc.: Rockville, USA.

Bybee, R.W. (1987). Science education and the Science-Technology-Society (S-T-S) theme. *Science Education,* 70, 667-683.

Cross, R.T. (1990). Science, Technology and Society: Social responsibility versus technological imperatives. *The Australian Science Teachers Journal*, 36 (3), 34-35.

Duschl, R.A. & Wright, E. (1989). A case of high school teachers' decision-making models for planning and teaching science. *Journal of Research in Science Teaching*, 26, 467-501.

Fensham, P. (1990). What will science education do about technology? The *Australian Science Teachers Journal*, 36, 9-21.

Fullan, M. & Stiegelbauer, S. (1991). *The New Meaning of Educational Change.* Teachers College Press, Columbia University: New York.

Gallagher, J.J. (1991). Prospective and practicing secondary school science teachers' knowledge and beliefs about the philosophy of science. *Science Education*, 75, 121-133.
Gesche, A. (1995). Beyond the promises of biotechnology. *Search*, 26, 145-147.
Heath, P.A. (1992). Organizing for STS teaching and learning: The doing of STS. *Theory Into Practice*, 31, 53-58.
Kimball, M. (1968). Understanding the nature of science: A comparison of scientists and science teachers. *Journal of Research in Science Teaching*, 5, 110-120.
Kuhn, T.S. (1970). *The Structure of Scientific Revolutions*. The University of Chicago Press: Chicago.
Lederman, N.G. (1986). Students' and teachers' understanding of the nature of science: A reassessment. *School Science and Mathematics*, 86, 91-99.
Lowe, I. (1993). *Making science teaching exciting: Teaching complex global issues*. In 44th Conference of the National Australian Science Teachers' Association: Sydney.
Lowe, I. (1995). Shaping a sustainable future. *Griffith Gazzette,* 9.
Lumpe, T., Haney, J.J. & Czerniak, C.M. (1998). Science teacher beliefs and intentions to implement Science-Technology-Society (STS) in the classroom. *Journal of Science Teacher Education*, 9(1), 1-24.
Masters, G.N. (1988). Partial credit models. In *Educational Research, Methodology and Measurement: An International Handbook*, Keeves, J.P. (ed). Pergamon Press: Oxford.
Merton, R.K. (1973). *The Sociology of Science: Theoretical and Empirical Investigations*. University of Chicago Press: Chicago.
National Board of Employment Education and Training. (1993). *Issues in Science and Technology Education A Survey of Factors which Lead to Underachievement*. Australian Government Publishing Service: Canberra:
National Board of Employment Education and Training. (1994). *Science and Technology Education: Foundation for the Future*. Australian Government Publishing Service: Canberra.
Norusis, M. J. (1990). *SPSS Base System. User's Guide*. SPSS: Chicago, Illinois.
Parker, L. H. (1992). Language in science education: Implications for teachers. *Australian Science Teachers Journal*, 38 (2), 26-32.
Parker, L.H., Rennie, L.J. & Harding, J. (1995). Gender equity. In *Improving Science Education*, Fraser, B.J. & Walberg, H.J. (eds). University of Chicago Press: Chicago.
Pomeroy, D. (1993). Implications of teachers' beliefs about the nature of science: Comparison of the beliefs of scientists, secondary science teachers and elementary teachers. *Science Education*, 77, 261-278.
Rennie, L.J. & Punch, K.F. (1991). The relationship between affect and achievement in science. *Journal of Research in Science Teaching*, 28, 193-209.
Rasch, G. (1960). *Probabilistic Models for some Intelligence and Attainment Tests*. Paedagogiske Institute: Copenhagen, Denmark.
Rubba, P.A., Bradford, C.S. & Harkness, W.J. (1996). A new scoring procedure for The Views on Science-Technology-Society instrument. *International Journal of Science Education*, 18, 387-400.
Rubba, P.A. & Harkness, W.L. (1993). Examination of preservice and in-service secondary science teachers' beliefs about Science-Technology-Society interactions. *Science Education*, 77, 407-431.
Tedman, D.K. (1998). *Science Technology and Society in Science Education*. PhD Thesis. The Flinders University, Adelaide.
Tedman, D.K. and Keeves, J.P. (2001) The development of scales to measure students', teachers' and scientists' views on STS. *International Education Journal*, 2 (1), 20-48. http://www.flinders.edu.au/education/iej

Thomas, I. (1987). Examining science in a social context. *The Australian Science Teachers Journal*, 33 (3), 46-53.

Thorndike, R.L. (1982). *Applied Psychometrics*. Houghton Mifflin Company: Boston.

Wright, B.D. (1988). Rasch measurement models. *In Educational Research Methodology and Measurement: An International Handbook*, Keeves, J.P. (ed) pp. 286-297. Pergamon Press: Oxford, England.

Yager, R.E. (1990a). STS: Thinking over the years - an overview of the past decade. *The Science Teacher*, March 1990, 52-55.

Yager, R. E. (1990b). The science/technology/society movement in the United States: Its origin, evolution and rationale. *Social Education*, 54, 198-201.

Ziman, J. (1980). *Teaching and Learning about Science and Society*. Cambridge University Press: Cambridge, UK.

Zoller, U., Donn, S., Wild, R. & Beckett, P. (1991). Students' versus their teachers' beliefs and positions on science/technology/society-oriented issues. *International Journal of Science Education*, 13, 25-36.

Chapter 14

ESTIMATING THE COMPLEXITY OF WORKPLACE REHABILITATION TASKS USING RASCH ANALYSIS

Ian Blackman
School of Nursing, Flinders University

Abstract: This paper explores the application of the Rasch model in developing and subsequently analysing data derived from a series of rating scales that measures the preparedness of participants to engage in workplace rehabilitation. Brief consideration is given to the relationship between effect and learning together with an overview of how the rating scales were developed in terms of their content and processes. Emphasis is then placed on how the principles of Rasch scaling can be applied to rating scale calibration and analysis. Data derived from the application of the Workplace Rehabilitation Scale are then examined for the evidence of differentiated item function (DIF).

Key words: partial credit, attitude measurement, DIF, rehabilitation

1. INTRODUCTION

This paper explores the application of the Rasch model in developing and subsequently analysing data derived from a series of rating scales that measures the preparedness of participants to engage in workplace rehabilitation. Brief consideration is given to the relationship between effect and learning together with an overview of how the rating scales were developed in terms of their content and processes. Emphasis is then placed on how the principles of Rasch scaling can be applied to rating scale calibration and analysis. Data derived from the application of the Workplace

S. Alagumalai et al. (eds.), Applied Rasch Measurement: A Book of Exemplars, 251–270.

Rehabilitation Scale are then examined for the evidence of differential item function (DIF).

1.1 Identifying relevant test items for the construction of a Workplace Rehabilitation Rating Scale

Useful educational information may be gained by asking participants to complete attitude rating scales at different times during their learning or employment. Their attitude is frequently measured this way for three different reasons:

1. to identify their attitudes and dispositions that underlie their thinking and which may can be either positive or negative;
2. to estimate the intensity of their attitude; and
3. to gauge the consistency of their attitude towards some belief/value.

In order to determine the desired scope of workplace rehabilitation training for managers, a survey was conducted using rating scales, specifically trying to capture the manager's attitudes towards their expected role and their preparedness to assist with vocational rehabilitation in their work sites. To complement those findings, a similar survey instrument was given to employees undertaking vocational rehabilitation. The study sought to ascertain if there were a difference between the perceived complexities of vocationally related rehabilitation tasks, as understood by workplace managers and their employees who were undertaking workplace rehabilitation. The data were generated by administering a 31 item, four point (Likert-type) rating scale to 352 staff members who were to rate their ability to perform numerous rehabilitation tasks.

By employing the partial credit model (part of the Rasch family), the research sought to discover if there was an underlying dimension of the work-related rehabilitation tasks and whether the ability to undertake workplace rehabilitation tasks were influenced by the status of the participants within in the work area: that is, either the injured employee or the workplace manager. Furthermore, the study sought to assess whether a scale of learning could be constructed, based on the estimation of the difficulty of the rehabilitation tasks based on the responses of the injured employees and workplace managers.

1.2 Developing the rating scale content: rehabilitation overview

In order to identify the numerous factors that influence vocational workplace rehabilitation and to inform the construction of the rating scale in the survey construction, a literature review was undertaken. This will be briefly alluded to next because the factors identified will become important in determining the validity of the scale that has been developed (see unidimensionality described below).

There has been much debate about the success of workplace rehabilitation since various Australian state governments enacted laws to influence the management of employees injured at work (Kenny, 1994; Fowler, Carrivick, Carrelo & McFarlane, 1996; Calzoni, 1997). The reasons for this are numerous but include such factors as confusion about how successful vocational rehabilitation can be measured, misunderstanding as to what are the purposes of workplace rehabilitation, poor utilisation of models that inform vocational rehabilitation (Cottone & Emener 1990; Reed, Fried &Rhoades, 1995) and resistance on the part of key players involved in vocational workplace rehabilitation (Kenny, 1995a; Rosenthal & Kosciulek, 1996; Chan, Shaw, McMahon, Koch & Strauser, 1997). Compounding this problem further is that, while workplace managers are assumed to be in the best position to oversee their employees generally, managers are ill-prepared to cater for the needs of injured employees in the workplace (Gates, Akabas & Kantrowitz, 1993; Kenny, 1995a). Industry restructuring, in which workplace managers are expected to take greater responsibility for larger numbers of employees, has put greater pressure on the workplace manager to cope with the needs of rehabilitating employees. Workplace managers who struggle with workplace rehabilitation and experience inadequate communication with stakeholders involved in the vocational rehabilitation workplace, may themselves become stressed, and this can result in workplace bullying (Gates et al., 1993; Kenny, 1995b; Dal-Yob, Taylor & Rubin, 1995; Garske, 1996; Calzoni, 1997; Sheehan, McCarthy & Kearns, 1998).

One vital mechanism that helps to facilitate the role of the manager in the rehabilitative process and simultaneously serves to promote a successful treatment plan of injured employees is vocational rehabilitation training for managers (Pati, 1985).

Based on the rehabilitation problems identified in the literature, 31 items or statements relating to different aspects of the rehabilitative process were identified for inclusion into a draft survey instrument for distribution and testing. Rehabilitation test items are summarised in Table 14-1 and the full questionnaires are included in the appendices.

Each of the 31 rehabilitation items were prescribed as statements followed by four ordered response options: namely, 1 a very simple task; 2 an easy task; 3 a hard task; and 4 a very difficult task.

Table 14-1. Overview of the content of the work-place rehabilitation questionnaire given to managers and rehabilitating employees for completion

Rehabilitation focus	**Questionnaire item number(s)**
Rehabilitation documentation	15, 17
Involvement with job retraining	5, 20
Staff acceptance of the presence of a rehabilitating employee in the work-place	6, 13, 27
Suitability of allocated return to work duties	1, 10, 23
Confidentiality of medically related information	2
Contact between manager and rehabilitating employee	3, 22,
Dealing with negative aspects about the workplace and rehabilitation	4, 21
Securing equipment to assist with workplace rehabilitation	7
Understanding legal requirements and entitlements related to rehabilitation	8, 14, 18
Communication with others outside the workplace (e.g. doctors, rehabilitation consultants, spouses, unions)	9, 11, 29, 30, 31
Budget adjustments related to work role changes	26
Gaining support from the workplace/organisation	12, 19, 24, 28
Dealing with language diversity in the workplace	16
Managing conflict as it relates to workplace rehabilitation	25

2. PITFALLS WITH THE CONSTRUCTION AND SUBSEQUENT ANALYSIS OF RATING SCALES

Two common assumptions are commonly made when rating scales are constructed and employed to measure some underlying construct (Bond & Fox, 2001, p. xvii). Firstly, it is assumed that equal scores indicate equality on the underlying construct. For example, if two participants, A and B, respond to the three items shown in Table 14-1 with scores of 2, 3 and 4 and the other with score of 4, 4 and 1 respectively, they would both have scores of nine. An implied assumption is that all items are of equal difficulty and therefore that raw scores may be added. Secondly, there is no mechanism for exploring the consistency of an individual's responses. The inclusion of inconsistent response patterns has been shown to increase the standard error of threshold estimates and to compress the threshold range during the instrument calibration. It is therefore desirable to use a method of analysis that can detect respondent inconsistency, that can provide estimates of item thresholds and individual trait estimates on a common interval scale, and that can provide standard errors of these estimates. The Rasch measurement model meets these requirements.

On the basis of 'sufficiency', the two respondents both with scores of 9, would receive the same estimate of trait ability under the Rasch model. What would differentiate them is the fit statistic. One would produce a reasonable fit statistic (>0.6 and <1.5 infit means square), while the other would probably have an infit means square of >1.7.

A second point is that where respondents do not respond to all items or where respondents are presented with different sets of items, the Rasch model is able to estimate trait score, taking into account the level of item difficulty. This is not possible using raw scores or classical test theory.

3. CALIBRATING THE WORKPLACE REHABILITATION SCALE

3.1 The partial credit model

The Workplace Rehabilitation Scale comprised 31 rehabilitation items that were used to survey effect on participants, values and vocational rehabilitation preparedness. Given that the scale used for the workplace rehabilitation questionnaire was polytomous in nature, the partial credit model was used for data collection and analysis. Unlike the rating scale model (which is a extension of the Rasch dichotomous model), the partial credit model has the advantage of not constraining threshold levels between items but allows threshold levels to vary across items. This model approach estimates if distances between response categories are constant for each rehabilitation item and if the options for each item vary in the number of response categories. The alternative approach (rating scale model) uses only one set of item thresholds estimates and it applies this to all the items that make up the Workplace Rehabilitation Scale (Bond & Fox, 2001, p. 204).

3.2 Unidimensionality

While the workplace rehabilitation survey instrument does not contain many items, some measurement of reliability of the instrument is warranted. Unidimensionality is an assumption of the Rasch model and is required to ensure that all test items used to construct the Workplace Rehabilitation Scale reflect the same underlying construct. If items are not seen to reflect a common rehabilitation construct, they need to be reviewed by the researcher and either modified or removed from the instrument (Hambenton, Swaminathan & Rogers, 1991; Linacre, 1995; Smith, 1996). This test for unidimensionality seeks to ensure test item validity. In order to test for unidimensionality of the rehabilitation data, the QUEST program (Adams, & Fox, 1996) was employed. The two main goodness of fit indices used in the analysis are the unweighted (or the outfit means square) and the weighted (or infit means square) index. Both are a form of chi-square ratios, which provide information about discrepancies between the predicted and observed data, particularly in terms of the size and the direction of the residuals for each rehabilitation item being estimated. Fit values will vary around a mean of zero and will either be positive or negative, depending whether the observed values show greater variation in responses than what was expected (greater variation showing a positive value and minimal variation as a negative value). In this way, the compatibility of the data obtained from the Workplace Rehabilitation Scale can be screened against the Rasch model requirement.

The outfit means square index is sensitive to outliers where the managers or the rehabilitating employees gained unexpectedly high scores on difficult rehabilitation items or achieved uncharacteristic low scores on easy rehabilitation items. One or two of these large outliers can cause the statistic to become very large as when a very able respondent views an easy rehabilitation item as difficult or when another respondent with low ability views a difficult rehabilitation item as being easy. With reference to Figures 14-1 and 14-2, which represent the fit model of all rehabilitation items completed by workplace managers and rehabilitating employees respectively, it can be seen that all questions fit the Rasch model with infit means square values not greater than 1.30 or less than 0.77 (Adams et al., 1996).

All rehabilitation items are seen as being compatible as the observed responses given by the participants are expected by the model's requirement for unidimensionality. Further analysis of the overall statistics as set out in Table 14-2 reveal more information about the reliability of the survey tool.

The mean for items for both cohorts is set at zero as part of the Rasch analysis. It can be seen that there is a greater item standard deviation for workplace manager responses for the items estimated than the employees. Item reliability for both cohorts is quite low especially for the workplace managers and negligible for rehabilitating employees. This index indicates the replicability of the items placed along the rehabilitation continuum, if these same items were given to another group of respondents with similar ability levels. This low index suggests that the employee group was more likely to be responding more erratically to some items in the scale than the manager group.

```
------------------------------------------------------------------------------------------
INFIT
 MNSQ     .56       .63       .71       .83      1.00      1.20      1.40      1.60      1.8
---------+---------+---------+---------+---------+---------+---------+---------+---------+
  1 Item 1                        .              |    *         .
  2 Item 2                        .              |  *           .
  3 Item 3                        .              |      *       .
  4 Item 4                        .              |       *      .
  5 Item 5                        .              |   *          .
  6 Item 6                        .              | *            .
  7 Item 7                        .              |  *           .
  8 Item 8                        .             *|              .
  9 Item 9                        .              *              .
 10 Item 10                       .              *              .
 11 Item 11                       .        *     |              .
 12 Item 12                       .       *      |              .
 13 Item 13                       .           *  |              .
 14 Item 14                       .         *    |              .
 15 Item 15                       .             *|              .
 16 Item 16                       .              |  *           .
 17 Item 17                       .       *      |              .
 18 Item 18                       .              | *            .
 19 Item 19                       .           *  |              .
 20 Item 20                       .             *|              .
 21 Item 21                       .   *          |              .
 22 Item 22                       .              |       *      .
 23 Item 23                       .              |*             .
 24 Item 24                       .              *              .
 25 Item 25                       .        *     |              .
 26 Item 26                       .              |           *  .
 27 Item 27                       .        *     |              .
 28 Item 28                       .      *       |              .
 29 Item 29                       .        *     |              .
 30 Item 30                       .          *   |              .
 31 Item 31                       .            * |              .
------------------------------------------------------------------------------------------
```

Figure 14-1. Fit indices for workplace managers' responses for all rehabilitation items

```
------------------------------------------------------------------------------------------
INFIT
 MNSQ     .56       .63       .71       .83      1.00      1.20      1.40      1.60      1.8
---------+---------+---------+---------+---------+---------+---------+---------+---------+
  1 Item 1                        .            * |              .
  2 Item 2                        .              |          *   .
  3 Item 3                        .              |*             .
  4 Item 4                        .              |  *           .
  5 Item 5                        .         *    |              .
  6 Item 6                        .     *        |              .
  7 Item 7                        .           *  |              .
  8 Item 8                        .              |   *          .
  9 Item 9                        .           *  |              .
 10 Item 10                       .              | *            .
 11 Item 11                       .              |*             .
 12 Item 12                       .      *       |              .
 13 Item 13                       .              |*             .
 14 Item 14                       .              |        *     .
 15 Item 15                       .              | *            .
 16 Item 16                       .     *        |              .
 17 Item 17                       .   *          |              .
 18 Item 18                       .             *|              .
 19 Item 19                       .           *  |              .
 20 Item 20                       .       *      |              .
 21 Item 21                       .    *         |              .
 22 Item 22                       .              |        *     .
 23 Item 23                       .              *              .
 24 Item 24                       .              |         *    .
 25 Item 25                       .        *     |              .
 26 Item 26                       .              |         *    .
 27 Item 27                       .              |  *           .
 28 Item 28                       .  *           |              .
 29 Item 29                       .              |         *    .
 30 Item 30                       .              |   *          .
 31 Item 31                       .      *       |              .
------------------------------------------------------------------------------------------
```

Figure 14-2. Fit indices for rehabilitating employees' responses for all rehabilitation items

Table 14-2. Summary statistics for rehabilitation items and persons

Participant	Index	Value
Workplace managers	Item mean	0.00
	Item standard deviation	0.51
	Item reliability	0.47
	Person mean	-0.59
	Person standard deviation	0.93
	Person reliability	0.89
Rehabilitating employee	Item mean	0.00
	Item standard deviation	0.27
	Item reliability	0.00
	Person mean	-0.18
	Person standard deviation	0.83
	Person reliability	0.89

When re-examining employee responses to the survey items, some responded to the survey quite differently from other participants. Some employees estimated all but one rehabilitation item as being very simplistic. Other respondents rated all rehabilitation items with extreme responses (rating with either 1 or a 4) only. A small number tended to rate rehabilitation items with numbers of 2 or 3 with no items being either very simple or very difficult. Item reliability in this case has alerted the researcher to the fact that participants may not be responding as accurately as possible or their rehabilitation experience in reality, is not consistent with the types of statements given in the survey.

While item reliability for the workplace managers is much more robust compared with the employee responses, their ratings need to be checked also for responses which show little or major variances. The person reliability indices for both groups are much higher than for the item index. Person reliability reflects the capacity to duplicate the ordering for persons alongside the rehabilitation scale, if this sample of persons were given another set of rehabilitation items measuring the same construct. The reliability indices for both groups are quite high. Table 14-3 and Table 14-4 show the fit statistics mentioned above, but also reveal threshold values for the rehabilitating employee and manager group respectively. These threshold values are all correctly ordered, ranging from negative values (most difficult items) through to zero (items of average difficulty), then they become positive values, which indicate item ease.

Table 14-3. Item estimates (thresholds) in input order for rehabilitating employees (n=80)

	ITEM NAME	SCORE	MAXSCR	THRESHOLD/S 1	2	3	INFT MNSQ	OUTFT MNSQ	INFT t	OUTFT t
1	Item 1	80	225	-.94	.60	1.44	.96	.91	-.2	-.4
				.41	.47	.50				
2	Item 2	84	225	-.78	.37	1.38	1.19	1.29	1.3	1.3
				.44	.43	.51				
3	Item 3	111	216	-1.30	-.19	.80	1.02	1.26	.2	1.3
				.45	.42	.43				
4	Item 4	104	216	-1.56	.01	.97	1.05	1.09	.4	.5
				.50	.44	.47				
5	Item 5	99	210	-1.13	-.02	.95	.92	.88	-.6	-.6
				.44	.44	.47				
6	Item 6	96	192	-1.41	-.18	1.23	.84	.84	-1.0	-.8
				.50	.45	.49				
7	Item 7	89	204	-1.34	.12	1.50	.94	.93	-.4	-.3
				.47	.46	.56				
8	Item 8	96	207	-1.38	.22	.92	1.08	1.14	.6	.8
				.50	.43	.45				
9	Item 9	79	195	-.91	.24	1.17	.94	.90	-.3	-.4
				.47	.44	.49				
10	Item 10	83	225	-.88	.41	1.51	1.04	.99	.3	.0
				.41	.45	.52				
11	Item 11	90	231	-.78	.25	1.22	1.03	1.01	.2	.1
				.42	.42	.48				
12	Item 12	106	228	-1.03	-.04	.96	.87	.83	-1.0	-.8
				.41	.43	.46				
13	Item 13	117	222	-1.59	-.17	.98	1.03	1.00	.3	.1
				.50	.43	.46				
14	Item 14	101	213	-1.34	.08	1.01	1.16	1.19	1.1	1.0
				.47	.43	.44				
15	Item 15	41	150	-.06	.49	1.72	1.03	1.38	.3	1.3
				.48	.53	.74				
16	Item 16	62	147	-1.38	.16	2.09	.85	.83	-.8	-.6
				.59	.53	.77				
17	Item 17	101	225	-1.69	.15	1.68	.82	.84	-1.2	-.8
				.50	.46	.54				
18	Item 18	114	219	-1.81	-.18	.97	.98	1.01	-.1	.1
				.53	.44	.47				
19	Item 19	88	186	-1.88	.25	1.09	.95	.89	-.3	-.5
				.59	.48	.49				
20	Item 20	80	189	-1.16	.20	1.27	.88	.90	-.7	-.4
				.47	.47	.52				
21	Item 21	110	228	-1.63	-.05	1.37	.84	.82	-1.1	-1.0
				.47	.43	.50				
22	Item 22	79	207	-1.25	.63	1.41	1.16	1.13	1.0	.6
				.47	.48	.56				
23	Item 23	82	201	-1.31	.32	1.40	1.01	.95	.1	-.2
				.47	.49	.53				
24	Item 24	117	207	-1.53	-.40	.62	1.18	1.15	1.2	.7
				.50	.44	.42				
25	Item 25	129	210	-1.94	-.38	.40	.89	.88	-.8	-.6
				.56	.44	.40				
26	Item 26	86	183	-1.13	-.25	1.55	1.17	1.23	1.1	1.0
				.50	.47	.55				
27	Item 27	89	201	-1.16	.31	.70	1.06	1.12	.5	.6
				.44	.44	.44				
28	Item 28	84	174	-1.25	.14	.57	.81	.83	-1.4	-.8
				.47	.45	.45				
29	Item 29	78	177	-.88	.25	.77	1.19	1.15	1.2	.7
				.47	.46	.46				
30	Item 30	91	195	-1.19	.14	.91	1.07	1.03	.5	.2
				.47	.45	.47				
31	Item 31	41	81	-.92	-.05	.91	.87	.82	-.6	-.5
				.70	.68	.70				
Mean				.00			.99	1.01	.0	.1
SD				.27			.12	.16	.8	.7

Table 14-4. Item estimates (thresholds) in input order for workplace managers (n=272)

	ITEM NAME	SCORE	MAXSCR	THRESHOLD/S 1	2	3	INFT MNSQ	OUTFT MNSQ	INFT t	OUTFT t
1	Item 1	346	798	-2.78	.04	2.06	1.11	1.14	1.3	1.3
				.31	.27	.38				
2	Item 2	149	538	-.44	1.70		1.05	1.03	.7	.3
				.25	.35					
3	Item 3	328	807	-2.22	.22	1.31	1.13	1.18	1.5	1.6
				.28	.25	.31				
4	Item 4	257	792	-2.00	.97	3.28	1.16	1.18	1.7	1.6
				.28	.31	.79				
5	Item 5	306	795	-2.75	.57	2.34	1.08	1.12	.8	1.0
				.31	.30	.45				
6	Item 6	380	801	-3.44	-.27	2.19	1.04	1.05	.5	.5
				.38	.25	.39				
7	Item 7	365	795	-2.63	-.26	1.81	1.07	1.18	.8	1.7
				.28	.26	.32				
8	Item 8	294	807	-2.41	.64	3.52	.97	.96	-.3	-.4
				.28	.28	.78				
9	Item 9	323	786	-2.88	.29	1.98	1.00	1.00	.1	.0
				.31	.28	.40				
10	Item 10	261	536	-2.25	1.27		1.00	1.00	.1	.0
				.28	.30					
11	Item 11	272	795	-2.31	.91	2.65	.89	.87	-1.2	-1.1
				.28	.31	.58				
12	Item 12	257	786	-1.94	.90	2.12	.88	.86	-1.2	-1.3
				.25	.33	.46				
13	Item 13	269	801	-2.19	.96	3.98	.94	.94	-.6	-.5
				.28	.31	1.07				
14	Item 14	373	807	-2.94	-.29	2.27	.90	.91	-1.2	-.9
				.34	.26	.40				
15	Item 15	329	804	-2.91	.33	2.34	.98	.99	-.1	-.1
				.34	.28	.44				
16	Item 16	319	768	-2.84	.21	2.46	1.05	1.05	.6	.5
				.31	.27	.48				
17	Item 17	315	798	-2.88	.48	2.24	.88	.85	-1.3	-1.3
				.31	.30	.45				
18	Item 18	226	780	-1.38	.93	2.17	1.05	1.00	.5	.1
				.25	.33	.48				
19	Item 19	212	783	-1.75	1.73	2.32	.95	.94	-.4	-.5
				.25	.51	.60				
20	Item 20	322	774	-2.59	.10	2.87	.98	.98	-.3	-.1
				.31	.27	.53				
21	Item 21	288	783	-2.78	.80	4.03	.82	.81	-2.0	-1.7
				.34	.31	1.06				
22	Item 22	419	774	-3.41	-.91	1.80	1.16	1.16	1.8	1.5
				.41	.25	.34				
23	Item 23	318	786	-2.75	.27	3.68	1.03	1.03	.4	.4
				.34	.25	.81				
24	Item 24	388	777	-3.22	-.59	2.16	1.00	1.01	.0	.2
				.38	.24	.39				
25	Item 25	373	777	-3.59	-.33	2.34	.90	.91	-1.2	-.9
				.41	.29	.45				
26	Item 26	446	747	-3.31	-1.10	.82	1.22	1.25	2.5	2.3
				.38	.26	.24				
27	Item 27	371	780	-3.44	-.22	2.00	.89	.89	-1.4	-1.1
				.38	.25	.37				
28	Item 28	312	777	-2.84	.44	2.03	.87	.85	-1.4	-1.3
				.31	.28	.39				
29	Item 29	291	771	-2.81	.69	2.35	.89	.88	-1.0	-1.0
				.34	.31	.52				
30	Item 30	287	753	-2.88	.65	2.34	.93	.91	-.7	-.7
				.34	.31	.49				
31	Item 31	301	750	-2.41	.24	2.06	.97	.98	-.3	-.2
				.28	.29	.40				
Mean				.00			.99	1.00	.0	.0
SD				.51			.10	.12	1.1	1.0

4. ITEM THRESHOLD VALUES

4.1 Rehabilitating employee ability estimates

Figure 14-3 reveals how rehabilitating employees rated the complexity of their workplace rehabilitation tasks. As mentioned earlier, the partial credit model assumes that threshold values will be different within each individual rehabilitation item itself and across all other rehabilitation survey items. The assumption of equidistance between the categories or thresholds of the rehabilitation rating scale is not held by the Rasch's partial credit model. The logit scale is plotted on the left of the histogram and item numbers to the right and beside each item number an additional dot point has been added (either a .1, .2 or a .3), which reflects the threshold value for that item. Low threshold values indicate less complexity of the rehabilitation task and higher threshold values reflecting greater task complexity.

A most difficult rehabilitation item for rehabilitating employees was Item 16 which reflected their ability to be understood in the workplace if English were not their first language. Note the thresholds for this item are well-spaced across the whole logit scale with the first threshold located at logit at –1.38, the second threshold at +0.16 and the last threshold at logit +2.09. This suggests that there is equal probability that rehabilitating employees will view this item as either being very easy or simple to complete, or as either an easy or difficult task to undertake, or view it as being hard to very difficult to negotiate. This distribution of threshold values for this item is not a typical pattern of threshold values across all the rehabilitation items.

Note how Item 15 (ensures that all appropriate rehabilitation documentation is submitted) has a different threshold configuration. The first threshold is located at logit -0.06, the second at +0.49 and the third threshold at +1.72. In this item, there is less probability that rehabilitating employees will score between the second and third thresholds, but are more likely to rate their own ability (and the task complexity) as being between the first and second thresholds. In contrast to this threshold pattern is Item 25 which explores the rehabilitating employee's ability to avoid doing things at work that could re-aggravate the original injury. Note the first threshold is located at logit -1.94, the second threshold at -0.38 and the last at +0.40. This item not only reveals a very small spread of logits between the three thresholds, but it is also viewed as a relatively easy task for rehabilitating employees to undertake as threshold values are well down the logit scale overall.

```
--------------------------------------------------------------------------------
                            |  16.3
                            |  15.3    17.3
                            |   7.3    10.3    26.3
                       X    |   1.3     2.3    21.3    22.3    23.3
                       X    |   6.3    11.3    20.3
                       X    |   2.3    19.3     9.3
 1.0                  XX    |   4.3     5.3     8.3    12.3    13.3    14.3    18.3
                       X    |   3.3    29.3                            31.3
                      XX    |  22.2    24.3    27.3
                    XXXX    |  15.2            28.3     1.2
                XXXXXXXX    |   2.2    10.2    25.3    23.2    27.2
                XXXXXXXX    |   8.2     9.2    11.2    19.2    20.2    29.2
               XXXXXXXXX    |   7.2    16.2    14.2    17.2    28.2    30.2
  .0               XXXXX    |  15.1     5.2    12.2     4.2    21.2    31.2
                  XXXXXX    |   3.2     6.2    13.2    18.2    26.2
                  XXXXXX    |  24.2            25.2
                   XXXXX    |
                       X    |
                  XXXXXX    |   2.1    11.1
-1.0                  XX    |   1.1     9.1    10.1    29.1            31.1
                       X    |   5.1    12.1    26.1
                     XXX    |  20.1    22.1    27.1    28.1    30.1
                            |   3.1    16.1     7.1     8.1    14.1     6.1    23.1
                            |   4.1    24.1
                     XXX    |  13.1    17.1    21.1
                            |  18.1
-2.0                        |  19.1            25.1
                            |
                       X    |
                            |
                            |
                            |
                            |
-3.0                        |
                            |
                            |
                            |
                            |
-4.0                        |
                       X    |
--------------------------------------------------------------------------------
 Each X represents 1 rehabilitating employee
```

Figure 14-3. Estimations of the complexity of workplace rehabilitation activities as rated by rehabilitating employees (n = 80)

4.2 Workplace managers' ability estimates

Referring to Item 21 on Figure 14-4, it can be seen that the threshold values are spread widely across the logit table and almost equidistant between the categories. This item differentiates well between the perceived ability levels of workplace managers. This pattern of threshold values differs significantly from Item 26, which explores the managers' capacity to deal with their workplace budget given the presence of a rehabilitating employee in their workplace. The first threshold is located at logit –3.31, the second at –1.10 and the third threshold at +0.82. The logit spread for this item is much smaller and the third threshold value is adjacent to the logit area that usually indicates average difficulty of an item for persons with average ability (logit area at 0). Item 19 is not discriminating well for manager ability or rehabilitation task complexity as shown by second and third thresholds levels at +1.73 and 2.32 logits, respectively. This item asks how complex is it for the workplace manager to communicate upwardly with his or her own supervisor if workplace rehabilitation difficulties occur. A 0.59 logit difference occurs between these thresholds, which does not strongly differentiate the item as being a hard or very difficult task to complete according to managers.

```
---------------------------------------------------------------------------------------
 4.0                         |    21.3
                             |    13.3
                             |    23.3
                             |     8.3
                             |     4.3
                             |
 3.0                         |
                             |    20.3
                             |    11.3
                             |    16.3
                             |     5.3    6.3   14.3   15.3   17.3   19.3   25.3
 2.0                         |     1.3   12.3   18.3   24.3   28.3   31.3
                             |     9.3   27.3
                             |     2.2    7.3   19.2   22.3
                             |
                          X  |     3.3   10.2
                         XX  |
 1.0                    XXX  |     4.2   11.2   13.2   18.2
                          X  |    12.2   21.2   26.3
                       XXXX  |     5.2    8.2   29.2   30.2
                    XXXXXXX  |    17.2   28.2
                XXXXXXXXXXX  |     3.2    9.2   15.2   16.2   23.2   31.2
 0.0             XXXXXXXXXX  |     1.2   20.2
                   XXXXXXXX  |
             XXXXXXXXXXXXXX  |     6.2    7.2   14.2   25.2   27.2
          XXXXXXXXXXXXXXXXX  |     2.1
                     XXXXXX  |    24.2
               XXXXXXXXXXXX  |
-1.0              XXXXXXXXX  |    22.2
                XXXXXXXXXXX  |    26.2
                       XXXX  |    18.1
                       XXXX  |
                          X  |    19.1
-2.0                    XXX  |     4.1   12.1
                      XXXXX  |
                        XXX  |     3.1   10.1   11.1   13.1
                             |     8.1   31.1
                          X  |     7.1   20.1
                          X  |     1.1    5.1    9.1   15.1   16.1   17.1   21.1
-3.0                      X  |    14.1
                          X  |    24.1
                             |     6.1   22.1   26.1   27.1
                             |    25.1
                             |
-4.0                         |
                          X  |
                             |
                          X  |
---------------------------------------------------------------------------------------
 Each X represents 2 workplace supervisors. (N=272)
```

Figure 14-4. Rehabilitation item estimates (thresholds) as rated by workplace managers (n=272)

5. DIFFERENTIAL ITEM FUNCTIONING

To further explore if the Workplace Rehabilitation Scale is working effectively, it would not be unreasonable to ascertain whether the responses given by workplace managers and rehabilitating employees could be possibly influenced by other factors or if some form of item bias exists in the Workplace Rehabilitation Scale. Differential item functioning exists to determine if different meanings occur between various members of groups undertaking the test or survey. In this study, the gender of the respondents was subjected to further analysis to determine if the sex of the respondents influenced their responses. This is done by comparing rehabilitation items across the two groups, with item difficulties for each group estimated separately and with item calibrations plotted against each other. With reference to Figures 14-5 and 14-6 it can be seen that some rehabilitation items differentiate for gender for workplace managers and rehabilitating employees.

```
------------------------------------------------------------------------------------------
                               Plot of standardised differences

      Easier for female workplace managers      Easier for male workplace managers

      -3           -2           -1            0            1            2            3
-------+------------+------------+------------+------------+------------+------------+
Item 1              .                         | *                       .
Item 2              .                         |       *                 .
Item 3              .                     *   |                         .
Item 4              .                         |      *                  .
Item 5              .                         |                 *       .
Item 6         *    .                         |                         .
Item 7              .                         |        *                .
Item 9              .                         |        *                .
Item 10           * .                         |                         .
Item 11             .                         |         *               .
Item 12             .                         |                     *   .
Item 14             .                 *       |                         .
Item 15             .                         |                    *    .
Item 16             .         *               |                         .
Item 17             .                         |         *               .
Item 18             .                         |                  *      .
Item 19             .                         |     *                   .
Item 22             .                         |                         .     *
Item 23             .         *               |                         .
Item 24             .  *                      |                         .
Item 25             .                         |         *               .
Item 26             .   *                     |                         .
Item 27             *                         |                         .
Item 28             .                         |         *               .
Item 29             .      *                  |                         .
Item 30             .                     *   |                         .
Item 31             .                         |       *                 .
==========================================================================================
```

Figure 14-5. Comparison of item estimates for groups female and male workplace managers

Significant items are located outside the two vertical lines of the graph, which reflect two or more standard deviations from the mean of the scores given by respondents. For female workplace managers, Items 6, 10 and 27 show this pattern with Item 22 estimated to be an easier rehabilitation Item for male workplace managers to undertake. It requires considerable understanding of the underlying construct in order to accurately ascertain what is implied by such possible bias. However, such items do need to be investigated closely. Items which were easier for female workplace managers to undertake all reflect the necessity to have strong 'people-oriented' skills and flexibility towards others in the workplace, whereas the one item that favoured male managers dealt with legal requirements for rehabilitation.

```
------------------------------------------------------------------------------------------
                                Plot of standardised differences

  Easier for female rehabilitating employees    Easier for male rehabilitating employees

      -3           -2           -1            0            1            2            3
-------+------------+------------+------------+------------+------------+------------+
Item 2              .               *         |                         .
Item 3              .             *           |                         .
Item 4              .                         |    *                    .
Item 5              .                         |      *                  .
Item 6              .              *          |                         .
Item 7              .                         |             *           .
Item 8              .                         |      *                  .
Item 10             .                       * |                         .
Item 11             .              *          |                         .
Item 12             .                         |        *                .
Item 15             .                         |                         .    *
Item 18             .                     *   |                         .
Item 19             .         *               |                         .
Item 20             .                *        |                         .
Item 21             .     *                   |                         .
Item 23             .                         |         *               .
Item 24             .                         |                         *
Item 25             .                         |*                        .
Item 26             .          *              |                         .
Item 28             .                        *|                         .
Item 29             .     *                   |                         .
Item 30             .                         |      *                  .
item 31             .                         |                *        .
==========================================================================================
```

Figure 14-6. Comparison of item estimates for groups female and male rehabilitating employees

With reference to Figure 14-6, it can be seen that two rehabilitation items differentiate in favour of male rehabilitating employees. Items 15 and 24 relate to the employees capacity to meet legal requirement for rehabilitation (ensuring adequate documentation) and participating in reviewing rehabilitation policy in the workplace.

6. CONCLUSION

It has been argued in this paper that Rasch analysis offers a great deal for the development and analysis of attitude scales, which in turn serves to give useful information to educators and rehabilitation planners about the readiness of rehabilitating employees to take on the learning tasks associated with workplace rehabilitation. There are limitations to using traditional analytical procedures to analyse rating scales which are overcome when Rasch scaling is used to measure item difficulty and abilities estimates of participants engaged in a learning process. By employing the partial credit model, the educational researcher is no longer constrained by the assumption that rating scale categories are static or uniformly estimated across each item. Instead, rating scales can be visualised as a continuum of participant ability which, when used on multiple occasions can be a valuable adjunct to see if learning has taken place. Instrument coherence can also be assessed in Rasch analysis by examining items for unidimensionality as indicated by their fit statistics and by looking for differential item functioning

7. REFERENCES

Adams, J.J. & Khoo, S. (1996) *Quest: The interactive test analysis system.* Australian Council for Educational Research. Camberwell. Victoria. Version 2.1.

Bond, T. & Fox, C. (2001) *Applying the Rasch model: fundamental measurement in then human sciences.* Lawrence Erbaum Associates, Publishers. New Jersey.

Calzoni, T. (1997) The client perspective: the missing link in work injury and rehabilitation studies. *Journal of Occupational Health and Safety of Australia and New Zealand,*13, 47-57.

Chan, F., Shaw, L., McMahon, B., Koch, L. & Strauser, D. (1997) A model for enhancing rehabilitation counsellor-consumer working relationship. *Rehabilitation Counselling Bulletin,* 41, 122-137.

Corrigan, P., Lickey, S., Campion, J. & Rashid, F. (2000) A short course in leadership skills for the rehabilitation team. *Journal of Rehabilitation,* 66, 56-58.

Cottone, R.R. & Emener, W.G. (1990) The psychomedical paradigm of vocational rehabilitation and its alternative. *Rehabilitation Counselling Bulletin,* 34, 91-102.

Dal-Yob, L., Taylor, D. W. & Rubin, S. E. (1995) An investigation of the importance of vocational evaluation information for the rehabilitation plan development. *Vocational Evaluation and Work Adjustment Bulletin,* 33-47.

Fabian, E. & Waugh, C. (2001) A job development efficacy scale for rehabilitation professionals. *Journal of Rehabilitation*, 67, 42-47.

Fowler, B., Carrivick, P., Carrelo, J. & McFarlane, C. (1996) The rehabilitation success rate: an organisational performance indicator. *International Journal of Rehabilitation Research,* 19, 341-343.

Garske, G. G. (1996) The relationship of self-esteem to levels of job satisfaction of vocational rehabilitation professionals. *Journal of Applied Rehabilitation Counselling*, 27,19-22.

Gates, L.B., Akabas, S.H. & Kantrowitz, W. (1993) Supervisor's role in successful job maintenance: a target for rehabilitation counsellor efforts. *Journal of Applied Rehabilitation Counselling,* 60-66.

Hambenton, R.K., Swaminathan, H. & Rogers, H.J. (1991) *Fundamentals Of Item Response Theory.* Sage Publications. Newbury Park.

Kenny, D. (1994) The relationship between worker's compensation and occupational rehabilitation. *Journal of Occupational Health and Safety of Australia and New*

Zealand, 10, 157-164.

Kenny, D.(1995a) Common themes, different perspectives: a systemic analysis of employer-employee experiences of occupational rehabilitation. *Rehabilitation Counselling Bulletin,* 39, 54-77.

Kenny, D. (1995b) Barriers to occupational rehabilitation: an exploratory study of long-term injured employees. *Journal of Occupational Health and Safety of Australia and New Zealand.* 11, 249-256.

Linacre, J.M. (1995) Prioritising misfit indicators. *Rasch Measurement Transactions*, 9, 422-423.

Pati, G.C. (1985) Economics of rehabilitation in the workplace. *Journal of Rehabilitation.* 22-30.

Reed, B.J., Fried, J.H. & Rhoades, B.J. (1996) Empowerment and assistive technology: the local resource team model. *Journal of Rehabilitation*, 30-35.

Rosenthal, D. & Kosciulek, J. (1996) Clinical judgement and bias due to client race or ethnicity: an overview for rehabilitation counsellors. *Journal of Applied Rehabilitation Counselling,* 27, 30-36.

Sheehan, M., McCarthy, P. & Kearns, D. (1998) Managerial styles during organisational restructuring: issues for health and safety practitioners. *Journal of Occupational Health and Safety of Australia and New Zealand,* 14, 31-37.

Smith, R. (1996) A comparison of methods for determining dimensionality in Rasch measurement. *Structural Equation Modelling,* 3, 25-40.

8. OUTPUT 14-1

Employee's role in rehabilitation: questionnaire

Some of these situations below may not directly apply to you when you were undergoing the rehabilitation process however, your opinions on these matters would still be equally valued.

Please insert a number in each of the boxes below which best indicates how you felt about your ability to get the following tasks done or situations resolved, when you were undergoing rehabilitation at your work place.

Rating Scale for all items:	1..................	2.....................	3.........................	4
	This was very easy for me to handle	This was easy for me to deal with	This was difficult for me to handle	This was very difficult for me to deal with

1. Actually doing the jobs that were given to me because of my limitation
2. Knowing that my employer would keep rehabilitation information about me confidential
3. Dealing with my supervisor on a daily basis when I was rehabilitating
4. Trying to stop feeling negative about my job/employer since being injured
5. Being taught or shown things related to my job that I had not done before my injury
6. Getting others around at work to understand my limitations while at work

7. Having extra equipment around to actually help me do the job when I got back to work
8. Knowing what I was entitled to as a rehabilitating worker returning to work
9. Attending the doctor with my supervisor/rehabilitation counsellor
10. Being able to take my time when doing new tasks when I got back to work
11. Interacting with a rehabilitation counsellor who was from outside the organisation
12. Communicating with management about the amount of work I had to do when I was at work to rehabilitate
13. Finding other people in the workplace who were able to help me
14. Dealing with the legal aspects related to rehabilitation that impact on workers like me
15. Ensuring that all documentation related to return to work, e.g. medical certificates etc. were given to the right people/right time
16. Being understood by others in the workplace: that is, my first language is NOT English
17. Developing return to work plans in consultation with my supervisor
18. Finding out just what were the rehabilitation policy and procedures around my workplace
19. Telling my supervisor/rehabilitation coordinator about any difficulties I was experiencing while at work rehabilitating
20. Doing the relevant training programs I needed to do to effectively do the job I was doing during rehabilitation
21. Making my supervisor understand the difficulties I had at work in the course of my return to work duties
22. Finding the time to regularly review my return to work program
23. Doing those duties I was allocated to do when other things that other workers did seemed more interesting
24. Participating or reviewing rehabilitation policy for use in my workplace
25. Avoiding the things at work that might be a bit risky and possibly re-hurt my original injury
26. Making budgetary adjustments in relation to changes in my income after the injury
27. Getting other workers to allow me to do their jobs which were safe and suitable for me
28. Getting worthwhile support from management which actually helped me to return to work
29. Involving unions as advocates for me in the workplace
30. Involving my spouse/partner about in my return to work duties during return to work review or conferences
31. Interacting with my organisation's claims office about my wages/costs etc.
32. How many years have you been engaged in your usual job?
33. How many staff apart from you is your supervisor responsible for?
34. Would you please indicate your gender as either F or M.

9. OUTPUT 14.2

Manager/supervisor's role in rehabilitation: questionnaire

The purpose of this questionnaire is to identify the supervisor's ongoing learning needs in relation to the management of an injured employee who is seeking a return to work. **Your contributions toward this survey are valued and all your opinions are confidential. Your name is not required.**

Would you please read the statements below and indicate in the box next to each item **how easy or difficult (rating from 1 to 4) you would or may find this task** in your role in supervising a worker who is undergoing rehabilitation in a work area you are responsible for.

Rating scale for all items:	**1..................**	**.2......................**	**.3......................**	**.4**
	A very easy task for me	**An easy task for me**	**A hard task for me**	**A very difficult task for me**

Supervisor's task in a rehabilitative context

1. Providing suitable duties for a rehabilitating worker in accordance with medical advice
2. Ensuring confidentiality of information/issues as they relate to a rehabilitating worker
3. Participating with a rehabilitating worker on a daily basis
4. Dealing with your own negative feelings towards a rehabilitating worker
5. Assisting in the retraining of a worker undergoing rehabilitation
6. Getting other members of your staff to accept the work limitations of the rehabilitating worker
7. Securing additional equipment and/or services that would assist a rehabilitating worker undertake work tasks
8. Understanding the entitlements of a rehabilitating worker who is seeking a return to work
9. Meeting with the rehabilitating worker and the treating doctor to negotiate suitable work roles
10. Allowing the rehabilitating worker to have flexibility in the duration of time taken to complete allocated tasks
11. Interacting with an external rehabilitation consultant about issues relating to a rehabilitating worker
12. Communicating with senior management about your own staffing needs, as a rehabilitation worker is placed in your area
13. Identifying others members of your staff who would be able to assist the rehabilitating worker at your worksite
14. Understanding the legal requirements that impact on your role when assisting a rehabilitating worker
15. Maintaining appropriate documentation as it relates to your management of a rehabilitating worker
16. Taking into account linguistic diversity of rehabilitating workers when planning a return to work
17. Developing return to work plans in consultation with a rehabilitating worker
18. Finding your workplace's rehabilitation policy and procedures to refer to
19. Reporting any difficulties that the rehabilitating workers are experiencing to your supervisor or rehab co-ordinator
20. Undertaking relevant training programs you need in order to effectively supervise a rehabilitating worker

21. Responding to difficulties that a rehabilitating worker reports to you in the course of their return to work duties
22. Finding an appropriate time to review return to work programs
23. Ensuring that the rehabilitating worker only undertakes duties that are specified and agreed
24. Participating in the construction or review of the rehabilitation policy used in your organisation
25. Managing the rehabilitating worker when he/she engages in 'risky' behaviour that may antagonise the original injury
26. Making budgetary adjustment to account for costs incurred in accommodating for a rehabilitating worker in your work area
27. Getting cooperation from other workers to take on alternative duties to allow the rehabilitating worker to undertake suitable and safe tasks
28. Obtaining active support from senior management that assists you in your role to facilitate a return to work for a rehabilitating worker
29. Interacting with union representatives who advocate for the rehabilitating worker
30. Responding to spouse's inquiries about the rehabilitating worker's workplace duties during return to work review conference
31. Liaising with your organisation's claims management staff about the rehabilitating worker's costs

OTHER QUESTIONS

32. How many years have you been engaged in a supervisory role?
33. How many staff would you be responsible for in your work place?

Lastly would you please indicate your gender as either F or M

Chapter 15

CREATING A SCALE AS A GENERAL MEASURE OF SATISFACTION FOR INFORMATION AND COMMUNICATION TECHNOLOGY USERS

I Gusti Ngurah Darmawan
Pendidikan Nasional University, Bali; Flinders University

Abstract: User satisfaction is considered to be one of the most widely used measures of information and communication technology (ICT) implementation success. Therefore, it is interesting to examine the possibility of creating a general measure of user satisfaction to allow for diversity among users and diversity in the ICT-related tasks they perform. The end user computing satisfaction instrument (EUCSI) developed by Doll and Torkzadeh (1988) was revised and used as a general measure of user satisfaction. The sample was 881 government employees selected from 144 organisations across all regions of Bali, Indonesia. The data were analysed with Rasch Unidimensional Models for Measurement (RUMM) software. All the items fitted the model reasonably well with the exception of two items which had the chi-square probability < 0.05 and one item which had disordered threshold values. The overall power of the test-of-fit was excellent.

Key words: user satisfaction, information and communication technology, Rasch measurement, government agency

1. INTRODUCTION

In an era of globalisation, innovations in information and communications technology (ICT) have had substantial influences on communities and businesses. The availability of cheaper and more powerful personal computers, combined with the capability of telecommunication infrastructures, has put increasing computing power into the hands of a much

S. Alagumalai et al. (eds.), Applied Rasch Measurement: A Book of Exemplars, 271–286.

greater number of people in organisations (Willcocks, 1994; Kraemer & Rischard, 1996; Dedrick, 1997). These people use ICT for a large variety of different tasks or applications: for example, word processing, spreadsheets, statistics, databases and communication (Harrison & Rainer, 1996). The large number of people using ICT in their work means that there is a need to measure the effectiveness of such usage.

User satisfaction is probably one of the most widely used single measures of information systems success (DeLone & McLean, 1992). User satisfaction reflects the interaction of ICT with users. User satisfaction is frequently employed to evaluate ICT implementation success. A number of researchers have found that user satisfaction with the organisation's information systems has a strong bearing on ICT utilisation by employees or that there is a significant relationship between user satisfaction and information and communication technology usage (Cheney, 1982; Baroudi, Olson, & Ives, 1986; Doll & Torkzadeh, 1991; Etezadi-Amoli & Farhoomand, 1996; Gelderman, 1998; Kim, Suh, & Lee, 1998; Al-Gahtani & King, 1999; Khalil & Elkordy, 1999). A general measure of user satisfaction is necessary to allow for diversity among users and diversity in the ICT-related tasks they perform.

2. END-USER COMPUTING SATISFACTION INSTRUMENT

User satisfaction is considered one of the most important measures of information systems success (Ives & Olson, 1984; DeLone & McLean, 1992). The structure and dimensionality of user satisfaction are important theoretical issues that have received considerable attention (Ives, Olson, & Baroudi 1983; Doll & Torkzadeh, 1988, 1991: Doll, Xia, & Torkzadeh 1994; Harrison & Rainer, 1996). The importance of developing standardized instruments for measuring user satisfaction has also been stressed by several researchers (Ives & Olson, 1984; DeLone & McLean, 1992).

Early research on user satisfaction developed measurement instruments that focused on general user satisfaction (e.g. Bailey & Pearson, 1983; Ives & Olson, 1984). However, several subsequent studies pointed out weakness in these instruments (for example, Galetta & Lederer, 1989).

In response to the perceived weakness in prior instruments, Doll and Torkzadeh (1988) developed a 12-item scale to measure user satisfaction which is called the End User Computing Satisfaction Instrument (EUCSI). Their scale is a measure of overall user satisfaction that includes a measure of the satisfaction of the extent to which computer applications meet the end-user's needs with regards to five factors: namely (a) content, (b) accuracy,

(c) format, (d) ease of use, and (e) timeliness. The use of these five factors and the 12-item instrument developed by Doll and Torkzadeh (1988) as a general measure of user satisfaction have been supported by Harrison and Rainer (1996).

Most of the previous research on user satisfaction focus on explaining what user satisfaction is by identifying its components, but the discussion usually suggests that user satisfaction may be a single construct. Substantive research studies use the Classical Test Theory and obtain a total score by summing items.

Classical Test Theory (CTT) involves the examination of a set of data in which scores can be decomposed into two components, a true score and an error score that are not linearly correlated (Keats, 1997). Under the CTT, the sums of scores on the items and the item difficulty are not calibrated on the same scale; the totals are strictly sample dependent. Therefore, CTT can not produce anything better than a ranking scale that will vary from sample to sample. The goal of a proper measurement scale for User Satisfaction can not be accomplished through Classical Test Theory.

The Rasch method, a procedure within item response theory (IRT), produces scale-free measures and sample-free item difficulties (Keeves & Alagumalai 1999). In Rasch measurement the differences between pairs of measures and pairs of item difficulty are expected to be relatively sample independent.

The use of the End User Computing Satisfaction Instrument (EUCSI) as a general measure of user satisfaction has several rationales. First, Doll and Torkzadeh (1988 p. 265) stated that 'the items ... were selected because they were closely related to each other'. Secondly, eight of Doll and Torkzadeh's 12 items use the term 'system'. Users could perceive the term 'system' to nonspecifically encompass all the computer-based information systems and applications that they might encounter.

The purpose of this paper is to examine the possibility of creating an interval type, unidimensional scale of User Satisfaction for computer-based information systems using the End User Computing Satisfaction Instrument (EUCSI) of Doll and Torkzadeh (1991) as a general measure of user satisfaction. In addition, it is also interesting to explore any differences between the sub-groups of respondents such as gender and the type of organisation where they work.

3. METHODS

This study employs the EUCSI as a general measure of user satisfaction. The second item of accuracy is modified to tap another aspect of accuracy. Data were collected by self-report questionnaires. Respondents were asked to answer to what extent they agreed to each of the 12 statements by considering all the computer-based information systems and applications they were currently using in their jobs (see Output 15-1). Their responses may vary from not at all (0), very low (1), low (2), moderate (3), high (4) and very high (5).

Data were analysed with the computer program Rasch Unidimensional Measurement Models (RUMM 2010) (Andrich, Lyne, Sheridan & Luo, 2000). A scale was created in which the User Satisfaction measures were calibrated on the same scale as the item difficulties.

4. SAMPLE

The data in this paper come from a study (Darmawan, 2001) focusing on the adoption and implementation of information and communication technology by local government in Bali, Indonesia. The legal basis for the current system of regional and local government in Indonesia is set out in Law No. 5 of 1974. Building on earlier legislation, this law separates governmental agencies at the local level into two categories (Devas, 1997):

1. decentralised agencies (decentralisation of responsibilities to autonomous provincial and local governments); and
2. deconcentrated agencies (deconcentration of activities to regional offices of central ministries at the local level).

In addition to these two types of governmental agencies, government owned enterprises also operate at the local level. These three types of government agencies, decentralised, deconcentrated, and state-owned enterprises, have distinctly different functions and strategies. It is believed that these differences affect attitudes toward the adoption of innovation (Lai & Guynes, 1997).

The number of agencies across all regions of Bali which participated in this study is 144. These agencies employed a total of 10 034 employees, of whom 1 427 (approximately 14%) used information technology in their daily duties. Of these, 881 employees participated in this study.

From the total of 881 respondents, 496 (56%) were male. Almost two-thirds of the government employees who participated in this survey (66.2%) had at least a tertiary diploma or a university degree. About 33 per cent had only completed their high school education. Almost one-third of them (33%)

had not completed any training. Most of the respondents had attended some sort of software training (67%). A small number of respondents (5%) had had the experience of attending hardware training. Even though almost two-thirds of respondents had experienced either software or hardware training, the levels of expertise of these respondents were still relatively low. Among the respondents most (93%) were computer operators. Only five per cent and two per cent had any experience as a programmer or a systems analyst respectively.

5. PSYCHOMETRIC CHARACTERISTICS OF THE USER SATISFACTION SCALE

As can be seen in Table 15-1, twelve items relating to the End User Computing Satisfaction have a good fit to the measurement model, indicating a strong agreement between all 881 persons to the difficulties of the items on the scale. However, there are two items that have the chi square probability < 0.05. Most of the item threshold values are ordered from low to high indicating that the persons have answered consistently and logically with the ordered response format used (except for Item 2, see also Table 15-3).

Table 15-1. Summary data of the reliabilities and fit statistics to the model for the 12-item EUCSI

Items with chi-square probability <0.05		2
Items with disordered threshold		1
Separation index		0.937
Item mean (SD)		0.000 (0.137)
Person mean (SD)		0.745 (1.439)
Item-trait interaction (chi-square)		63.364 (p = 0.068)
Item fit statistic	Mean	-1.212
	SD	1.538
Person fit statistic	Mean	-1.846
	SD	3.444
Power of test-of-fit		Excellent

Notes

1. The index of person separation is the proportion of observed variance that is considered true (94%) and is high.
2. The item and person fit statistics have an expectation of a mean near zero and a standard deviation near one, when the model fit the data.
3. The item-trait interaction test is a chi-square. The results indicate that there is a fair collective agreement between persons of differing User Satisfaction for all item difficulties.

The Index of Person Separability for the 12-item scale is 0.937. This means that the proportion of observed variance considered to be true is 94 per cent. The item-trait tests-of-fit indicates that the values of the item difficulties are consistent across the range of person measures. The power of the test-of-fit is excellent.

As stated earlier, two items (Item 10 and Item 12) have chi square probability values < 0.05 (see Table 15-2). According to Linacre (2003), 'as the number of degrees of freedom, that is, the sample size, increases, the power to detect small divergences increases, and ever smaller departures of the mean-square from 1.0 become statistically 'significant''. Since the chi square probability values are sensitive to sample size, it is not appropriate to judge the fit of the item solely on the chi square value.

In order to judge the fit of the model, the item characteristics curves were examined. The Item Characteristic Curves for the two items are presented in Figure 15-1 and Figure 15-2. There is no big discrepancy in these curves. The expected scores for the five groups formed are very close to the curves. Therefore, these two items can still be considered as having adequate fits.

Most of the item threshold values are ordered from the low to high except for Item 2. For Item 2, threshold 1 (-2.019) is slightly higher than threshold 2 (-2.207). Figure 15-3 shows the response probability curve for item 2. It can be seen in this figure that the probability curve for 0 cut the probability curve for 2 before it cut the probability curve for 1. As a comparison, an example of well-ordered threshold values is presented in Figure 15-4.

The item-person tests-of-fit (see Table 15-1) indicate that there is a good consistency of person and item response patterns. The User Satisfaction measures of the persons and the threshold values of the items are mapped on the same scale as presented in Figure 15-5. There is also another way of plotting the distribution of the User Satisfaction measures of the persons and the threshold values of the items as shown in Figure 15-6. In this study, the items are appropriately targeted against the User Satisfaction measures. That is, the range of item thresholds matches the range of User Satisfaction measures on the same scale. The item threshold values range from -3.022 to 4.473 and the User Satisfaction measures of the persons range from -3.300 to 5.657.

In Table 15-4, the items are listed based on the order of their difficulty. At one end, most employees probably would find it 'easy' to say that the information presented is clear (Item 8). It was expected that there would be some variation in each person's responses to this. At the other end, most employees would find it 'hard' to say that the information content meets their need (Item 2) and there would be some variation around this.

In regard to the five factors, namely (a) content, (b) accuracy, (c) format, (d) ease of use, and (e) timeliness, it seemed that most employees were highly satisfied with the format and the clarity of the output presented by the system. They seemed to be slightly less satisfied with the accuracy of the system followed by the timeliness of the information provided by the system and the ease of use of the system. The information content provided by the system seemed to be a factor that the employees felt least satisfied with.

Table 15-2. Location and probability of item fit for the End User Computing Satisfaction Instrument (12-item)

Item	Description	Location	SE	Residual	Chi Square	Probability
	Information content					
I0001	The system precisely provides the information I needs	0.111	0.05	-0.954	2.173	0.696
I0002	The information content meets my need	0.247	0.05	-2.236	6.696	0.130
I0003	The system provides reports that meet my needs	0.017	0.05	-3.453	6.230	0.161
I0004	The system provides sufficient information	0.041	0.05	0.065	1.831	0.760
	Information accuracy					
I0005	The system is accurate	-0.144	0.05	-3.320	2.624	0.612
I0006	The data is correctly/safely stored	-0.136	0.05	-2.715	3.481	0.467
	Information format					
I0007	The outputs are presented in a useful format	-0.137	0.05	-1.618	2.048	0.720
I0008	The information presented is clear	-0.213	0.05	-1.877	2.786	0.583
	Ease of use					
I0009	The system is user friendly	0.155	0.05	-0.030	4.030	0.386
I0010	The system is easy to learn	0.041	0.05	0.700	16.902	0.000*
	Timeliness					
I0011	I get the needed information in time	0.046	0.05	0.881	5.398	0.229
I0012	The system provides up-to-date information	-0.028	0.05	0.015	9.164	0.032*

Notes
* chi-square $p<0.05$

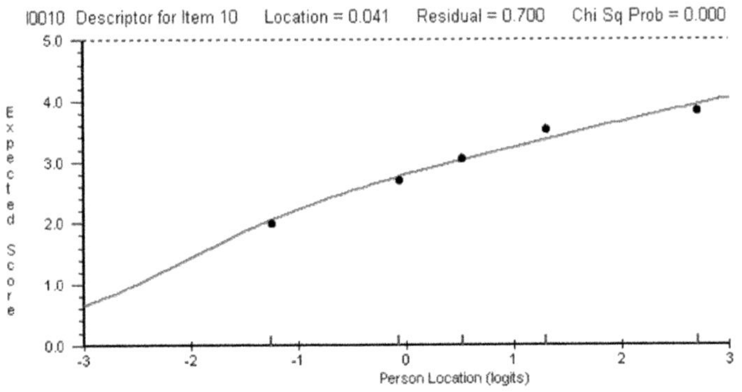

Figure 15-1. Item Characteristic Curve for Item 10

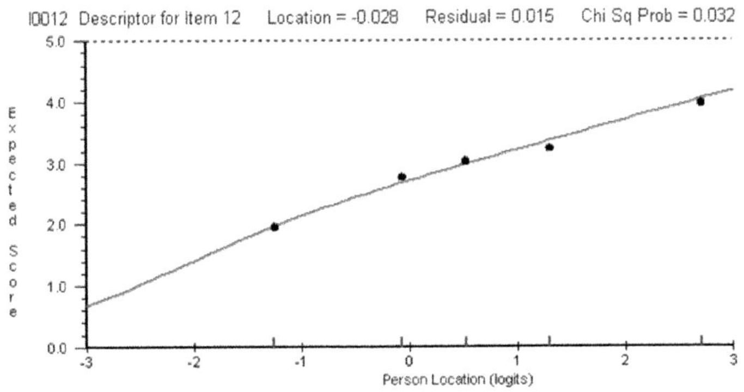

Figure 15-2. Item Characteristic Curve for Item 12

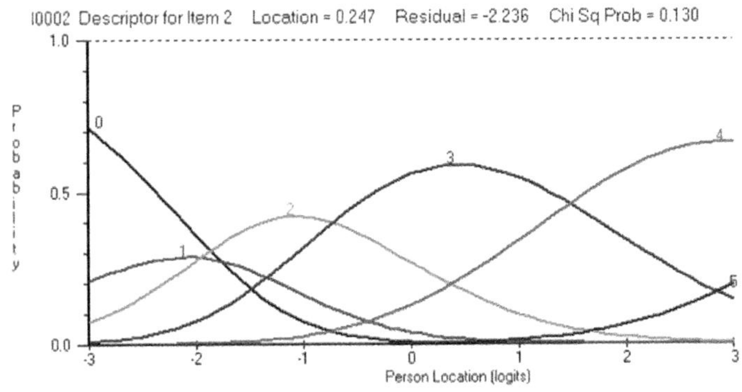

Figure 15-3. Response Probability Curve for Item 2

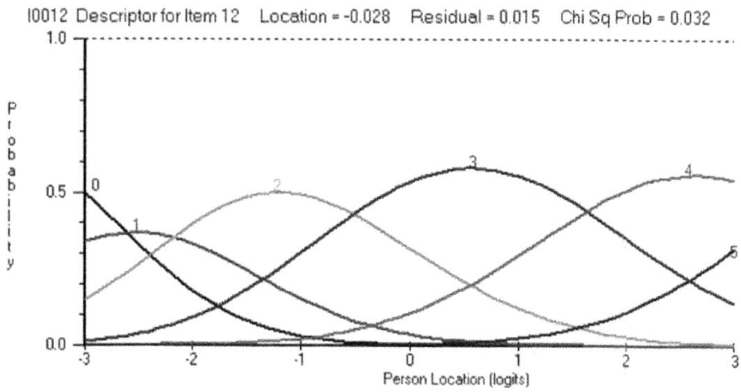

Figure 15-4. Response Probability Curve for Item 12

Table 15-3. Item threshold values

	Threshold				
Item	1	2	3	4	5
I0001	-2.103	-2.068	-0.735	1.333	3.573
I0002	-2.019	-2.207	-0.975	1.232	3.970
I0003	-2.260	-2.234	-0.762	1.472	3.783
I0004	-3.022	-2.207	-0.644	1.563	4.310
I0005	-2.595	-2.065	-0.694	1.371	3.982
I0006	-2.544	-2.104	-0.704	1.400	3.951
I0007	-2.859	-2.136	-0.807	1.329	4.473
I0008	-2.559	-2.112	-0.907	1.205	4.373
I0009	-2.993	-2.308	-0.575	1.733	4.143
I0010	-2.463	-2.345	-0.797	1.547	4.058
I0011	-2.329	-2.198	-0.685	1.513	3.699
I0012	-2.587	-2.149	-0.487	1.661	3.561

6. GENDER AND TYPE OF ORGANISATION DIFFERENCES

In addition to examining the overall fit of each item, it is also interesting to investigate whether the individual items in this instrument function in the same way for different groups of employees. RUMM software, which is used in this study, has the capability to undertake the differential item function (DIF) analysis. In DIFF analysis, the presence of item bias is checked and the significance of differences observed between different groups of employees is examined (Andrich et al., 2000). Gender (male and female) and type of organisation (centralised, decentralised, and enterprise type organisation) are the two person factors used in this study.

Male and female employees had a high agreement on most of the items except for item 2. Table 15-5 and Figure 15-7 indicate that male employees were less satisfied with the ability of the system to provide the contents that met their needs. The difference seems to be larger in less satisfied employees, and the difference is not obvious for highly satisfied employees.

```
-------------------------------------------------------------------------------------
LOCATION          PERSONS     ITEMS [uncentralised thresholds]
-------------------------------------------------------------------------------------
  6.0                       |
                            |
                            |
                            |
                            |
  5.0                     X |
                            |
                            |
                            |
                            | I0002.5  I0009.5  I0007.5  I0004.5
  4.0                     X | I0010.5  I0008.5
                            | I0003.5  I0006.5  I0005.5
                          X | I0001.5  I0011.5
                          X | I0012.5
                            |
  3.0                       |
                          X |
                      XXXXX |
                         XX |
                      XXXXX |
  2.0                  XXXX |
                       XXXX | I0009.4
                        XXX | I0004.4  I0012.4
                       XXXX | I0001.4  I0002.4  I0003.4  I0011.4  I0010.4
                       XXXX | I0005.4  I0006.4
  1.0                 XXXXX | I0007.4
                       XXXX | I0008.4
                    XXXXXXX |
         XXXXXXXXXXXXXXXXXX |
                        XXX |
  0.0                 XXXXX |
                       XXXX |
                     XXXXXX |
                        XXX | I0012.3  I0009.3
                         XX | I0010.3  I0003.3  I0002.3  I0011.3  I0001.3  I0004.3
 -1.0                   XXX | I0007.3  I0006.3  I0005.3
                       XXXX | I0008.3
                       XXXX |
                         XX |
                          X | I0002.1
 -2.0                     X | I0001.1  I0002.2  I0001.2
                          X | I0012.2  I0004.2  I0009.2  I0011.2
                          X | I0008.2  I0010.2  I0011.1  I0007.2  I0003.1  I0006.2  I0003.2  I0005.2
                            | I0010.1
                            | I0008.1  I0005.1  I0006.1  I0012.1
 -3.0                       | I0007.1  I0004.1  I0009.1
                            |
                            |
                            |
 -4.0                       |
-------------------------------------------------------------------------------------
           X = 8 Persons
```

Figure 15-5. Item threshold map

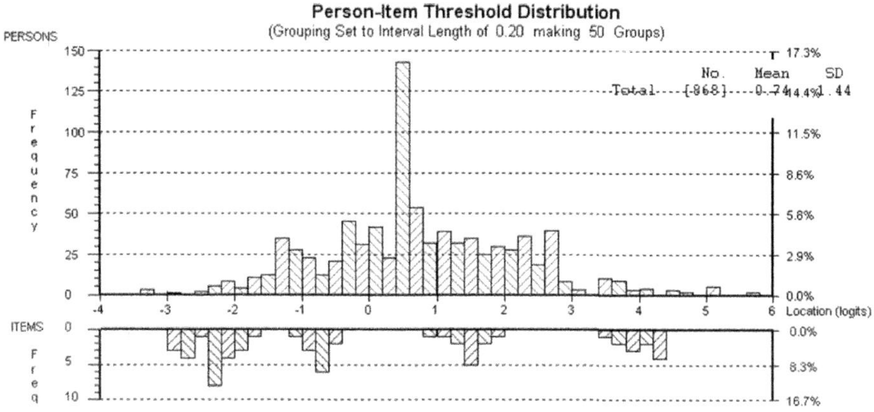

Figure 15-6. Person-Item Threshold Distribution

Table 15-4. Item difficulty order

Item	Description	Location	
I0002	The information content meets my needs	0.247	▲
I0009	The system is user friendly	0.155	
I0001	The system precisely provides the information I need	0.111	
I0011	I get the needed information in time	0.046	hard items
I0004	The system provides sufficient information	0.041	
I0010	The system is easy to learn	0.041	
I0003	The system provides reports that meet my needs	0.017	
I0012	The system provides up-to-date information	-0.028	
I0006	The data is correctly/safely stored	-0.136	
I0007	The outputs are presented in a useful format	-0.137	easy items
I0005	The system is accurate	-0.144	
I0008	The information presented is clear	-0.213	▼

Table 15-5. Analysis of variance for Item 2

SOURCE	S.S	M.S	DF	F-RATIO	Prob
BETWEEN	14.321	2.864	5		
gender	7.822	7.822	1	9.848	0.002
CInt	0.605	0.302	2	0.381	0.689
gender-by-CInt	5.894	2.947	2	3.710	0.024
WITHIN	684.698	0.794	862		
TOTAL	699.019	0.806	867		

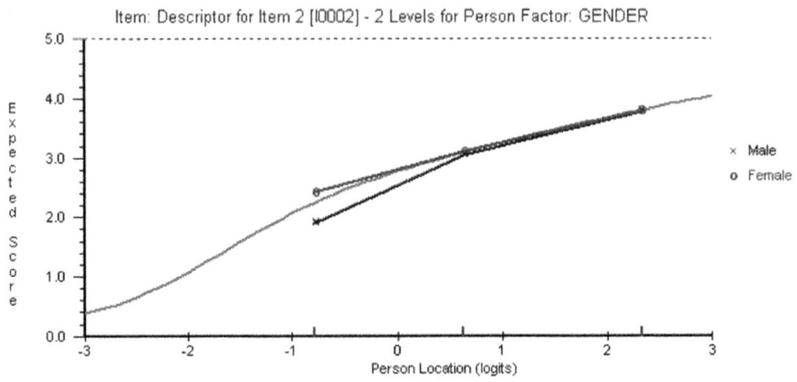

Figure 15-7. Item Characteristic Curve for Item 2 with male differences

Regarding the types of organisation, there is no big difference between employees in a decentralised, deconcentrated and enterprise type organisation. The only difference is observed in their responses to item 3 (see Table 15-6 and Figure 15-8). Employees in enterprise type organisations are more likely to feel satisfied with the level of sufficiency of the information provided by the system.

Table 15-6. Analysis of variance for Item 3

SOURCE	S.S	M.S	DF	F-RATIO	Prob
BETWEEN	24.098	3.012	8		
type	7.183	3.592	2	4.887	0.008
CInt	9.089	4.544	2	6.183	0.003
type-by-CInt	7.826	1.957	4	2.662	0.031
WITHIN	631.352	0.735	859		
TOTAL	655.450	0.756	867		

The values of the psychometric properties obtained from the analysis indicate that the 12-item instrument provides a reasonable fit to the data. These findings suggest the presence of a measure, User Satisfaction, which explained most of the observed variance. The reasonable fit of the model to the data supports the construct validity of the instrument developed by Doll and Torkzadeh (1991) when used as a general measure of user satisfaction.

The scale is created at the interval level of measurement. On the interval measurement scale, one unit on the scale represents the same magnitude ofthe trait or characteristic being measured across the whole range of the scale. Equal distance on the scale between measures of User Satisfaction corresponds to equal differences between the item difficulties on the scale. This scale does not have a true zero point of item difficulty or person measure. The zero value of the scale is usually set at the average level of item difficulty. The difficulty levels of the twelve items used in this study are mostly towards the average ability of the person. The item difficulty levels range from -0.213 to 0.247. This means, for example, that the majority of the persons found that they agreed to a moderate level to the items. Therefore, to improve the instrument, other items at both the easy end and hard end of the scale should be added. According to Keeves and Masters (1999), in order to get a meaningful representation of a characteristic being measured, there should be a sufficient range of item difficulties for that characteristic.

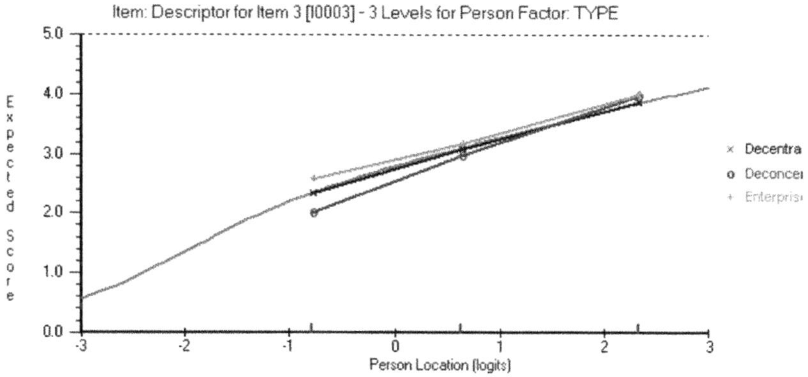

Figure 15-8. Item Characteristic Curve for item 2 with organisation type differences

7. DISCUSSION

The use of EUCSI as a general measure of satisfaction for information and communication technology users has an implication for management of ICT in organisations. The management of ICT is often examined in multiple levels: (a) the individual level, including the job-related ICT activities of the individual employees, and (b) the organisational level encompassing education, training, standards, security, controls, et cetera for the organisation as a whole (Rainer & Harrison, 1993). The use of the EUCSI as a general measure of user satisfaction can be particularly beneficial to ICT managers at the organisational level. This measure will allow managers to assess information and communication systems across departments, users and applications to gain an overall view of user satisfaction. The use of the EUCSI as a general measure does not contradict the original use of the instrument by Doll and Torkzadeh (1991), which measured application-specific computing satisfaction. Using the scale as a general measure as well as an application-specific measure could help the ICT manager gain a broader perspective of user satisfaction with the systems and applications across the organisation.

8. CONCLUSION

The Rasch model was useful in constructing a scale of User Satisfaction in using a computer-based system. The 12-item scale had desirable psychometric properties. The proportion of observed variance considered true was 94 per cent. The threshold values were ordered in correspondence with the ordering of the response category with a slight discrepancy for Item 2. Item difficulties and person measures are calibrated on the same scale. The scale is sample independent; the item difficulties do not depend on the sample of government employees used or on the opinions of the person who constructed the items. However, the person measures in this study are only relevant to the government agencies involved.

From DIF analyses, it was found that male and female employees had a high agreement on most of the items. The only difference was that male employees were less satisfied with the ability of the system to provide the contents that met their needs. The difference seemed to be larger in less satisfied employees, and the difference was not obvious for highly satisfied employees. Regarding the types of organisation, the only difference was observed in their responses to Item 3. Employees in enterprise type organisations were more likely to feel satisfied with the level of sufficiency of the information provided by the system.

Overall, ICT managers can use the EUCSI as a quick, easy to use, easy to understand indication of user satisfaction throughout an organisation. This measure of user satisfaction can give these managers an idea as to the effectiveness of the ICT resource in the organisation. The current form of the EUCSI may not be final. For example, the spread of the difficulty of the item need to be expanded. However, the instrument deserves additional validation studies.

9. OUTPUT 15-1

To what extent do the following statements describe your situation:

(0 = not at all, 1 = very low, 2 = low, 3 = moderate, 4 = high, 5 = very high)?

1. Information content
    (a) The system precisely provides the information I need
    (b) The information content meets my needs
    (c) The system provides reports that meet my needs
    (d) The system provides sufficient information
2. Information accuracy
    (a) The system is accurate
    (b) The data is correctly/safely stored
3. Information format
    (a) The outputs (e.g. reports) are presented in a useful format
    (b) The information presented is clear
4. Ease of use
    (a) The system is user friendly
    (b) The system is easy to learn
5. Timeliness
    (a) I get the needed information in time
    (b) The system provides up-to-date information

10. REFERENCES

Al-Gahtani, S., & King, M. (1999). Attitudes, satisfaction and usage: factors contributing to each in the acceptance of information technology. *Behaviour and Information Technology, 18*(4), 277-297.

Andrich, D., Lyne, A., Sheridan, B., Luo, G. (2000). RUMM 2010 Rasch Unidimensional Measurement Models [Computer Software]. Perth: RUMM Laboratory.

Bailey, J.E. & Pearson, S.W (1983) Development of a tool for measuring and analysing computer user satisfaction. *Management Science*, 24, 530-545.

Baroudi, J. J., Olson, M. H., & Ives, B. (1986). An Emperical Study of the Impact of User Involvement on System Usage and Information Satisfaction. *Communication of the ACM, 29*(3), 232-238.

Cheney, P. H. (1982). Organizational Characteristics and Information Systems: An Exploratory Investigation. *Academy of Management Journal, 25*(1), 170-184.

Darmawan, I. G. N. (2001). Adoption and implementation of information technology in Bali's local government: A comparison between single level path analyses using PLSPATH 3.01 and AMOS 4 and Multilevel Path Analyses using MPLUS 2.01. *International Education Journal, 2*(4): 100-125.

Delone, W. H., and Mclean, E. R. (1992). Information System Success: The Quest for The Dependent Variable. *Information System Research, 3*(1), 60-95.

Devas, N. (1997). Indonesia: what do we mean by decentralization? *Public Administration and Development, 17*, 351-367.

Doll, W. J. and Torkzadeh, G. (1988). The measurement of end-user computing satisfaction. *MIS Quarterly, 12*(3), 258-265.

Doll, W. J. and Torkzadeh, G. (1991). The measurement of end-user computing satisfaction: theoretical and methodological issues. *MIS Quarterly, 15*(1), 5-6.

Doll, W. J. Xia, W., and Torkzadeh, G. (1994). A confirmatory factor analysis of the end-user computing satisfaction instrument. *MIS Quarterly, 18(*4), 453-461.

Etezadi-Amoli, J., & Farhoomand, A. F. (1996). A structural model of end user computing satisfaction and user performance. *Information and Management, 30*(2), 65-73.

Galetta, D.F. & Lederer, A.L. (1989) Some cautions on the measurement of user information satisfaction. *Decision Sciences*, 20, 25-34.

Gelderman, M. (1998). The relation between user satisfaction, usage of information systems and performance. *Information and Management, 34*(1), 11-19.

Harrison, A. W. and R. Kelly Rainer, J. (1996). A General Measure of User Computing Satisfaction. *Computers in Human Behavior, 12*(1), 72-92.

Ives, B. and Olson, M. H. (1984). User Involvement and MIS Success: A Review of Research. *Management Science, 30*(5), 586-603.

Ives, B., Olson, M. H., and Baroudi, J. J. (1983). The Measurement of User Information Satisfaction. *Communication of The ACM, 26*, 785-793.

Keats, J. A. (1997). Classical Test Theory. In J. P. Keeves (Ed.), *Educational Research, Methodology, and Measurement: An International Handbook* (2nd ed., pp. 713-19). Oxford: Pergamon Press.

Keeves, J. P. and Alagumalai, S. (1999). New Approaches to Measurement. In G. N. Masters and J. P. Keeves (Eds.), *Advances in Measurement in Educational Research and Assessment* (pp. 23-42). Oxford: Pergamon.

Keeves, J. P. and Masters, G. N. (1999). Introduction. In G. N. Masters and J. P. Keeves (Eds.), *Advances in Measurement in Educational Research and Assessment* (pp. 1-19). Oxford: Pergamon.

Khalil, O. E. M., & Elkordy, M. M. (1999). The Relationship Between User Satisfaction and Systems Usage: Empirical Evidence from Egypt(1). *Journal of End User Computing, 11*(2), 21.

Kim, C., Suh, K., & Lee, J. (1998). Utilization and User Satisfaction in End User Computing: A Task Contingent Model. *Information Resources Management Journal, 11*(4), 11-24.

Kraemer, K. L. and Dedrick., J. (1997). Computing and public organizations. *Journal of Public Administration Research and Theory, 7*(1), 89-113.

Lai, V. S., & Guynes, J. L. (1997). An Assessment of the Influence of Organizational Characteristics on Information Technology Adoption Decision: A Descriminative Approach. *IEEE Transactions on Engineering Management, 44*(2), 146-157.

Linacre, J. M. (2003) Size vs. Significance: Standardized Chi-Square Fit Statistic. *Rasch Measurement Transactions, 17(1)*, 918.

Rischard, J.-F. (1996). Connecting Developing Countries to the Information Technology Revolution. *SAIS Review, Winter-Spring*, 93-107.

Willcocks, L. (1994). Managing Information Systems in UK Public Administration: Issues and Prospects. *Public Administration, 72*, 13-32.

Chapter 16

MULTIDIMENSIONAL ITEM RESPONSES: MULTIMETHOD-MULTITRAIT PERSPECTIVES

Mark Wilson and Machteld Hoskens
University of California, Berkeley

Abstract: In this paper we discuss complexities of measurement that can arise in a multidimensional situation. All of the complexities that can occur in a unidimensional situation, such as polytomous response formats, item dependence effects, and the modeling of rater effects such as harshness and variability, can occur, with a correspondingly greater degree of complexity, in the multidimensional case also. However, we will eschew these, and concentrate on issues that arise due to the inherent multidimensionality of the situation. First, we discuss the motivations for multidimensional measurement models, and illustrate them in the context of a state-wide science assessment involving both multiple choice items and performance tasks. We then describe the multidimensional measurement model (MRCML). This multidimensional model is then applied to the science assessment data set to illustrate two issues that arise in multidimensional measurement. The first issue is the question of whether one should (or perhaps, can) design items that relate to multiple dimensions. The second issue arises when there is more than one form of multidimensionality present in the item design: should one use just one of these dimensionalities, or some, or all? We conclude by discussing further issues yet to be addressed in the area of multidimensional measurement.

1. WHY SHOULD WE BE INTERESTED IN MULTIDIMENSIONAL MEASUREMENT MODELS?

A basic assumption of most item response models is that the set of items in a test measures one common latent trait (Hambleton & Murray, 1983;

S. Alagumalai et al. (eds.), Applied Rasch Measurement: A Book of Exemplars, 287–307.

Lord, 1980). This is the unidimensionality assumption. There are, however, at least two reasons why this assumption can be problematic. First, researchers have argued that the unidimensionality assumption is inappropriate for many standardised tests which are deliberately constructed from sub-components that are supposed to measure different traits (Ansley & Forsyth, 1985). Although it is often argued that item response models are robust to such violations of unidimensionality, particularly when the traits are highly correlated, this need not always be the case. For instance, in computerised adaptive testing, where examinees may take different combinations of test items, ability estimates may reflect different composites of abilities underlying performance and thus cannot be compared directly (Way, Ansley, and Forsyth, 1988). Further, when a test contains mutually exclusive subsets of items or when the underlying dimensions are not highly correlated, the use of a unidimensional model can bias parameter estimation, adaptive item selection, and ability estimation (Folk and Green, 1989).

Second, and perhaps more importantly, the demands of current assessment practice often go beyond single dimensional summaries of student abilities, achievements, understandings, or the like. Modern practice often requires the examination of single pieces of work from multiple perspectives; for example, it may be useful to code student responses not only for their accuracy or correctness but also for the strategy used in the performances or the conceptual understanding displayed by the performances.

The potential usefulness of multidimensional item response models has been recognised for many years and there has been considerable recent work on the development of multidimensional item response models and, in particular, on the consequences of applying unidimensional models to multidimensional data, both real and simulated (for example; Ackerman, 1992; Adams, Wilson & Wang, 1997; Andersen, 1985; Briggs & Wilson, 2003; Camilli, 1992; Embretson, 1991; Folk & Green, 1989; Glas, 1992; Kelderman and Rijkes, 1994; Luecht and Miller, 1992; Muthen, 1984; Muthen, 1987; Oshima & Miller, 1992; Reckase, 1985; Reckase & McKinley, 1991). Despite this activity it does appear that the application of multidimensional item response models in practical testing situations has been limited. This has probably been due to the statistical problems that have been involved in fitting such models and in the difficulty associated with the interpretation of the parameters of existing multidimensional item response models.

2. AN EXAMPLE: CLAS SCIENCE

In the 1993/4 school year, the California Learning Assessment System conducted a state-wide assessment of fifth grade science achievement (California State Department of Education, 1995). The tasks included both multiple choice items and performance tasks. Each student attempted 8 multiple choice items and 3 performance tasks, and over 100,000 students were tested. However, only one performance task was scored for each student. As one of our interests was in the performance tasks, a random sample of 3000 students' scripts for the performance tasks was retrieved and the other two tasks scored, by a small sample of the original raters. We shall use this smaller sample in our demonstration analyses below.

The structure of the item set is as follows. The multiple choice items were in six forms, with common items linking one half of the forms, and another set of linking items linking the other half. Below, we will show results for one of these halves. They were dichotomously scored in the usual way. There were three performance tasks, the same ones for each student. Each task was followed by five questions. These questions related to four "components" of science which had been developed by the CLAS test development committee. The scoring rubrics for the performance tasks were specific to each question. However there were some general guidelines they used for developing these rubrics. A score point of 1 is given to an attempted but inadequate response, scores 2 and 3 to acceptable and adequate responses, that to some extent may contain misconceptions; a score of 4 is given to a clear and dynamic response (one that goes beyond what is asked for).

The four components or process dimensions are given in Figure 16-1. Note that these have been developed to be substantively meaningful to science teachers--issues of psychometric structure have not figured in their development. At first glance, they certainly do not appear to have been developed with an intent of orthogonality. In fact, as they are often presented in contemporary curriculum documents as part of a single process, it would be reasonable to expect that they will be at least moderately inter-correlated. During the development process, each question for each task was assigned to at least one component (shown in Table 16-1). After development, the committee also made similar assignments for the multiple choice items. In doing so, they also noted that when there were multiple assignments, there was usually a "major" assignment and "minor" assignments. For example, question 2 of the second performance task noted in Table 16-1 (i.e., that is, item 2 in "PT2" in the Table) measures the first 3 components: subjects need to record the number of washers used (GD);

subjects need to use these data - as they have to make a comparison (UD); subjects need to use science concepts to explain why the number of washers used in both trials differs (US). But the major assignment is to US.

Table 16-1. Item assignments for the Between and Within models

		Between				Within			
Item Type	**Item No.**	**GD**	**UD**	**US**	**AS**	**GD**	**UD**	**US**	**AS**
MC	1	0	0	1	0	0	0	1	0
	2	0	0	1	1	0	0	1	0
	3	0	0	1	1	0	0	1	0
	4	0	0	1	0	0	0	1	0
	5	0	1	1	1	0	0	0	1
	6	0	1	1	1	0	0	1	0
	7	1	0	1	1	1	0	0	0
	8	0	1	1	1	0	0	1	0
	9	0	0	1	1	0	0	0	1
	10	0	0	1	1	0	0	0	1
	11	0	0	1	1	0	0	0	1
	12	0	0	1	1	0	0	1	0
PT1	1	1	0	0	0	1	0	0	0
	2	0	1	0	0	0	1	0	0
	3	0	0	1	1	0	0	1	0
	4	0	1	0	0	0	1	0	0
	5	0	0	1	1	0	0	0	1
PT2	1	0	1	1	0	0	0	1	0
	2	1	1	1	0	0	0	1	0
	3	0	1	1	0	0	0	1	0
	4	0	0	1	1	0	0	0	1
	5	0	0	1	1	0	0	0	1
PT3	1	1	1	0	0	0	1	0	0
	2	0	1	1	0	0	1	0	0
	3	0	1	1	0	0	1	0	0
	4	0	1	1	0	0	1	0	0
	5	0	0	1	1	0	0	0	1

<u>Data generation/organization (GD)</u>
- gather data
- record / organize data
- observe and describe

<u>Using and discussing data (UD)</u>
- test and describe what happens
- compare / group and describe
- draw conclusions based on own data
- explain / discuss results
- make prediction based on data
- provide rationale to support prediction

<u>Using science concepts to explain data and conclusions (US)</u>
- drawing on prior knowledge to help make decisions and to explain data and conclusions

<u>Applying science beyond immediate task (AS)</u>
- using what you know about a situation and your own data, make an application
- extension of concept
- extend ideas beyond scope of investigation
- generalize and infer principles about real world
- transfer information and apply
- apply outside of classroom to real-life situation

Figure 16-1. The four components of CLAS science

Note that the use of these components was not based on an explicit wish to report those components to the public, in fact, only one score (a "sort of" total score)[1] was reported for each student. The immediate purpose of the components was more a matter of expressing to teachers and students the importance of the components. However, their explicit inclusion in the information about the test also foreshadows their possible use in reporting in future years. This ascription of items to dimensions raises a couple of

[1] The process of combining the scores on the multiple choice items and the performance tasks involved a committee of subject matter specialists who assigned score levels to specific pairs of scores

interesting issues: (a) are the assignments defining psychometrically distinct dimensions, and (b) are the multiple assignments actually making a useful contribution?

The committee also developed a content categorization of the items (both performance and multiple choice), referred to in their documents as the "Big Ideas", roughly the traditional high-school discipline-based divisions of science: Earth Sciences, Physical Sciences, and Life Sciences. These assignments were also carried out after the development had taken place.

3. A MULTIDIMENSIONAL ITEM RESPONSE MODEL: MRCML

Considering the CLAS Science context that has just been described, we can see that multidimensional ideas are being expressed by the item developers. To match this thinking, we also need multidimensional measurement and analysis models. Even if the formal aim of the item developers is a single score, multidimensional measurement models will be needed to properly diagnose empirical problems with the items. The usual multidimensional models suffer from two drawbacks, however. First, the psychometric development has focused on dichotomously scored items, so that most of the existing models and computer programs cannot be applied to multidimensional polytomously scored items like those that generally arise in performance assessments such as those used by CLAS Science. Second, the limited flexibility of existing models and computer programs does not match the complexity of real testing situations which may involve structural features like those of CLAS Science: Raters and item-sampling.

The RCML model has been described in detail in earlier papers (Adams & Wilson, 1996; Wilson and Wang, 1996), so we can use that development (and the same notation) to simply note the additional features of the Multidimensional RCML (MRCML; Adams, Wilson & Wang, 1997; Wang, 1994; Wang & Wilson, 1996). We assume that a set of D traits underlie the individuals' responses. The D latent traits define a D-dimensional latent space and the individuals' positions in the D-dimensional latent space are represented by the vector $\theta = (\theta_1, \theta_2, \ldots, \theta_D)$. The scoring function of response category *k* in item *i* now corresponds to a *D* by 1 column vector rather than a scalar as in the RCML model. A response in category *k* in dimension *d* of item *i* is scored b_{ikd}. The scores across *D* dimensions can be collected into a column vector $\mathbf{b}_{ik} = (b_{ik1}, b_{ik2}, \ldots, b_{ikD})$, and then again be collected into the scoring sub-matrix for item *i*, $\mathbf{B}_i = (\mathbf{b}_{i1}, \mathbf{b}_{i2}, \ldots, \mathbf{b}_{iD})$, and then collected into a scoring matrix $\mathbf{B} = (\mathbf{B}'_1, \mathbf{B}'_2, \ldots, \mathbf{B}'_I)$ for the whole test. If the item parameter vector, ξ, and the design matrix, **A**, are defined as they were in the RCML model the probability of a response in category k of item i is modelled as:

$$\Pr(\mathbf{X}_{ij}=1;\mathbf{A},\mathbf{B},\xi\,|\,\boldsymbol{\theta}) = \frac{\exp\left(\mathbf{b}_{ij}\theta+\mathbf{a}'_{ij}\xi\right)}{\sum_{k=1}^{K_i}\exp(\mathbf{b}_{ik}\theta+\mathbf{a}'_{ik}\xi)} . \quad (1)$$

And for a response vector we have;

$$\Pr(\mathbf{X}=\mathbf{x}\,|\,\boldsymbol{\theta}) = \Omega(\boldsymbol{\theta},\xi)\exp\{\mathbf{x}'(\mathbf{B}\boldsymbol{\theta}+\mathbf{A}\xi)\}, \quad (2)$$

with

$$\Omega(\boldsymbol{\theta},\xi)=\left[\sum_{\mathbf{z}\in V}\exp\{\mathbf{z}'(\mathbf{B}\boldsymbol{\theta}+\mathbf{A}\xi)\}\right]^{-1} . \quad (3)$$

The difference between the RCML model and the MRCML model is that the ability parameter is a scalar, θ, in the former, and a D by 1 column vector, $\boldsymbol{\theta}$, in the latter. Likewise, the scoring function of response k to item i is a scalar, b_{ik}, in the former, whereas it is a D by 1 column vector, $\mathbf{b}_{ik}$, in the latter.

As an example of how a model is specified with the design matrices, consider a test with one four response category question and design matrices

$$\mathbf{A}=\begin{bmatrix}1&0&0\\1&1&0\\1&1&1\end{bmatrix} \text{ and } \mathbf{B}=\begin{bmatrix}1&0&0\\1&1&0\\1&1&1\end{bmatrix} .$$

Substituting these matrices in (1) gives

$$\begin{aligned}\Pr(\mathbf{X}_{11}=1;\mathbf{A},\mathbf{B},\xi\,|\,\boldsymbol{\theta})&=\exp(\theta_1+\xi_1)/D\\ \Pr(\mathbf{X}_{12}=1;\mathbf{A},\mathbf{B},\xi\,|\,\boldsymbol{\theta})&=\exp(\theta_1+\theta_2+\xi_1+\xi_2)/D\\ \Pr(\mathbf{X}_{13}=1;\mathbf{A},\mathbf{B},\xi\,|\,\boldsymbol{\theta})&=\exp(\theta_1+\theta_2+\theta_3+\xi_1+\xi_2+\xi_3)/D\end{aligned} \quad (4)$$

where

$$D = \exp(\theta_1 + \xi_1) + \exp(\theta_1 + \theta_2 + \xi_1 + \xi_2) + \exp(\theta_1 + \theta_2 + \theta_3 + \xi_1 + \xi_2 + \xi_3)$$

which is a multidimensional partial credit model (note that we have not imposed the constraints that would be necessary for the identification of this model).

Note that a different way to express this model would be to consider the log of the ratio of the probabilities of each score and the preceding one:

$$\phi_{12} = \log\left[\frac{\Pr(\mathbf{X}_{12} = 1; \mathbf{A}, \mathbf{B}, \boldsymbol{\xi} \mid \boldsymbol{\theta})}{\Pr(\mathbf{X}_{11} = 1; \mathbf{A}, \mathbf{B}, \boldsymbol{\xi} \mid \boldsymbol{\theta})}\right] = \theta_2 + \xi_2$$

$$\phi_{23} = \log\left[\frac{\Pr(\mathbf{X}_{13} = 1; \mathbf{A}, \mathbf{B}, \boldsymbol{\xi} \mid \boldsymbol{\theta})}{\Pr(\mathbf{X}_{12} = 1; \mathbf{A}, \mathbf{B}, \boldsymbol{\xi} \mid \boldsymbol{\theta})}\right] = \theta_3 + \xi$$ [redacted]

In combination with the first equation in (4), this gives a somewhat more compact expression to the model, and shows that this multidimensional partial credit model parameterizes each step on a different dimension.

In this example, each step is associated with a different dimension. This is a somewhat unusual assumption, and has been chosen especially to illustrate something of the flexibility of the MRCML model. The more usual scenario is that all the steps of a polytomous item would be seen as associated with the same dimension, but that different items may be associated with different dimensions. This is the case with the models used in the examples in the next section.

The analyses in this paper were carried out with the *ACERConquest* software (Wu, Adams & Wilson, 1998), which estimates all models specifiable under the MRCML framework, and some beyond.

4. FIRST ISSUE: WITHIN VERSUS BETWEEN MULTIDIMENSIONALITY

To assist in the discussion of different types of multidimensional models and tests we have introduced the notions of *within* and *between item multidimensionality* (Adams, Wilson & Wang, 1997; Wang, 1994; Wang & Wilson, in 1996). A test is regarded as multidimensional *between* item if it is made up of several unidimensional sub-scales. A test is considered multi-dimensional *within* item when at least one of the items relate to more than one latent dimension.

The Multidimensional Between-Item Model. Tests that contain several sub-scales each measuring related, but supposedly distinct, latent dimensions are very commonly encountered in practice. In such tests each item belongs to only one particular sub-scale and there are no items in common across the sub-scales. In the past, item response modelling of such tests has proceeded by either (a) applying a unidimensional model to each of the scales separately (which Davey and Hirsch (1991) call the *consecutive* approach) or by ignoring the multidimensionality and treating the test as unidimensional. Both of these methods have weaknesses that make them less desirable than undertaking a joint, multidimensional, calibration. The unidimensional approach is clearly not optimal when the dimensions are not highly correlated, and would generally only be considered when the reported outcome is to be a single score. In the consecutive approach, while it is possible to examine the relationships between the separately measured latent ability dimensions, such analyses must take due consideration of the measurement error associated with the dimensions—particularly when the sub-scales are short. Another shortcoming of the consecutive approach is its failure to utilise all of the data that is available. See Adams, Wilson & Wang (1997) and Wang (1994) for empirical examples illustrating his point.

The advantage of a model like the MRCML with data of this type is that; (1) it explicitly recognises the test developers' intended structure, (2) it provides direct estimates of the relations between the latent dimensions and (3) it draws upon the (often strong) relationship between the latent dimensions to produce more accurate parameters estimates and individual measurements.

The Multidimensional Within-Item Model. If the set of items in a test measure more than one latent dimension and some of the items require abilities from more than one of the dimensions then we say the test has within item multidimensionality. The distinction between the within and between item multidimensional models is illustrated in Figure 16-2. When we consider the design matrix **A** and the score matrix **B** in the MRCML model, the distinction between a Within and a Between model has fairly simple expression:

(a) in terms of the design matrix, Between models are always decomposable into block matrices that reflect the item structure, whereas Within models are not;

(b) in terms of the score matrix, for Between models, each item scores on only one dimension, whereas for Within models, an item may score on more than one dimension.

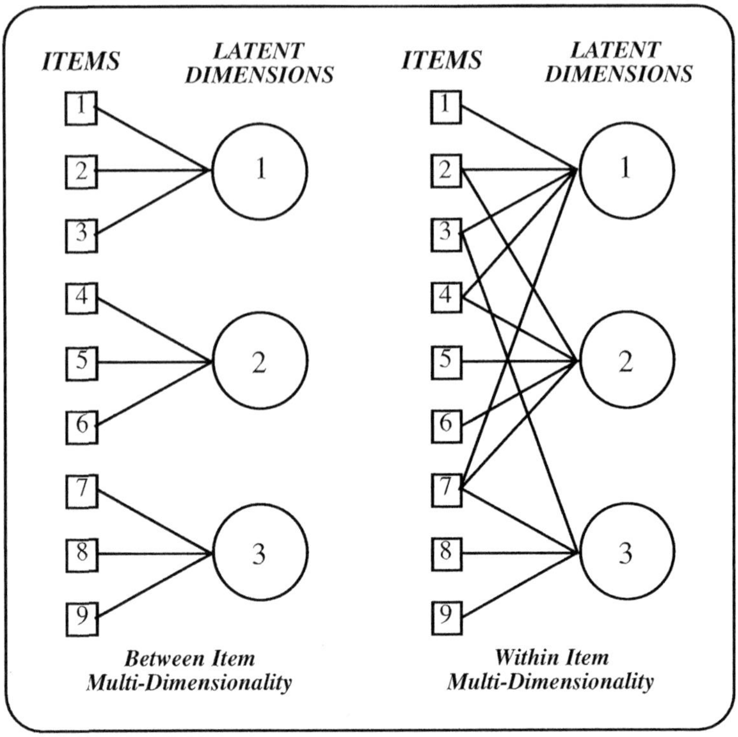

Figure 16-2. A Graphic Depiction of Within and Between Item Multi-dimensionality

5. RETURN TO THE CLAS SCIENCE EXAMPLE

In the CLAS Science context, this distinction between Within and Between models corresponds to an important design issue: Should we design items that relate primarily to a single component, or should we design ones that relate to several simultaneously? Of course, one may not always be able to make this choice, but, nevertheless, there will be situations where designers have such freedom, so one would like to know something about how to go about answering the question. Note that this issue has not been so prominent in the use of multiple choice items. This is probably because such items can be attempted quite quickly. This allows the designers to have (relatively) lots of items, so that one need only worry about primary assignment. However, when designing performance tasks, one is immediately confronted with the problem that they take quite a long time for students to respond in a reasonable way. Hence designers of performance tasks are subject to the temptation to use their (few) items in multiple ways.

In order to examine this issue in the context of CLAS Science, we took one segment of the data corresponding to students who had responded to one linked set of the multiple choice forms. This corresponded to approximately half of the data. We analysed this data first with all of the assignments made by the design committee (the "Within" model), and then with only those assignments deemed primary by the Committee (the "Between" model). The actual assignments for the two models are given in Table 16-1 (Note that the values in the Table correspond to entries in the score matrix for the multiple choice items, but they are only indicators of choice for the performance items). For the multiple choice items, this corresponds to using a simple Rasch model; for the performance items, we used a partial credit item model. Neither of the sets of assignments is balanced across components.

The deviances (-2Xloglikelihood) of the two models were: 56206.5 for the Between and 56678.6 for the Within. As they both have the same number of parameters, 67, this corresponds to a difference in Akaike's Information Index (AIC: Akaike, 1977) of 472.1 in the favour of the Between model. This leads one to prefer the Between model. Although, the AIC does not have a corresponding significance test, it is worth noting that the difference is not trivial.

We can further illustrate the differences by summarising the fit of the two models. We used a classical Pearson χ^2 Fit statistic based on aggregated score groups. To obtain the aggregated score groups we used an unweighted overall sum score (matches sum over sufficient statistics for the 4 dimensional between-item multidimensional solution) and a weighted overall sum score (to match the sum over sufficient statistics for the 4 dimensional within-item multidimensional solution). The aggregation of score groups we carried out for each of the 3 forms separately. Each of the fit statistics is based on a contingency table with 12 cells: 3 (aggregated score groups) x 4 (response categories) cells for the performance task items; 6 (aggregated score groups) x 2 (response categories) cells for multiple choice items. We only included subjects for which there were no missing data--as only then were sum scores comparable. Hence, for items that are included in all three forms, we obtained 3 sets of fit statistics (See Table 16-2 for fit statistics for all items by forms). We also summarised these fit statistics in the form of the mean values of the ratios of the χ^2 to its corresponding degrees of freedom: This is shown in Table 16-3. The general picture that we gather from these results is:

(a) whether weighted or unweighted sum scores are used to form score groups hardly makes a difference (they are highly correlated, .95 or higher);

(b) usually the fit is better for the between-item multidimensional solution;

(c) overall fit seems acceptable.

Using the unweighted fit statistics for comparative purposes, for Form 1 under the Within model, 5 of the 21 items have statistics that are significant at the .05 level, and only 3 for the Between model. For Form 2, 7 of the 27 items have statistics that are significant at the .05 level under the Within model; only 1 statistic is significant under the Between model. For Form 3 this number is 6 and 1, respectively.

Table 16-2. Weighted and unweigheted fit statistics for the components models

	Form 1				Form 2				Form 3			
	unweighted		weighted		unweighted		weighted		unweighted		weighted	
item	bet	with	bet	with	bet	with	bet	with	bet	with	bet	with
01	4.96	9.18	6.93	6.30	8.06	6.13	11.63	6.10	1.59	10.50	4.63	7.12
02	5.00	5.76	2.91	2.59	6.94	5.50	7.90	6.15	11.28	11.92	8.69	10.44
03	10.40	11.89	11.58	12.15	3.81	3.39	7.32	6.63	1.38	1.70	1.03	2.50
04	15.76	22.45	16.81	16.16								
05	29.62	63.30	19.22	50.28								
06	8.74	12.31	9.01	8.59								
07					14.47	94.33	9.66	78.82				
08					13.38	13.00	10.29	5.32	3.59	10.30	5.22	7.28
09									16.75	7.32	14.94	5.39
10									4.97	4.56	3.74	3.29
11									11.00	6.19	6.26	3.35
12									3.99	4.20	2.08	2.17
13	6.63	24.34	5.08	17.76	13.03	32.03	8.60	23.98	10.17	21.19	5.70	13.47
14	16.02	20.59	25.63	24.30	11.93	34.77	9.15	25.04	6.36	23.31	6.16	15.96
15	19.98	20.89	20.90	21.29	14.06	12.87	10.77	9.77	13.46	13.28	6.34	6.44
16	4.66	17.16	9.36	16.71	12.05	37.80	5.42	22.59	9.04	23.73	7.38	16.96
17	12.84	18.14	10.72	15.37	18.78	31.93	16.69	29.20	7.39	11.80	7.84	11.73
18	28.87	16.55	29.53	15.15	9.91	9.86	12.21	8.69	8.86	6.39	10.09	6.59
19	13.80	4.01	22.74	6.37	9.68	9.24	15.23	7.77	13.47	3.41	21.68	5.62
20	15.05	8.96	19.08	11.51	9.10	11.22	8.63	12.94	9.41	8.46	8.80	7.54
21	21.79	33.48	21.78	32.40	9.33	16.70	7.92	15.66	18.35	25.75	16.25	22.53
22	18.56	30.22	15.94	26.71	9.71	20.33	10.49	20.39	21.08	30.08	19.72	28.01
23	12.01	18.50	16.19	22.37	5.10	8.73	4.91	10.25	7.82	12.50	6.89	11.08
24	11.84	16.34	15.75	21.68	10.05	13.41	10.37	12.59	10.98	12.47	9.83	10.76
25	15.37		19.08		20.94		20.88		20.82		19.70	
26	15.05	13.34	17.30	14.31	21.24	27.05	22.55	26.27	16.55	16.09	17.51	15.49
27	14.10	18.84	16.83	19.20	17.16	26.75	14.44	21.94	19.51	25.15	18.65	22.77

Table 16-3. Summary of the Fit Statistics

		Unweighted		**Weighted**	
Form	**Item Type**	**Between**	**Within**	**Between**	**Within**
1	pt	1.26	1.56	1.48	1.58
1	mc	1.03	1.73	0.92	1.33
2	pt	1.07	1.74	0.99	1.47
2	mc	0.78	2.04	0.78	1.72
3	pt	1.07	1.39	1.01	1.16
3	mc	0.57	0.59	0.49	0.43

Looking now at the Between model, the latent[2] correlations among the four components are given in Table 16-4. As one might have expcected, these correlations are quite high.. One can ask a number of interesting questions, based on these results. One question of some practical significance is: Could one make the model simpler by collapsing the more highly correlated components (say, AS and UD, or AS and US)? What we need to do , in order to investigate this, is to test whether the correlation between these pairs of dimensions is 1.0. That can be achieved by reducing the dimensionality (assigning items from the original two dimensions to just one dimension), and testing for a significant difference in the fit between the two models. Taking the first of these, we find that the resulting three dimensional model has AIC=57059.3, a worse fit than either of the four dimensional models fitted above. Thus, at least in a statistical significance sense, collapsing dimensions seems to not be advisable.

Table 16-4. Estimated Correlations among the Components

	UD	US	AS
GD	.76	.75	.70
UD		.75	.87
US			.86

We can use these fit statistics to focus attention on particular items. The highest fit statistic is obtained for item 7, which is the only multiple-choice item measuring dimension 1 (GD). Under the within-item multidimensional model it is also assumed to measure dimensions 3 (US) and 4 (AS). This item is badly fitting under the within-item multidimensional model, but not under the between-item multidimensional model. When examining the residuals, it can be seen that the item is under-discriminating, that is, with increasing ability the proportion of subjects having the item correct increases less than expected, and this effect is much more marked under the within-item multidimensional model. This is displayed in Figure 3, where these proportions are compared to expected proportions, for Within and Between models, (for the statistics) for US (the relationships do not change substantively among the dimensions). Item 7 (which is the third item of form 5) is a tricky question. It actually tests subjects' knowledge of the concept of 'scientific observation'. Only one alternative (the correct one) is a descriptive statement, the other 3 give explanations of subjects' observations--and hence are wrong, as subjects are asked to pick the statement that best describes the observations. The item investigates

[2] We use the term *latent*, because they are the directly-estimated correlations in the MRCML model--and may differ from correlations calulated in other ways, such as correlating the raw scores, or even the estimated person ablities

subjects' knowledge of what is science rather than a particular scientific content. The other items measuring dimension 1, items 13, 19 and 23 actually ask subjects to carry out observations, i.e., write down characteristics of the objects under investigation. Thus, the lower empirical discrimination for this item probably corresponds to this item having a weak relationship with the specific dimension defined by the other items.

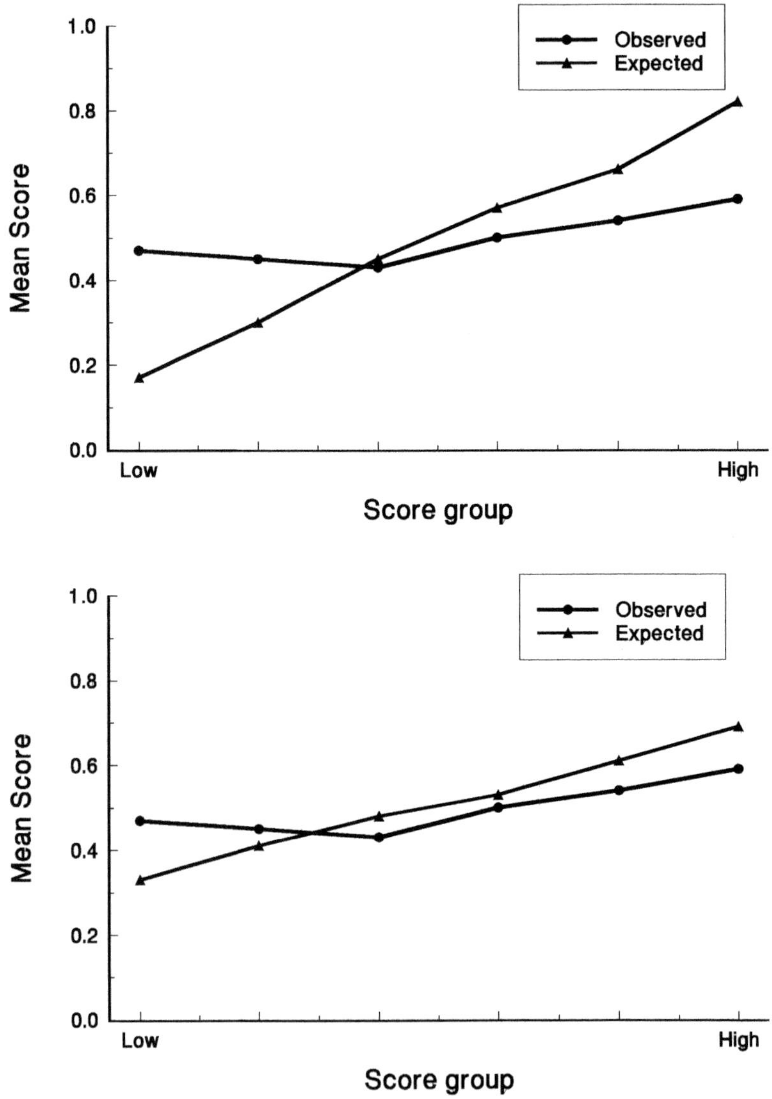

Figure 16-3. . Residuals for item 7 under the within (top) and between (bottom) item multidimensional solutions

We repeated these analyses for the other half of the data (different students and multiple choice items, same performance tasks), and found that the results were essentially the same (i.e., the numerical results differed somewhat, but the interpretations did not change). All in all, it seems that the Between model is noticeably better for this data set. This may seem counter-intuitive to some developers who might wish that by making more assignments from an item to different components, one must be squeezing more information out of the student responses. There are two ways to express why this is not necessarily so. One way to think of it is that there really is only a certain amount of information in the data set to begin with, so that adding "links" will not improve the situation once that information has been exhausted. Another is to see that by adding these assignments, at some point, one will not be adding information to each dimension, but, indeed, one will be making it more difficult to find the right orientation for each component. That is, at a certain point, more links may make the components "fuzzier". In this case, the designers have not improved their model by adding the Within-Item assignments, but have in fact, made it worse.

6. SECOND ISSUE: DIFFERENT DIMENSIONALITIES--MODES VERSUS CONTENT

Having considered the possibilities of using components as a way of looking at the CLAS Science data, we can also look at the other two possibilities for dimensionalities raised above. The other two dimensionalities are (a) the "Big Ideas" content analysis of the items, and (b) the distinction between the two modes of assessment items--multiple choice items and performance tasks. The difference between the item modes has been described above. As mentioned above, the three "Big Ideas" of CLAS Science are: Earth Sciences, Physical Sciences, and Life Sciences. The issue that arises is: Should we be considering only one, or some, or all, such dimensionalities in specifying our model? The MRCML model allows us to use much the same kind of modeling as we used to illustrate the situation for the components.

This issue can be specified in the following way for the CLAS Science context: Is the relationship among the "Big Ideas" the same when represented by the multiple choice items as it is when represented by the performance tasks. Or, to put it another way, does the way we gather the data (item modes) alter the relationships among the cognitive structures in which we are interested? This is a fundamental issue of invariance that we need to understand in order (a) to know how to deploy different item modes in instrument design, and (b) to use the resulting item sets in an efficient and meaningful way. As such, it represents one of the major challenges for the next decade of applied psychometrics in education, because mixed item modes are coming to be seen as one of the major strategies of instrument design, especially for achievement tests (cf., Wilson & Wang, 1996).

In order to examine this issue we constructed several different MRCML models:

(a) a unidimensional model (UN), where all items are associated with a single dimension;
(b) a two dimensional model, based on the item modes (MO);
(c) a three dimensional mode, based on the "Big Ideas" (BI); and
(d) a six dimensional model based on the cross-product of the item mode and the big idea models (MOBI) (i.e., think of it as having a "Big Ideas" model (BI) within each item mode (MO)).

We fitted these four models with the same data as before, and came up with the results illustrated in Figure 4. This Figure shows likelihood ratio tests conducted between the hierarchical pairs of models (i.e., pairs for which one model is a submodel of the other). Thus, the UN model is a submodel of each of the alternative models MO and BI, and each of them is a submodel of MOBI, but MO and BI are not related in this way.

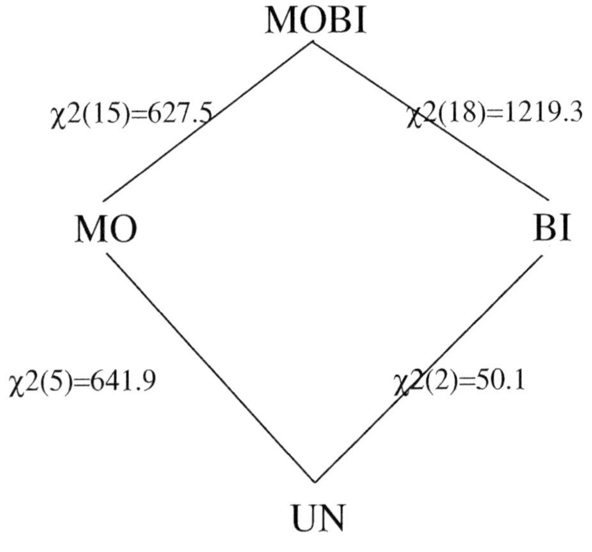

Figure 16-4. Relationships among the item mode and "Big Idea" models

Each of these tests is statistically significant at a standard α=.05 level of statistical significance. Thus, we can say that both of the multidimensional models (MO and BI) fit better than the unidimensional model (UN), and that the model that combines both item modes and "Big Ideas" (MOBI) fits better than the models that use only one of them (MO and BI). To get an idea of the relative success of these two, we cannot use a likelihood ratio test, as they are not hierarchical. Instead we can, as above, use the AIC: For MO, AIC=1680.6, and for BI, AIC=1094.8. Hence, the "Big Ideas" model is a relatively better fit than the modes model. All in all, the fit analyses confirm that there are statistically significant differences in the way that the "Big Ideas" are psychometrically represented by the multiple choice items and the performance task questions. Some idea of the effect of these differences can be gained by considering the latent correlations among the three "Big Ideas" within each mode, which are shown in Table 16-5.

Table 16-5. Correlations among and between multiple choice items and performance tasks[3]

	Earth Sciences	Physical Sciences	Life Sciences
Multiple Choice			
Earth Sciences	-	.64	.64
Physical Sciences	50°	-	.76
Life Sciences	50°	40°	-
Performance Tasks			
Earth Sciences	-	.69	.77
Physical Sciences	46°	-	.59
Life Sciences	40°	54°	-
MC to PT			
correlation	.65	.63	.54
angle	50°	51°	57°

Table 16-5 shows that the pattern of relationships among the "Big Ideas" in the two modes is somewhat different, sufficiently so in a technical sense to give the fit results mentioned above. But the differences, between 50° and 46°, 50° and 40°, and 40° and 54°, are probably not so great that a substantive expert would remark upon them. There is also a considerable "mode effect" that is fairly constant across the three "Big Ideas", ranging from a correlation of .54 to .65. This is consistent with, though somewhat higher than, similar comparisons for CLAS Mathematics (Wilson & Wang, 1995).

[3] In the two matrices, correlations are above the diagonal and their corresponding angles are below.

7. DISCUSSION AND CONCLUSIONS

The results reported above have supported a number of observations relevant to multidimensional item response modeling: (a) more links from items to dimensions (or "components") do not necessarily make for a better model; (b) different content dimensions may exhibit themselves in different ways under examination through alternate measurement modes (although, in this case, the difference seemed small from a substantive point of view). The evidence presented arises in just one specific context, CLAS Science, and so cannot be formally generalised beyond that. But, the results are suggestive of some possible interpretations as follows.

1. In designing questions related to performance assessments and other complex item response modes, developers have a choice of strategy between developing questions (or other micro observational contexts), that are either sharply focussed on a specific dimension, or ones that straddle two or more dimensions. When such performance tasks are expensive (in development, student, and/or scoring costs), there is a temptation to pursue the latter strategy. These results suggest that test developers, when left to their own devices, may be poor judges of the allocation of their items to multiple underlying dimensions.
2. One alternative that has been suggested for use in large-scale assessments, where both coverage of many topics is wanted, and also use of time-expensive modes such as performance assessment are valued, is to mix different assessment modes into what has been called an "Assessment Net" (Wilson, 1994; Wilson & Adams, 1995). Use of such a strategy is dependent upon knowledge of the relationship between the mode of assessment (i.e., "method"), and the dimension being measured (i.e., "trait"). The MRCML approach allows modeling of this situation, and examination of the consistency with which the modes assess the dimensions. The present results are suggestive of the possibility of substantive consistency, even in the presence of technically-detectable levels of discrepancy.

Both of these sorts of investigation are rather new in the item response modeling framework. There is pressing need for the development of research designs, appropriate modeling strategies, focussed fit statistics and diagnostic techniques, and an extensive research literature, in this area.

As examples of the use of the MRCML model, the models described and estimated above have proved useful for illustrating certain types of analyses. They are not, however, indicative of the full range of application of the model. For example, in the CLAS context, we ignored two issues that could well have an important impact on the measurement situation. First, each performance assessment question is embedded in a performance task, and hence, there is a certain possibility that each of the five sets of qucstions should be considered an item "bundle", and the dependence formally modeled. This has been demonstrated in the unidimensional case (Wilson & Adams, 1994), but use of a within-item multidimensional model is also now a possibility. This opens up a number of intriguing possibilities that await further research, such as whether the bundle parameters and the question parameters lie on the same dimension. Second, the performance task questions were each rated by a specific rater, many of whom rated sufficient student scripts to be formally modeled. This has also been carried out in the unidimensional context (Wilson & Wang, 1995), but has not so far been considered in a multidimensional setting. Here too, new possibilities arise, and new challenges, both conceptual and technical can be discerned, such as the dimensionality of the raters (c.f., the discussion above about the dimensionality of the items). The MRCML model formally includes the possibilities of modeling both these complexities, simultaneously, if need be. Whether real data sets will be rich and deep enough to support such models, and whether we will find it such complex models interpretable, remain to be seen.

8. REFERENCES

Ackerman, T. A. (1992). A didactic explanation of item bias, item impact, and item validity from a multidimensional perspective. *Journal of Educational Measurement, 29,* 67-91.

Adams, R. J., & Wilson, M. (1996). Formulating the Rasch model as a mixed coefficients multinomial logit. In G. Engelhard and M. Wilson, (Eds.), *Objective measurement: Theory into Practice.* Vol III. Norwood, NJ: Ablex.

Adams, R. J., & Wilson, M. (1996, April). Multi-level modeling of complex item responses in multiple dimensions: Why bother? Paper presented at the annual meeting ofthe American Educational Research Association, New York.

Adams, R. J., Wilson, M., & Wang, W. (1997). The multidimensional random coefficients multinomial logit model. *Applied Psychological Measurement, 21*(1), 1-23.

Akaike, H. (1977). On entropy maximisation principle. In P. R. Krischnaiah (Ed.), *Applications of statistics.* New York: North Holland.

Andersen, E. B. (1985). Estimating latent correlations between repeated testings. *Psychometrika, 50,* 3-16.

Ansley, T. N., & Forsyth, R. A. (1985). An examination of the characteristics of unidimensional IRT parameter estimates derived from two-dimensional data. *Applied Psychological Measurement, 9,* 37-48.

Briggs, D. & Wilson, M. (2003). An introduction to multidimensional measurement using Rasch models. *Journal of Applied Measurement, 4*(1), 87-100.

California Department of Education. (1995). *A sampler of science assessment.* Sacramento, CA: Author.

Camilli, G. (1992). A conceptual analysis of differential item functioning in terms of a multidimensional item response model. *Applied Psychological Measurement, 16,* 129-147.

Davey, T. & Hirsch, T. M. (1991). *Concurrent and consecutive estimates of examinee ability profiles.* Paper presented at the Annual Meeting of the Psychometric Society, New Brunswick, NJ.

Embretson, S. E. (1991). A multidimensional latent trait model for measuring learning and change. *Psychometrika, 56,* 495-515.

Folk, V. G., & Green, B. F. (1989). Adaptive estimation when the unidimensionality assumption of IRT is violated. *Applied Psychological Measurement, 13,* 373-389.

Glas, C.A.W. (1989). Contributions to Estimating and Testing Rasch Models. Doctoral Dissertation. Univeristy of Twente

Glas, C. A. W. (1992). A Rasch model with a multivariate distribution of ability. In M. Wilson (Ed). *Objective measurement: Theory into Practice.* Vol. 1. *Norwood,* NJ: Ablex Publishing Corporation.

Hambleton, R. K., & Murray, L. N. (1983). Some goodness of fit investigations for item response models. In R. K. Hambleton (Ed.), *Applications of item response theory.* Vancouver BC: Educational Research Institute of British Columbia.

Kelderman, H., & Rijkes, C. P. M. (1994). Loglinear multidimensional IRT models for polytomously scored items. *Psychometrika, 59,* 149-176.

Lord, F. M. (1980). *Applications of Item Response Theory to Practical Testing Problems.* Hillsdale, NJ: Erlbaum.

Luecht, R. M., & Miller, R. (1992). Unidimensional calibrations and interpretations of composite traits for multidimensional tests. *Applied Psychological Measurement, 16, 3,* 279-293.

Muthen, B. (1984). A general structural equation model with dichotomous, ordered categorical and continuous latent variable indicators. *Psychometrika, 49,* 115-132.

Muthen, B. (1987). *LISCOMP: Analysis of linear structural equations using a comprehensive measurement model. User's guide.* Mooresville, IN: Scientific Software.

Oshima, T. C., & Miller, M. D. (1992). Multidimensionality and item bias in item response theory. *Applied Psychological Measurement, 16, 3,* 237-248.

Rasch, G. (1960, 1980).*Probabilistic Models for Some Intelligent and Attainment Tests.* Copenhagen: Danmarks Paedogogiske Institut.

Reckase, M. D. (1985). The difficulty of test items that measure more than one ability. *Applied Psychological Measurement, 9,* 401-412.

Reckase, M. D., & McKinley, R. L. (1991). The discriminating power of items that measure more than one dimension. *Applied Psychological Measurement, 15,* 361-373.

Wang, W. (1994). Implementation and application of the multidimensional random coefficients multinomial logit. Unpublished doctoral dissertation. University of California, Berkeley.

Wang, W., & Wilson, M. (1996). Comparing open-ended items and performance-based items using item response modeling. In, G. Engelhard & M. Wilson (Eds.), *Objective measurement III: Theory into practice.* Norwood, NJ: Ablex.

Way, W. D., Ansley, T. N., & Forsyth, R. A. (1988). The comparative effects of compensatory and non-compensatory two-dimensional data on unidimensional IRT estimation. *Applied Psychological Measurement, 12,* 239-252.

Wilson, M. (1994). Community of judgement: A teacher-centered approach to educational accountability. In, Office of Technology Assessment (Ed.), *Issues in Educational Accountability.* Washington, D.C.: Office of Technology Assessment, United States Congress.

Wilson, M. , & Adams, R. J. (1995). Rasch models for item bundles. *Psychometrika, 60,* 181-198.

Wilson, M., & Adams R.J. (1996). Evaluating progress with alternative assessments: A model for Chapter 1. In M.B. Kane (Ed.), *Implementing performance assessment: Promise, problems and challenges*. Hillsdale, NJ: Erlbaum.

Wilson, M.R. & Wang, W.C. (1995). Complex composites: Issues that arise in combining different models of assessment. *Applied Psychological Measurement, 19*(1), 51-72.

Wu, M., Adams, R.J., & Wilson, M. (1998). ACER*ConQuest* [computer program]. Hawthorn, Australia: ACER.

Chapter 17

INFORMATION FUNCTIONS FOR THE GENERAL DICHOTOMOUS UNFOLDING MODEL

Guanzhong Luo and David Andrich
Murdoch University, Australia

Abstract: Although models for the unfolding response processes are single-peaked, their information functions are generally twin peaked though in rare exceptions may be single-peaked. This contrasts with the models for cumulative response process which are monotonic and for which the information function is always single-peaked. In addition, in the cumulative models, the information is a maximum when the person and item locations are identical, whereas for most unfolding models, the information is minimum at this point. The general unfolding model (Luo, 1998, 2000) for dichotomous responses, of which all proposed probabilistic unfolding models are special cases, makes explicit two item parameters, one the location of the item, the other the latitude of acceptance which defines the thresholds between which the positive response is more likely than the negative response. The current paper carries on further studies of this general model, particularly the information function of the general model. First, the information function of this general unfolding model is resolved into two components, one related to the latitude of acceptance, the other related only to the distance between the person and item locations. The component relative to the latitude of acceptance has a maximum value at the affective thresholds, but is moderated by the operational function. Second, the contrasts between the information functions for unfolding and cumulative models is reconciled by showing that the key points for maximising the information is where the probability of the positive and negative responses are equal, which is the threshold where the person and item locations are identical in the cumulative models and are the two thresholds which define the latitude of acceptance in the unfolding models. As a result of the explication of these relationships, it is shown that some single peaked response functions have no defined information when the person is at the location of the item.

Key words: attitude measurement, information functions, operational functions, unfolding models, single-peaked functions, item response theory.

S. Alagumalai et al. (eds.), Applied Rasch Measurement: A Book of Exemplars, 309–328.

1. INTRODUCTION

Fisher's information function (Fisher, 1956) provided a measure of the intrinsic accuracy of a distribution (Rao, 1973). Birnbaum (1968) derived item and test information functions in the field of psychometrics with essentially the same definition, although the concept was investigated much earlier in Lord's definition of test discrimination power (Lord, 1952, 1953, Baker 1992). In the context of item response theory, Samijima (1969; 1977; 1993) studied information functions that are related to this definition but commencing with a different expression.

In educational measurement, the measurement models are usually concerned with the substantive areas of achievement, performance and the like, in which the probability of a positive response as a function of the location on the continuum is monotonically increasing. The response process is referred to as cumulative and it has an ideal direction – the greater the value the better. Nicewander (1993) related the information functions of these measurement models to Cronbach's reliability coefficient. The information function has been used widely for computer adaptive testing (CAT) and automated test construction (Theunissen, 1985).

In many case of attitude measurement, however, the relevant models for the response process have an *ideal point* rather than an ideal direction. The process is referred to as unfolding, and the corresponding response function is single-peaked rather than monotonic (Coombs, 1964, 1977). The last two decades or so of 20th century provided a rapid development of single-peaked probabilistic unfolding models (e.g. DeSarbo & Hoffman, 1986; Andrich, 1988; 1995, Böckenholt & Böckenholt, 1990, Hoijtink, 1990; 1991, Andrich & Luo, 1993, Verhelst & Verstralen, 1993). Various specific forms for unidimensional unfolding have been proposed. Among them, some frequently referenced unfolding models for dichotomous responses include the Simple Square Logistic Model (SSLM, Andrich, 1988), the PARELLA model (Hoijtink, 1990) and the Hyperbolic Cosine Model (HCM, Andrich & Luo, 1993). Abstracted from these specific unidimensional unfolding models, Luo (1998) proposed the general form of unfolding models for dichotomous responses, which was further extended into the general form of unfolding model for polytomous responses (Luo, 2001). One important feature of these two general forms for unfolding models is that the latitude of acceptance, also referred to as the unit of an item, is explicitly parameterized. Andrich (1996) provided a framework of the unidimensional measurement using either monotonic or single-peaked response functions. This framework permits a clearer understanding of the distinctions and relationships between these two types of responses and models. The result presented there attempted to provide a closure to the line of development in social measurement begun by Thurstone in the 1920s (Thurstone, 1927, 1928) in which he more or less explicitly applied both the monotonic and single-peaked response processes and developed further by Likert (1932), Guttman (1950), Coombs (1964) and Rasch (1961). In addition, it implied that some results from cumulative models could be recognized and applied with unfolding models. The particular concern in the current paper is the understanding and application of the information functions of unfolding models. Though the information functions have already been used in the context of unfolding to conduct computerized adaptive test and optimal fixed length test design (Roberts, Lin & Laughlin, 1999; Laughlin & Roberts, 1999), further studies on their properties and structure with a comparison to their counterparts within cumulative models are worthwhile.

The purpose of the paper is to present some characteristics of information functions for single-peaked, dichotomous unfolding models. In doing so, the central role played by the unit of an item in unfolding models, which is the range in which the relative probability of a positive response is greater than the negative response, is revealed.

The starting point for the discussion of information in this paper is the definition by Samejima (1969)

$$I(\theta) = E[-\frac{\partial^2}{\partial\theta^2}\log P\{X = x \mid \theta\}] = \frac{[p'(\theta)]^2}{p(\theta)q(\theta)}; \tag{1}$$

where X is a dichotomous random variable, $p(\theta) = P\{X = 1 \mid \theta\}$ and $q(\theta) = 1 - p(\theta) = P\{X = 0 \mid \theta\}$. This is not directly Fisher's original definition given as

$$I(\theta) = E[(\frac{\partial}{\partial\theta}\log P\{X = x \mid \theta\})^2] \tag{2}$$

Although it is implied in the psychometric literature that these definitions are equivalent, there seems no explicit proof of this equivalence in the common psychometric literature (e.g. Baker, 1992). In addition, there is an emphasis on IRT models which are generally specified and monotonically increasing, giving the possible impression that this equivalence is confined to the cumulative models. Therefore, for completeness, and because the unfolding models have unusual properties in their information functions, the Appendix shows that this definition is identical to Fisher's for dichotomous response models irrespective of whether or not the models are cumulative or unfolding.

The rest of the paper is presented as follows. First, the information function defined above is reviewed and derived for the general unfolding model. Second, in order to understand the structure of information functions in unfolding models, the information is resolved into two components, one which has its maximum value when the person-item distance equals the value of the item's unit, and another which is independent of the item's unit, but dependent on the distance between the locations of the person and the item. Third, special cases of unfolding models are considered in which the features at the location and thresholds of items are considered. Finally, a comparison is made between the cumulative and unfolding models regarding the location of the maximum information, and it is shown that this occurs as a function of the point in which the probabilities of a positive and negative response are equal in both kinds of models. In unfolding models, however, the location of the maximum information is moderated by a second component that is a function of the person-item distance.

2. FACTORISATION OF THE INFORMATION FUNCTION FOR THE GENERAL UNFOLDING MODEL

Figure 17-1 shows the general form of the probabilistic unfolding models (Luo, 1998) for dichotomous responses, which takes the mathematical form

$$\pi_{ni} = \Pr\{X_{ni} = 1 \mid \beta_n, \delta_i, \rho_i\} = \frac{\Psi(\rho_i)}{\Psi(\rho_i) + \Psi(\beta_n - \delta_i)}; \qquad (3)$$

where β_n is the location parameter for person *n*, δ_i is the location parameter for item *i* and $\rho_i \geq 0$ characterises the range in the which the probability of a positive response is greater than its complement. The parameter ρ_i is in fact parameterises the *latitude of acceptance*, an important concept in attitude measurement (Sherif & Sherif, 1967). Hereafter ρ_i is called the item unit and the two points on the continuum that define the unit are termed thresholds. A remarkable feature of this general form is that the function Ψ of the unit parameter ρ_i and the distance $\beta_n - \delta_i$ between the location and the item is the same. The function Ψ, which defines the form of the response model, is termed the *operational function.* It has the following properties:

(P1) Non-negative: $\Psi(t) \geq 0$ for any real t;

(P2) Monotonic in the positive domain: $\Psi(t_1) > \Psi(t_2)$ for any $t_1 > t_2 > 0$; and

(P3) Ψ is an even function (symmetric about the origin): $\Psi(t) = \Psi(-t)$ for any real t.

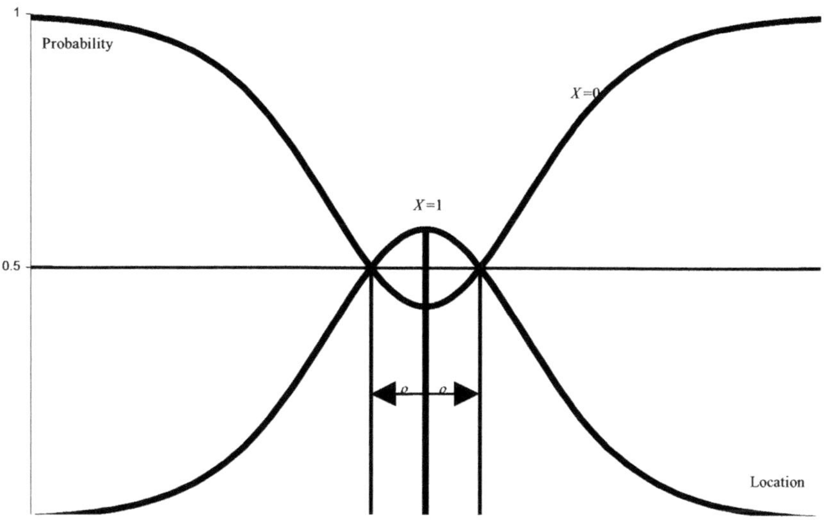

Figure 17-1. The general form of the probabilistic unfolding models for dichotomous responses

The form of Eq. (3) and Figure 17-1 show that the unit ρ_i is a structural parameter of unfolding models. Figure 17-2 shows the functions of the positive responses for the SSLM, the PARELLA model and the HCM for a value of ρ_i = 1.0 (the specific expressions of these models are given in equations 14, 16 and 18).

In this section, the information function is presented as the mathematical expectation of the negative of the second derivative of the log-likelihood function. This formulation (Birnbaum, 1968; Samejima, 1969; Baker, 1992) is more familiar in psychometrics than is Fisher's original definition, which is considered in a later section. In addition, we focus on the information for the location of a person parameter with the values of item parameters are given.

Generally, the information function is obtained as part of obtaining the maximum likelihood estimate (MLE) of the location parameter β_n. Under various specific/general unfolding models, various algorithms for estimating person parameters with given item parameters are similar (Andrich, 1988;

Hoijtink, 1990, 1991; Andrich & Luo, 1993; Verhelst & Verstralen, 1993; Luo, Andrich & Styles, 1998; Luo, 2000). In general, given the values of all item parameters $\{\delta_i, \rho_i; i = 1, \ldots, I\}$ consider the likelihood function

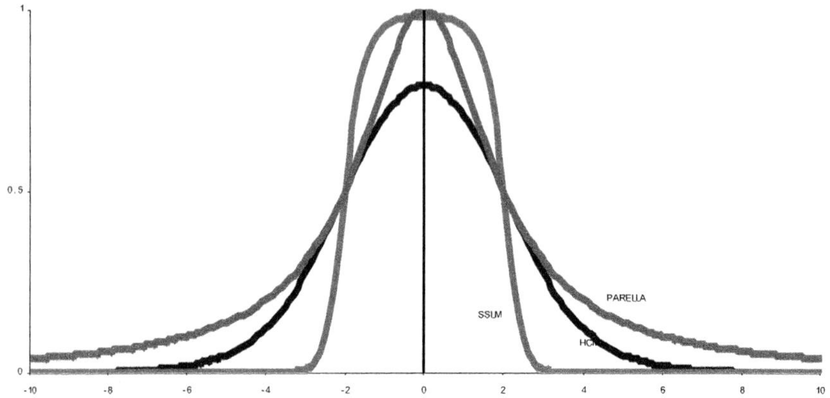

Figure 17-2. Three specific probabilistic unfolding models for dichotomous responses

$$L = \prod_i \frac{[\Psi(\rho_i)]^{x_{ni}}[\Psi(\beta_n - \delta_i)]^{1-x_{ni}}}{\Psi(\rho_i) + \Psi(\beta_n - \delta_i)}$$

$$= \frac{\prod_i [\Psi(\rho_i)]^{x_{ni}} \prod_i [\Psi(\beta_n - \delta_i)]^{1-x_{ni}}}{\prod_i [\Psi(\rho_i) + \Psi(\beta_n - \delta_i)]}. \tag{4}$$

Its logarithm is given by

$$\log L = \sum_i x_{ni} \log \Psi(\rho_i) + \sum_i (1 - x_{ni}) \log \Psi(\beta_n - \delta_i) - \sum_i \log[\Psi(\rho_i) + \Psi(\beta_n - \delta_i)]. \tag{5}$$

Differentiating $\log L$ with respect to β_n leads to the solution equations as

$$-\sum_i \Delta(\beta_n - \delta_i)(x_{ni} - \pi_{ni}) = 0. \tag{6}$$

where

$$\pi_{ni} = P\{x_{ni} = 1 \mid \beta_n, \delta_i, \rho_i\} = \frac{\Psi(\rho_i)}{\Psi(\rho_i) + \Psi(\beta_n - \delta_i)};$$

$$\Delta(t) = \frac{\partial \log \Psi(t)}{\partial t} = \frac{\partial \Psi(t)/\partial t}{\Psi(t)}, \tag{7}$$

The expectation of the second derivative is given by

$$\begin{aligned}
E[\frac{\partial^2 \log L}{\partial \beta_n^2}] &= E[-\sum_{i=1}^{I}\{[\frac{\partial}{\partial \beta_n}\Delta(\beta_n - \delta_i)](x_{ni} - \pi_{ni}) + \Delta(\beta_n - \delta_i)\frac{\partial}{\partial \beta_n}[(x_{ni} - \pi_{ni})]\}] \\
&= -\sum_{i=1}^{I}\{\frac{\partial}{\partial \beta_n}\Delta(\beta_n - \delta_i)]E[(x_{ni} - \pi_{ni})] + \Delta(\beta_n - \delta_i)E[\frac{\partial}{\partial \beta_n}(x_{ni} - \pi_{ni})]\} \\
&= -\sum_{i=1}^{I}\Delta(\beta_n - \delta_i)E[\frac{\partial}{\partial \beta_n}(x_{ni} - \pi_{ni})] \\
&= \sum_{i=1}^{I}\Delta(\beta_n - \delta_i)\frac{\partial}{\partial \beta_n}\pi_{ni}
\end{aligned} \tag{8}$$

Because

$$\begin{aligned}
\frac{\partial \pi_{ni}}{\partial \beta_n} &= \frac{\partial}{\partial \beta_n}[\frac{\Psi(\rho_i)}{\Psi(\rho_i) + \Psi(\beta_n - \delta_i)}] \\
&= \Psi(\rho_i)\frac{-\frac{\partial}{\partial \beta_n}\Psi(\beta_n - \delta_i)}{[\Psi(\rho_i) + \Psi(\beta_n - \delta_i)]^2} \\
&= \frac{\Psi(\rho_i)}{[\Psi(\rho_i) + \Psi(\beta_n - \delta_i)]}\frac{\Psi(\beta_n - \delta_i)}{[\Psi(\rho_i) + \Psi(\beta_n - \delta_i)]}\frac{-\frac{\partial}{\partial \beta_n}\Psi(\beta_n - \delta_i)}{\Psi(\beta_n - \delta_i)} \\
&= (-1)\pi_{ni}(1 - \pi_{ni})\frac{\frac{\partial}{\partial \beta_n}\Psi(\beta_n - \delta_i)}{\Psi(\beta_n - \delta_i)} \\
&= (-1)\pi_{ni}(1 - \pi_{ni})\Delta(\beta_n - \delta_i)
\end{aligned} \tag{9}$$

Substituting Eq. (8) into Eq. (7) gives

$$E[\frac{\partial^2 \log L}{\partial \beta_n^2}] = \sum_{i=1}^{I} \Delta(\beta_n - \delta_i)\frac{\partial}{\partial \beta_n}\pi_{ni}$$

$$= \sum_{i=1}^{I} \Delta(\beta_n - \delta_i)(-1)\pi_{ni}(1-\pi_{ni})\Delta(\beta_n - \delta_i) \quad (10)$$

$$= -\sum_{i=1}^{I} \pi_{ni}(1-\pi_{ni})\Delta^2(\beta_n - \delta_i)$$

That is, transposing (-1),

$$E[-\frac{\partial^2 \log L}{\partial \beta_n}] = \sum_{i=1}^{I} \pi_{ni}(1-\pi_{ni})\Delta^2(\beta_n - \delta_i). \quad (11)$$

Following Samejima (1969, 1977,1993) for any one item i, denote the *item information function* with respect to the estimate of β_n as the term within the summation on the right-hand side of Equation (11), that is,

$$\begin{aligned} I_{ni} &= \pi_{ni}(1-\pi_{ni})\Delta^2(\beta_n - \delta_i) \\ &= \frac{\Psi(\rho_i)}{\Psi(\rho_i)+\Psi(\beta_n-\delta_i)}\frac{1}{\Psi(\rho_i)+\Psi(\beta_n-\delta_i)}\Delta^2(\beta_n-\delta_i) \end{aligned} \quad (12)$$

This definition is consistent with information being additive across items. It is evident that the variables in Equation (12) are the person-item distance $\beta_n - \delta_i$ and the item unit ρ_i. Let

$$f(\pi_{ni}) = \pi_{ni}(1-\pi_{ni}), \quad (13)$$

which has the maximum value when $\pi_{ni} = 0.5$. This occurs when the person–item distance is the same as the item unit, $|\beta_n - \delta_i| = \rho_i$, that is where the positive response and the negative response functions intersect.

Using this definition of Eq. (13), Eq. (12) can be written as

$$I_{ni} = f(\pi_{ni})\Delta^2(\beta_n - \delta_i) \tag{14}$$

Equation (13) shows the factorisation of the item information function I_{ni} into the two components, $f(\pi_{ni})$, which has the maximum value when the person-item distance equals the value of the item unit; and the second component $\Delta^2(\beta_n - \delta_i)$, which is independent of the item unit ρ_i but is a function of the person-item distance.

The following examination of specific models shows that the item information functions for the SSLM (Andrich, 1988), the PARELLA model (Hoijtink, 1990) and the HCM (Andrich and Luo, 1993) with variable units (Luo, 1998), derived separately in those papers, are special cases of Eq. (14) or its equivalence, Eq. (12). In the graphs illustrating the factorisation of the information functions (Figures 17-3, 17-4 and 17-5), the value of $\rho_i = 1$. The same patterns arise for other values of ρ_i.

(1) The SSLM with a variable unit ρ_i.

The SSLM with a variable unit is defined as

$$\pi_{ni} = \frac{\exp(\rho_i^2)}{\exp(\rho_i^2) + \exp[(\beta_n - \delta_i)^2]}; \tag{15}$$

Therefore,

$$\Delta(\beta - \delta_i) = \frac{d}{d\beta}\log\exp(\beta - \delta_i)^2$$
$$= \frac{d}{d\beta}(\beta - \delta_i)^2 = 2(\beta - \delta_i) \quad . \tag{16}$$

giving

$$I_{ni} = \pi_{ni}(1 - \pi_{ni})4(\beta - \delta_i)^2 \tag{17}$$

Figure 17-3 shows the components of Eq. (17) as well as the information function. The first component gives the twin peaks to the information function and has a maximum value at the thresholds defining the unit. The second component takes the value of 0 at $\beta = \delta_i$.

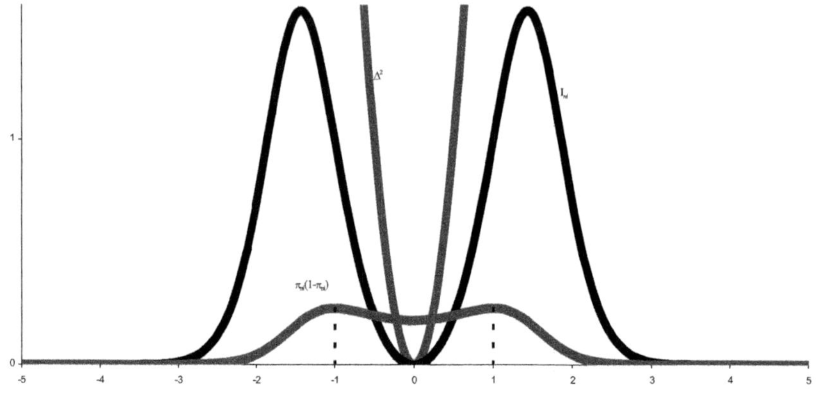

Figure 17-3. Item information function for the SSLM

(2) The HCM with a variable unit ρ_i.

The HCM with a variable unit is defined as

$$\pi_{ni} = \frac{\cosh(\rho_i)}{\cosh(\rho_i) + \cosh(\beta_n - \delta_i)} . \tag{18}$$

Therefore

$$\Delta(t) = \frac{\frac{d}{dt}\cosh(\beta - \delta_i)}{\cosh(\beta - \delta)_i} = \frac{\sinh(\beta - \delta_i)}{\cosh(\beta - \delta_i)} = \tanh(\beta - \delta_i), \tag{19}$$

giving

$$I_{ni} = \pi_{ni}(1 - \pi_{ni})\tanh^2(\beta - \delta_i) . \tag{20}$$

Figure 17-4 shows the components of equation (20) as well as the information function. Again, the first component of the information function gives it twin peaks, and the second component takes the value of 0 at $\beta = \delta_i$.

(3) The PARELLA model with a variable unit ρ_i.

The PARELLA model with a variable unit is defined as

$$\pi_{ni} = \frac{\rho_i^2}{\rho_i^2 + (\beta_n - \delta_i)^2} \; . \tag{21}$$

This model has the special feature that the unit parameter ρ_i is a scale parameter (Luo, 1998). Therefore, unlike the HCM and the SSLM, this parameter is not a property of the data independently of the scale. This has consequences for the information function.

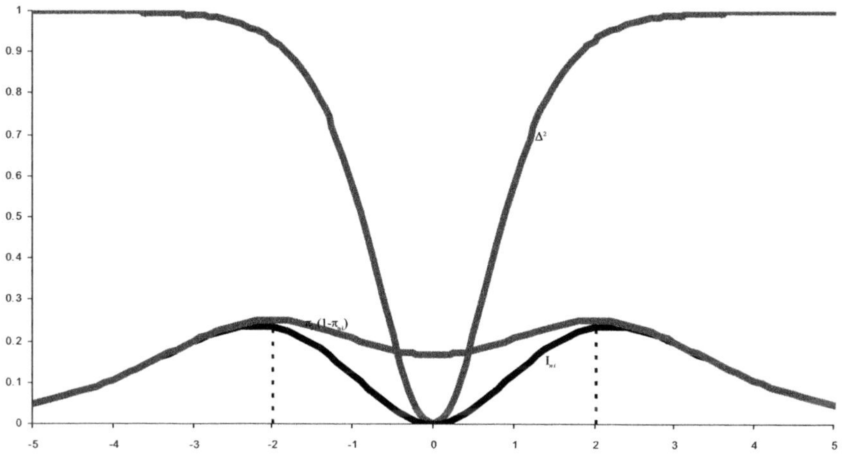

Figure 17-4. Item information function for the HCM

From (16),

$$\Delta(\beta - \delta_i) = \frac{\frac{d}{d\beta}(\beta - \delta_i)^2}{(\beta - \delta_i)^2} = \frac{2\beta - \delta_i}{(\beta - \delta_i)^2} = \frac{2}{\beta - \delta_i} \; ; \tag{22}$$

Therefore

$$\begin{aligned} I_{ni} &= f(\pi_{ni})\Delta^2(\beta_n - \delta_i) \\ &= \frac{\rho_i^2}{\rho_i^2 + (\beta_n - \delta_i)^2} \cdot \frac{(\beta_n - \delta_i)^2}{\rho_i^2 + (\beta_n - \delta_i)^2} \cdot \frac{4}{(\beta_n - \delta_i)^2} \\ &= \frac{4\rho_i^2}{[\rho_i^2 + (\beta_n - \delta_i)^2]^2} \end{aligned} \tag{23}$$

Unlike the SSLM and the HCM, when $\beta_n = \delta_i$, the information is a maximum, $4/\rho_i^2$. For example, if the arbitrary unit is given a value of 1, then the maximum information has the value 4 for that scale. In addition, and again unlike the SSLM and the HCM, the information function of the PARELLA model above is single-peaked. Figure 17-5 shows the two components of the information function and the information function for the case that $\rho_i = 1$. It is noted, however, that in the general form of the PARELLA (Hoitink, 1990); $\pi_{ni} = 1/(1 + (\beta_n - \delta_i)^{2\gamma})$, the information function can be single peaked or twin-peaked, depending on the value of the structural parameter γ: it is twin-peaked when $\gamma > 1$ but single-peaked when $\gamma \leq 1$.

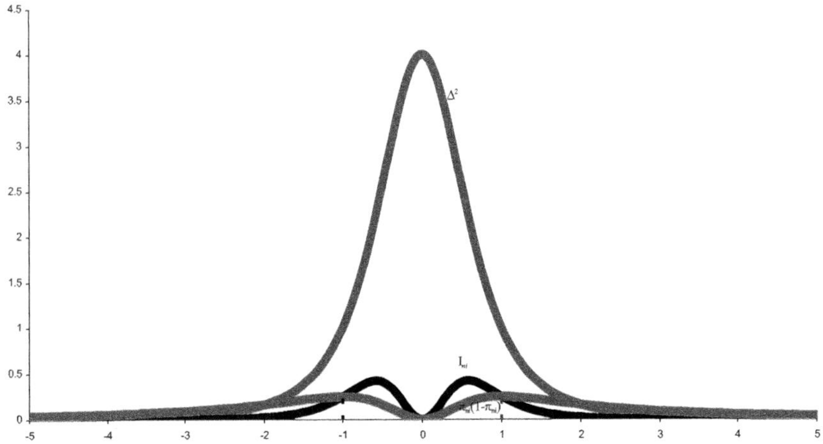

Figure 17-5. Item information function for the PARELLA when $\gamma \leq 1$.

It can be seen from the examples above that the point at which the item information function is a maximum deviates from the threshold points because of the effect of the component $\Delta^2(\beta_n - \delta_i)$ in Equation (12). This component, though independent of the unit, depends on the distance between the person and the location and the operational function. If the operation function Ψ in Equation (3) is so chosen so that it is a constant, in particular say $[\Delta(\beta-\delta_i)]^2 \equiv 1$, then it leads to an unfolding model in which the item information function has the maximum information at the threshold points $\{|\beta_n - \delta_i| = \rho_i\}$.

(4) Absolute Logistic Model (ALM) with a variable unit ρ_i.

Note how relatively simple it is to specify a new model given the general form of Eq. (3): All that is required is that the function of the distance between the location of the person and the item satisfies the straightforward properties of being positive (P1), monotonic in the positive domain (P2), and symmetrical about the origin (P3). The absolute function satisfies this property.

Let the operational function $\Psi(\beta-\delta_i) = \exp(|\beta-\delta_i|)$. Then

$$\pi_{ni} = \frac{\exp(\rho_i)}{\exp(\rho_i) + \exp[|\beta_n - \delta_i|]}; \quad (24)$$

$$\Delta(\beta-\delta_i) = \frac{\frac{d}{d\beta}\exp(|\beta-\delta_i|)}{\exp(|\beta-\delta_i|)} = \frac{d}{d\beta}|\beta-\delta_i| = \begin{cases} -1, & \beta-\delta_i < 0 \\ 1, & \beta-\delta_i > 0 \end{cases}. \quad (25)$$

Therefore, $[\Delta^2(\beta-\delta_i)] \equiv 1$ except for $\beta-\delta_i = 0$ and the value of $[\Delta^2(0)]$ is undefined. It is noted that no matter how small ε and irrespective of whether it is positive or negative, $[\Delta^2(\varepsilon)] \equiv 1$. That is, $\lim_{\varepsilon\to 0}[\Delta^2(\varepsilon)] = 1$. Therefore, without losing generality, we can define

$$[\Delta^2(0)] = 1. \quad (26)$$

Then

$$
\begin{aligned}
I_{ni} &= \pi_{ni}(1-\pi_{ni}) \\
&= \frac{\exp(\rho_i)}{\exp(\rho_i)+\exp[|\beta_n-\delta_i|]} \cdot \frac{\exp[|\beta_n-\delta_i|]}{\exp(\rho_i)+\exp[|\beta_n-\delta_i|]} \\
&= \frac{\exp(\rho_i+|\beta_n-\delta_i|)}{\{\exp(\rho_i)+\exp[|\beta_n-\delta_i|]\}^2}
\end{aligned}
\tag{27}
$$

Figure 17-6 shows the probabilistic function of the ALM for the value $\rho_i = 1$ and Figure 17-7 shows the corresponding information function. Note the discontinuity of the response function and the information function at $\beta = \delta_i$. Thus although this model has the attractive feature that the information is a maximum at the thresholds, it has a discontinuity at the location of the item. Whether that makes it impractical is yet to be determined.

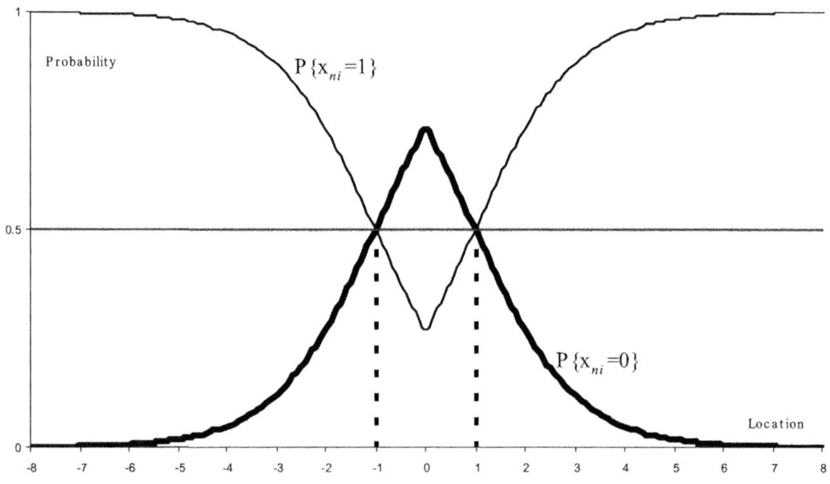

Figure 17-6. Probabilistic function of the Absolute Logistic Model (ALM)

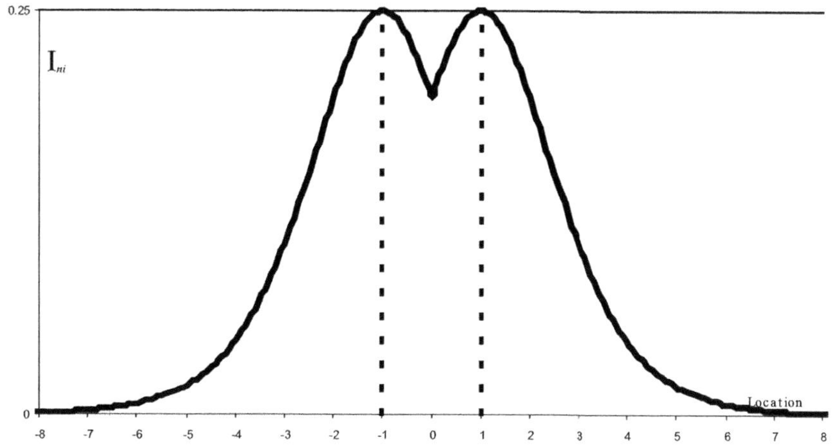

Figure 17-7. Item information function for the Absolute Logistic Model (ALM)

3. COMPARISON OF THE INFORMATION FUNCTIONS FOR CUMULATIVE AND UNFOLDING DICHOTOMOUS MODELS

The most commonly considered models with monotonic response functions are the Rasch (1961) model and the two and three parameter logistic models (Birnbaum, 1968). The maximum of the information is obtained at $\theta = \beta_n - \delta_i = 0$ for the Rasch model and the two-parameter logistic model, but not for three parameter logistic model (Birnbaum, 1968). These models respectively take the forms

$$P\{X_{ni} = x\} = \frac{1}{\lambda_{ni}} \exp\{(\beta_n - \delta_i)x\}, \tag{28}$$

$$P\{X_{ni} = x\} = \frac{1}{\lambda_{ni}} \exp\{\alpha_i(\beta_n - \delta_i)x\}, \tag{29}$$

and

$$P\{X_{ni} = x\} = \gamma_i + (1-\gamma_i)\frac{1}{\lambda_{ni}}\exp\{\alpha_i(\beta_n - \delta_i)x\}; \tag{30}$$

where λ_{ni} is a normalising factor, α_i is the discrimination of item *i*, and γ_i is a guessing parameter for item *i*. The reason the last model does not take its maximum information at the item location is that it is not symmetric, having a lower asymptote different from 0.

To reinforce the connection between the cumulative and unfolding models, note that in the monotonic response function the value of the item location is a threshold in the sense that when the person location is less than this value, then the probability for a positive response is less than 0.5, and vice versa (Bock and Jones, 1968). Thus it can be seen that in the Rasch model, and the two parameter logistic model, information is a maximum at their respective item thresholds.

The unfolding models have two item parameters – the item location and the item unit. As noted earlier, the item unit is defined by the two thresholds, and it is at these thresholds that one of the components of the information function is a maximum. The behaviours of the information functions for the dichotomous and cumulative and unfolding models are thus reconciled in that in both the information increases as the response function gets closer to the thresholds at which the probability of the response in one of the two categories is equal, that is, least certain. The exact point of the maximum information in the single-peaked functions depends then on the operational function. We defined a new operational function, the absolute logisitic, in which the information was indeed a maximum at the thresholds.

4. SUMMARY

Information functions are central in understanding the range in which a scale may be useful. In monotonic models, the information function with respect to an item for a person has the convenient property that in general, when a person is located at the same place as an item, then the item gives maximum information about the person. In contrast, in single-peaked models, the information function with respect to an item for a person has the inconvenient property that when a person is located at the same place as an item, then in general, the item gives a minimum, which may be zero, information about the person. This paper reconciles this apparent difference by showing that the location for maximising information in both the monotonic and unfolding models is where the probability of positive and negative responses are equal, that is, at the thresholds. However, in the case of the single peaked response models, in contrast to monotonic ones, there are two qualifications to this general feature. First, there are two thresholds at which the positive and negative responses arc equally likely – these define the range in which the positive response is more likely. Therefore the information function is generally twin-peaked. Second, unfolding models in general also involve a function of the person-item location. The information function can be resolved into a component defined by each, and it is the form of this function which moderates the maximum value of the information function so that its maximum is not at the thresholds. By constraining this component, a new model which gives maximum information at the thresholds is derived. However, this model has the inconvenient property that it is discontinuous at the location of the item.

5. REFERENCES

Andrich, D. (1988). The application of an unfolding model of the PIRT type to the measurement of attitude. *Applied Psychological Measurement,* 12, 33-51.

Andrich, D. (1995). Hyperbolic cosine latent trait models for unfolding direct responses and pairwise preference. *Applied Psychological Measurement,* 19, 269-290.

Andrich, D. (1996). A hyperbolic cosine latent trait model for unfolding polytomous responses: Reconciling Thurstone and Likert methodologies. *British Journal of Mathematical and Statistical Psychology,* 49, 347-365.

Baker, F. B. (1992) *Item response theory: parameter estimation techniques*. New York: Marcel Dekker.

Birbaum, A. (1968). Some latent trait models and their use in inferring an examinee's ability. In Lord, F. M. and Novick, M. R. (eds.), *Statistical theories of mental test scores* (pp. 397-472). Reading, MA:Addison-Wesley.

Bock, R.D. & Jones, L.V. (1968). *The measurement and prediction of judgement and choice.* San Francisco: Holden Day.

Böckenholt, U. & Böckenholt, I. (1990). Modeling individual differences in unfolding preference data: A restricted latent class approach. *Applied Psychological Measuremen,* 14, 257-266.

DeSarbo, W. S. (1986). Simple and weighted unfolding threshold models for the spatial representation of binary choice data. *Applied Psychological Measuremen,* 10, 247-264.

Fisher, R. A. (1956). *Statistical methods and scientific inference.* Edinburgh: Oliver and Boyd.

Hoijtink, H. (1990). *PARELLA: Measurement of latent traits by proximity items.* University of Groningen, The Ncthrlands.

Hoijtink, H. (1991). The measurement of latent traits by proximity items. *Psychometrika.* 57. 383-397.

Laughlin,J. E. & Roberts, J. S. (1999). Optimal fixed length test designs for attitude measures using graded agreement response scales. Paper presented at the annual meeting of Psychometric Society, Kansas University.

Lord, F. M. (1952). A theory of test scores. *Psychometric Monograph, No. 7.*

Lord, F. M. (1953). An application of confidence intervals and of maximum likelihood to the estimation of an examinee's ability. *Psychometrika.* 18. 57-75.

Luo, G. (1998). A general formulation for unidimensional unfolding and pairwise preference models: Making explicit the latitude of acceptance. *Journal of Mathematical Psychology*, 42, 400-417.

Luo, G. (2000). The JML estimation procedure of the HCM for single stimulus responses. *Applied Psychological Measurement*, 24, 33-49.

Luo, G. (2001). A class of probabilistic unfolding models for polytomous responses. *Journal of Mathematical Psychology*, 45, 224-248.

Luo,G., Andrich, D. & Styles, I. (1998). The JML estimation of the generalized unfolding model incorporating the latitude of acceptance parameter. *Australian Journal of Psychology*, 50, 187-198.

Nicewander, W. A. (1993). Some relationships between the information function of IRT and the signal/noise and reliability coefficient of classical test theory. *Psychometrika.* 58. 134-141.

Rao, C. R. (1973). *Linear statistical inference and its application (2nd edition).* New York: Wiley & sons.

Rasch, G. (1961). On general laws and the meaning of measurement in psychology. In J. Neyman (Ed.). *Proceedings of the Fourth Berkeley Symposium on Mathematical Statistics and Probability. IV*, 321-334. Berkeley CA: University of California Press.

Roberts, J. S., Lin, Y. & Laughlin, J. E. (1999). Computerized adaptive testing with the generalized graded unfolding model. Paper presented at the annual meeting of Psychometric Society, Kansas University.

Samejima, F. (1969). Estimation of latent ability using a response pattern of graded scores. *Psychometrika Monograph*, 1969, 34 (4, Part 2).

Samejima, F. (1977). A method of estimating item characteristic functions using the maximum likelihood estimate of ability. *Psychometrika*, 42, 163-191.

Samejima, F. (1993). An approximation for the bias function of the maximum likelihood estimate of a latent variable for the general case when the item responses are discrete. *Psychometrika*, 58, 115-138.

Sherif, M. and Sherif, C. W. (1967). *Attitude, Ego-involvement and Change*. New York: Wiley.

Theunissen, T. J. J. M. (1985). Binary programming and test design. *Psychometrika*, 50, 411-420.

Verhelst, H. D. and Verstralen, H. H. F. M. (1993). A stochastic unfolding model derived from partial credit model. *Kwantitatieve Methoden*. 42, 93-108.

6. APPENDIX

6.1 Relationship of the information function used in this paper to the original expression of Fisher (1956) for any dichotomous random variable

In general, for random variable X with probabilistic distribution $P\{X = x \mid \theta\}$, the information function defined by Fisher (1956) is the function

$$I(\theta) = E[(\tfrac{\partial}{\partial\theta}\log P\{X = x \mid \theta\})^2] \quad \text{(A1)}$$

In comparing Equation (21) with the derivation of Equation (9) in the paper, it can be seen that emphasis in the two expressions is different.

Lemma 1. Let X be a dichotomous random variable, $X = 0, 1$. Let

$$p(\theta) = P\{X = 1 \mid \theta\} \quad \text{(A2)}$$

and let

$$q(\theta) = 1 - p(\theta) = P\{X = 0 \mid \theta\}; \quad \text{(A3)}$$

then

$$E[(\tfrac{\partial}{\partial\theta}\log P\{X = x \mid \theta\})^2] = E[-\tfrac{\partial^2}{\partial\theta^2}\log P\{X = x \mid \theta\}] = \frac{[p'(\theta)]^2}{p(\theta)q(\theta)}. \quad \text{(A4)}$$

Proof. It is evident that

$$\frac{dp(\theta)}{d\theta} + \frac{dq(\theta)}{d\theta} \equiv p'(\theta) + q'(\theta) = 0;$$

$$[p'(\theta)]^2 - [q'(\theta)]^2 = [p'(\theta) + q'(\theta)][p'(\theta) - q'(\theta)] = 0;$$

$$\frac{d^2 p(\theta)}{d\theta^2} + \frac{d^2 q(\theta)}{d\theta^2} \equiv p''(\theta) + q''(\theta) = 0.$$

Therefore,

$$
\begin{aligned}
&E[(\tfrac{\partial}{\partial\theta}\log P\{X=x\,|\,\theta\})^2] \\
&=[\frac{p'(\theta)}{p(\theta)}]^2 p(\theta)+[\frac{q'(\theta)}{q(\theta)}]^2 q(\theta) \\
&=\frac{q(\theta)[p'(\theta)]^2+p(\theta)[q'(\theta)]^2}{p(\theta)q(\theta)} \\
&=\frac{[p'(\theta)]^2-p(\theta)[p'(\theta)]^2+p(\theta)[q'(\theta)]^2}{p(\theta)q(\theta)} \\
&=\frac{[p'(\theta)]^2-p(\theta)\{[p'(\theta)]^2-[q'(\theta)]^2\}}{p(\theta)q(\theta)} \\
&=\frac{[p'(\theta)]^2}{p(\theta)q(\theta)};
\end{aligned}
$$

and

$$
\begin{aligned}
&E[-\tfrac{\partial^2}{\partial\theta^2}\log P\{X=x\,|\,\theta\}] \\
&=[-\tfrac{\partial^2}{\partial\theta^2}\log p(\theta)]\cdot p(\theta)+[-\tfrac{\partial^2}{\partial\theta^2}\log q(\theta)]\cdot q(\theta) \\
&=[-\frac{d}{d\theta}\frac{p'(\theta)}{p(\theta)}]\cdot p(\theta)+[-\frac{d}{d\theta}\frac{q'(\theta)}{q(\theta)}]\cdot q(\theta) \\
&=\frac{[p'(\theta)]^2-p(\theta)p''(\theta)}{p(\theta)}+\frac{[q'(\theta)]^2-q(\theta)q''(\theta)}{q(\theta)} \\
&=\frac{q(\theta)[p'(\theta)]^2-p(\theta)q(\theta)p''(\theta)+p(\theta)[q'(\theta)]^2-p(\theta)q(\theta)q''(\theta)}{p(\theta)q(\theta)} \\
&=\frac{q(\theta)[p'(\theta)]^2+p(\theta)[q'(\theta)]^2-p(\theta)q(\theta)[p''(\theta)+q''(\theta)]}{p(\theta)q(\theta)} \\
&=\frac{q(\theta)[p'(\theta)]^2+p(\theta)[q'(\theta)]^2}{p(\theta)q(\theta)} \\
&=\frac{[1-p(\theta)][p'(\theta)]^2+p(\theta)[q'(\theta)]^2}{p(\theta)q(\theta)} \\
&=\frac{[p'(\theta)]^2-p(\theta)\{[p'(\theta)]^2-[q'(\theta)]^2\}}{p(\theta)q(\theta)} \\
&=\frac{[p'(\theta)]^2}{p(\theta)q(\theta)}.
\end{aligned}
$$

Chapter 18

PAST, PRESENT AND FUTURE: AN IDIOSYNCRATIC VIEW OF RASCH MEASUREMENT

Trevor G. Bond
School of Education, James Cook University

Abstract: This chapter traces the developments in Rasch measurement, and its corresponding refinement in both its application and programs to compute pertinent item and person parameters. The underlying principles of conjoint measurement are discussed, and its implications for education and research in social sciences are highlighted.

Key words: test design, fit, growth in thinking, item difficulty, item estimates, person ability, person estimates, unidimensional, multidimensional, latent trait

1. SERENDIPITY

How was it that my first personal use of Rasch analysis was in London in front of a BBC 2 micro-computer using a program called *PC-Credit* (Masters & Wilson, 1988) which sat on a five and a half inch floppy disc? Data was typed in live – there was no memory to which write a data file. Hit the wrong key, once in 35 items for 160 cases and, "Poof!" – all gone. Mutter another naughty word and start again. Since then I have had access to even earlier Rasch software – Ben Wright inadvertently passed on a Mac version of Mscale to me when he saved an output file from Bigsteps onto a Mac formatted disc. I had bumped into David Andrich as well as Geoff Masters and Mark Wilson at AARE from time to time. I had heard already about the Rasch model because when I wrote to ACER about the possibility of publishing my test of formal operational thinking, I was advised to get

S. Alagumalai et al. (eds.), Applied Rasch Measurement: A Book of Exemplars, 329–341.

some evidence from Rasch analysis to help make my case. (I believe a certain John Keeves was the head at ACER at that time.) Well sure, we might all live in the same country, but Melbourne is close to 3000 km from Townsville while flying to Perth then took most of a day – and most of a week's pay as well.

Masters and Wilson sat in on my AARE presentation at Hobart in 1984 – I was trying to make sense of BLOT scaling using Ordering Theory (Bart & Airasian, 1974; Airasian, Bart & Greaney, 1975). They tried to tell me that the Rasch model was the answer to my prayers. Well, at least I had two more than otherwise in the audience, or were they trying to get just one more in the audience for the Rasch workshop they were running later in the conference? So two years later as I was planning my first sabbatical at King's College (University of London), to work on formal operational thinking tests, Michael Shayer suggested I bring a Rasch program for the data analysis. He knew a fellow called Andrich at Murdoch, so why didn't I just drop round and ask for something that would run on his BBC2? Well Geoff had already spent half a day with me running analysis of more than 900 BLOT cases and said the test looked pretty good: something about estimates, and logits, and something else about fit. Yeah, whatever. Very nice and thanks.

Geoff Masters said that he had a version of *PC-Credit* (Masters & Wilson, 1988) somewhere that should run on the BBC machine – he would post it to me c/- Shayer, but he couldn't recall why the name Shayer rang a bell. Never mind. We even had an email link between Townsville and London (via bitnet and mainframe terminals) in '87! So, back to line one – that's how it all started. Masters later remembered he was impressed reading Shayer's *Towards a Science of Science Teaching* (Shayer & Adey, 1981) that he picked up by chance at the University of Chicago bookshop a year or two earlier. Ah, serendipity. PC-Credit provided a rather basic item-person map, item and person estimates and precision details as well as fit statistics as v, q and t. That's van der Wollenberg's q by the way. I know because I tracked down an article (Wollenberg, 1982) and read about it, then figured that every body else in psychometrics knew about the t distribution, so I would have an easier row to hoe if I just talked about that. What could be easier: $p<.05$=success; $p>.05$=failure. Thirty-four successes out of 35 on the BLOT; 12/12 on the PRT; 47/47 for a joint analysis - a little bit of common person equating thrown in, and it was a very successful sabbatical. This Rasch analysis is obviously pretty good stuff. It found out (finally) how good my *BLOT* test was!

1.1 Does everything fit?

As an independent learner with *Best Test Design* and *Rating Scale Analysis* to one side, I had learnt a simple rule: fit was everything. Good fit meant good items – misfit mean problem items. My only 'problem' was that in our Piagetian research at James Cook, I was not able to find misfitting items. It did help that *Quest* (Adams & Khoo, 1993) gave me a table four sorts of fit indicators – infit and outfit, both untransformed and standardized - as well as a fit graph with a pair of tram-tracks (two parallel lines) which always contained the asterisks that represented the items. I could always find some evidence that said the test fitted. Given my focus on indicators of Piagetian cognitive development, I rarely looked at the fit statistics for the persons – it was the underlying construct that counted, and the construct was represented by the items (what wonderful naïveté).

But maybe the fit statistics didn't really work properly. While the development of the *BLOT* and Shayer's Pendulum task (Piagetian Reasoning Task– *PRTIII*; Shayer, 1978), had benefited from the earlier intelligent use of true-score statistics, our innovative Partial Credit analyses of Piagetian interview transcripts used Rasch analysis for the first (and practically final) analysis of the data set. No misfit there, either. I started to wonder, "Will all our items always fit? More or less first up?" That hardly seemed right. Perhaps the fit stats are there to give users (me, in particular) a false sense of security. After all, everyone knew that you couldn't make good quantitative indicators for Piagetian cognitive development –American developmentalists had made the empirical disproof of Piagetian theory almost an art form. And you can be sure that I didn't want to look too closely in case all the attention Rasch-based research was getting at Jean Piaget Society meetings in the US was based on over-ambitious claims supported by fit statistics that wouldn't really separate a single latent trait from a mish-mash of responses.

In Chicago for a JPS meeting, I dragged prominent Piagetian, Terrance Brown down to Judd Hall where we discussed with Ben Wright the results that would eventually be published (as Bond, 1995a,b and Bond & Bunting, 1995) in the *Archives de Psychologie* – the very Geneva-based journal over which Piaget himself had earlier reigned as editor for decades. Terry knew Ben from years earlier when he worked at the Chicago medical school. Terry and I had met in Geneva at a Piaget conference and I knew Ben through the Australian links that tied Rasch measurement to Chicago (just see all those Australian names scattered thoughout this *festschrift* to John Keeves). We argued about the data, the results, the possible interpretations and most particularly, how Piagetian epistemological theory was implicated in the research and what sort of impact such results could have for Piagetian theory more broadly. I mentioned casually, *en passant*, that, from Ben's comments, these results then seemed to be more than just pretty good. And, then I revealed that each really was close to a first attempt at Rasch analysis of Piagetian derived tasks and tests. I hestitated, then ventured to ask Ben, if, in his experience, Rasch measurement was usually so straight forward.

Well, apparently not. Others often work years to develop a good test. Some give up after their devotion to the task produces a mere handful of questions that come up to the standard required by the Rasch model. In other areas successful Rasch-based tests are put together by whole teams of researchers dedicated to the task. Our results seemed the exception, rather than the rule – the fit statistics did discriminate against items that other researchers thought should be in their tests. When we started discussing the development of the Piagetian testing procedures used in the research, I dragged my dog-eared and annotated copy of *The Growth of Logical Thinking from Childhood to Adolescence* (*GLT*, Inhelder & Piaget, 1958) out of my brief case and outlined how the logical operational structures outlined on pages 293-329 became the starting point for the development of *Bond's Logical Operations Test* (Bond, 1976/1995) and how the ideas from *GLT*'s chapter four had been incorporated in the *PRTIII* developed by the King's team. (Shayer, 1976). I described how Erin Bunting had sweated drops of blood combing the same chapter over and over to develop her performance criteria (Bond & Bunting, 1995, pp.236-237; Bond & Fox, 2001, pp.94-96) for analysing interview transcripts garnered from administering the pendulum problem in free-flowing, semi-structured investigatory sessions conducted with individual high school students. Erin developed 45 performance criteria across 18 aspects of children's performances – ranging from failing to order correctly the pendulum weights or string lengths (typical of preschoolers' interaction with 'the swinging thing') to logically excluding the effect of pushes of varying force on the bob (a rare enough event in high school science majors) - criteria of such subtle and detailed richness had not been assembled for this task before (*e.g.* Kuhn, & Brannock, 1977; Somerville, 1974). In each case where we had doubts we recurred to the original French text and consulted the 50 year old original transcripts for the task held in the *Archives Jean Piaget* in Geneva.

Ben was quick to see the implication – we had the benefit of the ground-breaking work of one of the greatest minds of the twentieth century as the basis for our research (see Papert, 1999 in *Time. The Century's Greatest Minds*). We had nearly sixty full length books, almost 600 journal articles and a library of secondary research and critique to guide our efforts (Fondation Archives Jean Piaget, 1989). With a life time of Piaget's epistemological theory and a whole team's empirical research to guide us, how could we expect less than the level of success we had obtained – even first up? In contrast, many test developers had to sit around listening to the expert panel opining about the nature of the latent trait being tested, and the experts often had very little time for the empirical evidence that appeared to disconfirm their own prejudices. Our research at James Cook University has continued in that tradition: Find the Piaget text that gives chapter and verse of the theoretical description and empirical investigation of an interesting aspect of children's cognitive development. Take what Piaget says therein very seriously. Tear the appropriate chapter apart searching for every little nuance of description of the Genevan children's task performances from half a century ago. Encapsulate them into data-coding procedures and do your darnedest to reproduce as faithfully as possible the very essence of Piaget's insights. Decide when you are ready to put your efforts to the final test before typing "estimate <return>" as the command line. Be ready to report the misfits.

Our research results at James Cook University (see Bond, 2001; Bond, 2003; Endler & Bond, 2001) convince me that the thoughtful application of Georg Rasch's models for measurement to a powerful substantive theory such as that of Jean Piaget can lead to high quality measurement in quite an efficient manner. No wonder I publicly and privately subscribe to the maxim of Piaget's chief *collaborateur*, Bärbel Inhelder, "If you want to get ahead, get a theory." (Karmiloff-Smith & Inhelder, 1975)

2. FIT: COMPARISONS BETWEEN TWO MATRICES

While we expect that there will be variations in item difficulty estimates and person ability estimates representing the extent of a developmental continuum, our primary focus tends to be on the extent to which the relationships between the data could be held to imply one latent trait (Piagetian cognitive development), rather than many. To that end I encourage my research students to do all the revision necessary to their data coding before those data meet the Rasch software for the first time. Regardless of what *ex post facto* revisions are made as a result of the Rasch evidence, the original data analysis results must be recorded and explained in each research student's dissertation. In light of this focus, the summaries fit statistics and then the fit details for that item, and eventually, for each person are paramount. Of course, some revisions to scoring criteria, item wording and the like might follow, but each student's conceptualisation of the application of Piagetian theory to empirical practice is 'on the line' when the first Rasch analysis of the new data file is executed. Quite a step from the more pragmatic data analysis practices often reported.

Of course, the key to the question of data fit to the Rasch model's requirements for measurement lies in the comparison of two matrices. The first is the actual person-item matrix of 1s and 0s (in the case of dichotomous responses); that is, the data file that is submitted for analysis. The raw scores for each person and each item are the sufficient statistics for estimating item difficulties and person abilities. Those raw scores (in fact, the actual score/possible score decimal fractions) are iterated until the convergence criterion is reached yielding the array of person and item estimates (in logits) which provides a parsimonious account of the item and person performances in the data. These estimates for items and persons are then used to calculate the expected response probabilities based on those estimates: if the Rasch model could explain a data set collected with persons of those abilities interacting with items of those difficulties what would this (second) resultant item/person matrix look like?

That is the basis for our Rasch model fit comparison: the *actual* data matrix of 1s and 0s provides the information for the item/person estimations; those item person estimates are used to calculate the *expected* response probabilities for each item/person interaction. If we remove the information accounted by the model (*i.e.* the *expected* probabilities matrix) from the information collected with these items from these persons (*i.e.* the *actual* data matrix), is the matrix of *residual* information (actual – expected = residual) for any item or person too large to ignore? Well that should be easy, except . . . Except, there is always a residual - in every item/person cell. The actual data matrix (the data file) has entries of 1s or 0s (or sometimes 'blank'). The Rasch expected probability matrix always has a decimal fraction - never 1 or 0. That's the essence of a probabilistic model, of course. Even the cleverest child *might* miss the easiest item, and the child at the other end of the scale might have heard the answer to the hardest item on the way to school that very morning. That there must always be some fraction left over for every residual cell is not a concept that comes easily to beginners. Surely, if the person responds as predicted by the Rasch model, it should mean that the person scores 1 for the appropriate items and 0 for the rest. After all, a person must score either 1 or 0; right or wrong (for dichotomous items).

Having acknowledged that something *must* always be left over, the question is, "How much is ok?" Is the *actual* sufficiently like the *expected* that we can assume that the benefits of the Rasch measurement model do apply for this instantiation of the latent trait (*i.e.* this matrix of item / person interactions.) When the summary of the residuals is too large for an item (or a person), we infer that the item (or person) has *actually* behaved more erratically (unpredictably) than the model *expected* for an item (or a person) at that estimated location. If it is an erratic item, and we have plenty of items, we tend to dump the item. We don't often seem struck by the incongruity of dumping poorly performing items until we think seriously of applying the same principle to the misfitting persons.

For our JCU research students, dumping a poorly performing item is more akin to performing an amputation. Every developmental indicator went into the test or the scoring schedule for the task, because a genuine clever person (Prof. Piaget, himself) said it should be in there . . . and the research student was clever enough to be able to develop an instantiation of that indicator in the task. The response to misfit then is not to dump the item, but to attempt to find out what went wrong. That's why I oblige my students to be sure that their best efforts are reflected in the data set before the software runs for the first time. Those results must be reported and explained. Erratic performances by the children need the same sort of theory-driven attention – and often reveal aspects of children's development that were suspected / known by the child's teacher but waiting to be discovered empirically by the research candidate. It is often both practically and theoretically useful to suspend judgement on those misfitting items (by temporarily omitting them from a reanalysis) to see if person performance fit indicators improve. While we easily presume that the improved (items omitted) scale then works better, we cannot be sure that is the case until we re-administer the scale without those items. We have a parallel issue with creating Rasch measures from rating scales. A new instrument with three Likert-style response options might not produce the same measurement characteristics as were discovered when five response categories were collapsed into three during the previous Rasch analysis.

As a developmentalist, I have rarely been concerned when the residuals that remained were too small: less than –2.0 as t or z in the standardized form, or much less than .8 as mean squares. It seemed quite fine to me that performance on a cognitive developmental task was less stochastic than Rasch allowed – that success on items would turn quickly to failure when the cognitive developmental engine had reached its limits. But I have learned not to be too tolerant of items, in particular, which are too Guttman-like. It is likely that a number of the indicators in Piagetian schedules are the logical precursors of later abilities; those abilities incorporate the earlier pre-requisites into more comprehensive, logically more sophisticated developmental levels. This seems to be a direct violation of the Rasch model's requirement for local independence of items. We might try to reorganize those indicators into partial credit bundle format as recommended by Mark Wilson.

We are now much more aware of the limitations of our usual misfit indicators. A test made up of two discrete sub-tests (*e.g.* language and maths) might fit better than a straight test of either trait. Mean square statistics indicate the *size* of mis-fit while the transformed versions (z or t) indicate the probability of that much misfit. While transformed fit stats are easy to interpret from a classical test background ($p<.05$ is acceptable, while $p>.05$ is not), we know they tend to be too accepting for small samples but too rejecting for large N. Moreover, the clamour for indicators of effect size in educational and psychological research reminds us to focus more on significance in substantive rather than statistical terms. While many of us just mouth (and use) oft-repeated 'rules-of-thumb' for rejecting item and person performances due to misfit, Richard Smith (*e.g.* 1991, 2000) has studied our usual residual fit statistics over a long period to remind us of exactly what sorts of decisions we are making when we use each fit statistic. Refereeing papers for European journals, in particular, reveals to me that our Rasch colleagues in Europe usually require a broader range of fit indicators than we have in the US and Australia, and are often more stringent in the application of misfit cut-offs. *Winsteps* (Linacre & Wright, 2000) software now gives us the option of looking at the factor structure of the residual matrix – *i.e.* after the Rasch model latent trait (or factor) has been removed from the data – to help us to infer whether a single dimension or more than one might be implied by the residual distribution. *RUMM* (Rummlab, 2003) can divide the person sample into ability sub-groups and plot the mean location of each sub-group against the item characteristic curve for each item, providing a graphical indication of the amount of misfit and its location (on average). The *ConQuest* (Adams, Wu & Wilson, 1998) authors allow us to examine whether a unidimensional or multidimensional latent trait structure might make a more parsimonious account of test data – a property that has been well exploited by the *PISA* international educational achievement comparisons.

3. CONJOINT MEASUREMENT

In the terms in which most of us want to analyse and report our data and tests, we probably have enough techniques and advice on how do build useful scales for the latent traits that interest us and how to interpret the person measures - as long as the stakes are not too high. Of course, being involved in high stakes testing should make all of us a little more circumspect about the decisions we make. But, that's why we adhere to the

Rasch model and eschew other less demanding models (even other IRT models) for our research. But if we had ever been satisfied with the status quo in research in the human sciences, the Rasch model would merely be a good idea, not a demanding measurement model that we go well out of our ways to satisfy.

I had been attracted to the Rasch model for rather pragmatic reasons – I had been told that it was appropriate for the developmental data that were the focus of my research work and that it would answer the questions I had when other techniques clearly could not. It was only later, as I wanted to defend the use of the Rasch model and then to recommend it to others that I became interested in the issues of measurement and philosophy (and scientific measurement, in particular). How fortunate for me that I had received such good advice, all those years ago. Seems the Rasch model had much more to recommend it that could have possibly been obvious to a novice like me. The interest in philosophy and scientific knowledge has plagued me for a long time, however. My poor second-year developmental psychology teacher education students were required to consider whether the world we know really exists as such (empiricism) or whether it is a construction of the human mind that comes to know it (rationalism). Bacon and Locke *v.* Descartes and Kant. Poor students.

It's now exactly quarter of a century since Perline, Wright and Wainer (1979) outlined how Rasch measurement might be close to the holy grail of genuine scientific measurement in the social sciences - additive conjoint measurement as propounded by R. Duncan Luce (*e.g.* Luce & Tukey, 1964), David Andrich's succinct SAGE paperback (Andrich, 1988) even quietly (and not unwittingly) invoked the title, *Rasch models for measurement* (*sic*). In 1992, however, Norman Cliff decried the much awaited impact of Luce's work as 'the revolution that never happened', although, in 1996, Luce was writing about the 'ongoing dialogue between empirical science and measurement theory'. To me, the 'dialogue' between mathematical psychologists and the end-users of data analysis software has been like the parallel play that Piaget described in pre-schoolers: they talk (and play) in each other's company rather that *to* and *with* each other. Discussion amongst Rasch practitioners at conferences and online revealed that we thought we had something that no-one else had in the social sciences – additive conjoint measurement – a new kind of fundamental scientific measurement. We had long ago carefully and deliberately resiled from the S. S. Stevens (1946) view that some sort of measurement was possible with four levels of data, nominal, ordinal, interval and ratio; a view, we held, that allowed psychometricians to pose (unwarrantedly) as scientists. In moments of exuberance we even mentioned Rasch measurement and ratio scale in the same sentence.

4. OUR CURRENT CHALLENGES

Some of us were awakened from our self-induced torpor, by the lack of recognition for the unique role of Rasch measurement in the social sciences in Joel Michell's *Measurement in Psychology* (1999). Michell rehearsed many of the same criticisms of psychometrics that we knew by heart, but he did not come to our conclusions. In a review of his book for *JAM* (Bond, 2001) I took him to task for that omission. Well, in due course we come back to the issue of fit to the model. Luce's prescriptions for additive conjoint measurement require that the data matrix satisfies an hierarchical sequence of cancellation axioms. And while this might be demonstrated with the matrix of Rasch expected values – and not with values derived from 2- and 3- PL IRT models (*e.g.* Karabatsos, 1999; 2000), having an actual data matrix which does fit the Rasch model *does not* address the cancellation issues countenanced in Luce's axioms. If our goals had been just a little more modest and pragmatic, this revelation would not have alarmed us at all. But, while we might have issues of fit more or less covered for all practical purposes, by setting our measurement goals (impossibly?) high, the issue of model fit will haunt Rasch practitioners well into this millennium – like the albatross hung around the neck of Coleridge's ancient mariner, perhaps.

5. NOT JUST SERENDIPITY

John Keeves's address to the Rasch Measurement Special Interest Group of AERA in Chicago, 1997 prompted my recollections of the long and influential career that John has had in Australia – and internationally. Although I was not mentored into Rasch measurement directly by John, it was those he had mentored who directly influenced my adoption of Rasch models for measurement. In a discussion with him in the middle of this year, I pointed out that, for me, the success of our book (Bond & Fox, 2001) had come somewhat as a surprise. In spite of what some seem to think, I have never considered myself to be an expert in Rasch measurement in the way that others like Wright, Andrich, Linacre, Wilson, Adams and Masters have contributed to the model. To me the Rasch model has always been the means – and never the end-purpose – of our developmental and educational research at James Cook University. Of course the Rasch model is a very worthy object of research in its own right – and I value the work of the Rasch theoreticians very highly. But I think it is not just serendipity and geography that brings me the chance to write this chapter. John Keeves and I share an orientation to the *development* of knowledge in children; that children's development and their school achievement move in consonance. We both hold that the Rasch model provides the techniques whereby both cognitive development and school achievement might be faithfully measured and the relationships between them more clearly revealed. Professor John Keeves has contributed significantly to my research past and our research future.

6. REFERENCES

Adams, R.J., & Khoo, S.T. (1993). *Quest: The interactive test analysis system* [computer software]. Camberwell, Victoria: Australian Council for Educational Research.

Adams, R.J., Wu, M.L. & Wilson, M.R. (1998) *ConQuest: Generalised item response modelling software* [Computer software]. Camberwell: Australian Council for Australian Research.

Andrich, D. (1988). *Rasch models for measurement.* Newbury Park, CA: Sage.

Airasian, P. W., Bart, W. M. & Greaney, B. J. (1975) The Analysis of a Propositional Logic Game by Ordering Theory. *Child Study Journal,* 5, 1, 13-24.

Bart, W. M. & Airasian, P. W. (1974) Determination of the Ordering Among Seven Piagetian Tasks by an Ordering Theoretic Method, *Journal of Educational Psychology, 66,* 2, 277-284.

Bond, T.G. (1976/1995). *BLOT - Bond's logical operations test.* Townsville: James Cook University.

Bond, T.G. (1995a). Piaget and measurement I: The twain really do meet. *Archives de Psychologie,* 63, 71-87.

Bond, T.G. (1995b). Piaget and measurement II: Empirical validation of the Piagetian model. *Archives de Psychologie, 63,* 155-185.

Bond, T.G. & Bunting, E. (1995). Piaget and measurement III: Reassessing the *méthode clinique. Archives de Psychologie, 63,* 231-255.

Bond, T.G. (2001a) Book Review 'Measurement in Psychology: A Critical History of a Methodological Concept'. *Journal of Applied Measurement, 2*(1), 96-100.

Bond, T. G. (2001b). Ready for school? Ready for learning? An empirical contribution to a perennial debate. *The Australian Educational and Developmental Psychologist, 18*(1), 77-80.

Bond, T.G. (2003) Relationships between cognitive development and school achievement: A Rasch measurement approach, In R. F. Waugh (Ed.), *On the forefront of educational psychology.* New York: Nova Science Publishers (pp.37-46).

Bond, T.G. & Fox, C. M. (2001) *Applying the Rasch model: Fundamental measurement in the human sciences.* Mahwah, N.J.: Erlbaum.

Cliff, N. (1992). Abstract measurement theory and the revolution that never happened. *Psychological Science, 3*(3), 186 - 190.

Endler, L.C. & Bond, T.G. (2001). Cognitive development in a secondary science setting. *Research in Science Education, 30*(4), 403-416.

Fondation Archives Jean Piaget (1989) *Bibliographie Jean Piaget*. Genève: Fondation Archives Jean Piaget.

Inhelder, B., & Piaget, J. (1958). *The growth of logical thinking from childhood to adolescence* (A. Parsons & S. Milgram, Trans.). London: Routledge & Kegan Paul. (Original work published in 1955).

Karabatsos, G. (1999, April). Rasch vs. two- and three-parameter logistic models from the perspective of conjoint measurement theory. Paper presented at the Annual Meeting of the American Education Research Association, Montreal, Canada.

Karabatsos, G. (1999, July). Axiomatic measurement theory as a basis for model selection in item-response theory. Paper presented at the 32nd Annual Conference for the Society for Mathematical Psychology, Santa Cruz, CA.

Karabatsos, G. (2000). A critique of Rasch residual fit statistics. *Journal of Applied Measurement, 1*(2), 152 - 176.

Karmiloff-Smith, A. & Inhelder, B. (1975) If you want to get ahead, get a theory. *Cognition, 3*(3)195-212.

Keeves, J.P. (1997, March). International practice in Rasch measurement, with particular reference to longitudinal research studies. Invited paper presented at the Annual Meeting of the Rasch Measurement Special Interest Group, American Educational Research Association, Chicago.

Kuhn, D. & Brannock, J. (1977) Development of the Isolation of Variables Scheme in Experimental and "Natural Experiment" Contexts. *Developmental Psychology, 13*, 1, 9-14.

Linacre, J.M., & Wright, B.D. (2000). *WINSTEPS: Multiple-choice, rating scale, and partial credit Rasch analysis* [computer software]. Chicago: MESA Press.

Luce, R.D., & Tukey, J.W. (1964). Simultaneous conjoint measurement: A new type of fundamental measurement. *Journal of Mathematical Psychology, 1*(1), 1 - 27.

Masters, G.N. (1984). DICOT: Analyzing classroom tests with the Rasch model. *Educational and Psychological Measurement, 44*(1), 145 - 150.

Masters, G. N. & Wilson, M. R. (1988). *PC-CREDIT* (Computer Program). Melbourne: University of Melbourne, Centre for the Study of Higher Education.

Michell, J. (1999). *Measurement in psychology: Critical history of a methodological concept.* New York: Cambridge University Press.

Papert, S. (1999). Jean Piaget. *Time. The Century's Greatest Minds*. (March 29, 1999. No. 13, 74-75&78).

Perline, R., Wright, B.D., & Wainer, H. (1979). The Rasch model as additive conjoint measurement. *Applied Psychological Measurement, 3*(2), 237 - 255.

Rasch, G. (1960). *Probabilistic models for some intelligence and attainment tests.* Copenhagen: Danmarks Paedagogiske Institut.

Shayer, M. (1976) The Pendulum Problem. *British Journal of Educational Psychology,* 46, 85-87.

Shayer, M. & Adey, P. (1981) *Towards a Science of Science Teaching.* London: Heinemann.

Smith, R.M. (1991a). The distributional properties of Rasch item fit statistics. *Educational and Psychological Measurement, 51*, 541 - 565.

Smith, R.M. (2000). Fit analysis in latent trait measurement models. *Journal of Applied Measurement, 1*(2), 199 - 218.

Somerville, S. C. (1974) The Pendulum Problem: Patterns of Performance Defining Developmental Stages. *British Journal of Educational Psychology, 44,* 266-281.

Stevens, S.S. (1946). On the theory of scales of measurement. *Science, 103,* 677 - 680.

Wollenberg, A.L. (1982). Two new test statistics for the Rasch model. *Psychometrika, 47*, 2, 123-140.

Epilogue

OUR EXPERIENCES AND CONCLUSION

Sivakumar Alagumalai, David D Curtis and Njora Hungi
Flinders University

1. INFLUENCES ON RESEARCH AND METHODOLOGIES

John, through his insights into the Rasch model and support for collaborative work with fellow researchers, has guided our common understanding of measurement, the probabilistic nature of social events and objectivity in research. John has indicated in a number of his publications "the response of a person to a particular item or task is never one of certainty" (Keeves and Alagumalai, 1999, p.25). Parallel arguments have been expressed by leading measurement experts like Ben Wright, David Andrich, Geoff Masters, Luo Guanzhong, Mark Wilson and Trevor Bond.

> People *construct* their social world and there are creative aspects to human action, but this freedom will always be constrained by the structures within which people live. Because behaviour is not simply determined we cannot achieve deterministic explanations. However, because behaviour is constrained we can achieve probabilistic explanations. We can say that a given factor will increase the likelihood of a given outcome but there will never be certainty about outcomes. Despite the probabilistic nature of causal statements in the social sciences, much popular ideological and political discourse translates these into deterministic statements. (de Vaus, 2001, p.5)

We argue that any form of research, be it in the social sciences, education or psychology, needs to transcend popular beliefs and subjective ideologies. There are methodological similarities between objectivity in psychosocial

S. Alagumalai et al. (eds.), Applied Rasch Measurement: A Book of Exemplars, 343–346.

measurement and geometrical measurement, as all measurement is indirect, as abstract ideas and thinking are never observable in and of themselves (Fisher, 2000). Fisher (2000) further argues that any interpretation of the world, including those of hermeneutics (interpretation theory) is relevant to science.

In examining the probabilistic nature of social processes and, in particular, processes in education, we have come to understand better the importance of knowledge and justification, and thus epistemology. "Knowledge and justification represent positive values in the life of every reasonable person" (Audi, 2003, p.10). We note the importance of objective measurement and associated techniques in getting the research design and the instrument refined before making inferences.

> Well-developed concepts of knowledge and justification can play the role of ideals in human life; positively, we can try to achieve knowledge and justification in subjects that concerns us; negatively, we can refrain from forming beliefs where we think we lack justification, and we can avoid claiming knowledge where we think we can at best hypothesize. (Audi, 2003, p.11)

We have been privileged in our interactions with experts in this area to gain insights and an understanding of both the qualitative and quantitative aspects of what we are seeking to understand and explain. The pursuits of carefully conceived research studies undertaken by colleagues in this book highlight the approach that "data are acquired to subscribe to aspects of validity and to conform to the chosen model" (Andrich, 1989, p.15). Non-conformity of data is not an end point, but initiates further review of why it is so in light of current conception and knowledge.

2. OBJECTIVITY REVISITED

Two conditions are necessary to achieve the objectivity discussed above. "First, the calibration of measuring instruments must be independent of those subjects that happened to be used for calibration. Second, the measurement of objects must be independent of the instrument that happens to be used for measuring" (Wright and Stone, 1979, p.xii). Hence, object-free instrument calibration and instrument-free object measurement are crucial for generalisation beyond the instrument and for comparison. All contributors to this book have clearly highlighted the sample-free item calibration and the result is a test-free person measurement before proceeding to answer their major research questions in their respective studies.

In most of the studies, the variable in question began as a general idea of what was to be measured. This general idea was then given form by the test items. Thus, these items became operational definition (Wright and Stone, 1979, p.2) of the variable under examination. An understanding of the Rasch model and its application in various fields has helped this process of variable conception and refinement.

The Rasch model has enabled us to understand better the world in questioning deterministic judgements, the implications of using a flawed 'rubber-ruler' for measurement and the problems associated with using raw-scores. It has forced us to rethink how variables are conceived, the concept of dimensionality and why items and persons 'misfit' in a model.

The collaborative process of examining fellow Rasch users' work, participating in discussion groups and conferences, have provided us useful techniques in addressing issues of equating, biases, rater consistencies, separability of items and persons, bandwidth and fidelity. It has empowered us to consider research from the micro- to the macro-levels, from critically examining distractors of items and scales in a questionnaire to addressing causality and interactions in hierarchical models.

3. CONCLUSION

Most, if not all, the contributors have expanded their research interests beyond their contributions to this book. An insight into the Rasch model has challenged us to explore other disciplines and research paradigms, especially in the areas of cognition and cognitive neuroscience. Online testing, in particular adaptive testing and adaptive surveys, and the use of educational objects and simulations in education, are being examined in the light of the Rasch model. Conceptualising and developing criteria for stages and levels of learning are being examined carefully, with a view to gauging learning and to understand diversity in learning.

We are receptive to the emerging models and applications of the Rasch principles (see *Multidimensional item responses: Multimethod-multitrait perspectives* and *Information functions for the general dichotomous unfolding model*), and the challenge of making simple some of the axioms and assumptions. The exemplars in this book are contributions from beginning researchers who had been introduced to the Rasch model, and we testify to its usefulness. For some of us, our journey into objective measurement and the use of the Rasch model has just started, and we strive to continue the interests and passion of John in the use of the Rasch model beyond education and the social sciences.

> The advantages of using Rasch scaling to improve measurement are: (1) developing an interval scale; (2) the ease with which equating can be carried out; (3) that item bias can be detected; (4) that persons, items and raters are brought to a common scale that is independent of the specific situation in which the data were collected; and (5) estimates of error are calculated for each individual person, item and rater, rather than for the instrument as a whole. These advances in educational measurement have the capacity of helping to consolidate and extend the recent developments in theory in the field of education. (Keeves and Alagumalai, 1998, p.1242)

It is our firm belief that "what you cannot measure, you cannot manage and thus cannot change!"

4. REFERENCES

Andrich, D. (1989). Distinctions between assumptions and requirements in measurement in social science. In Keats, J.A., Taft, R.A., & Heath, S.H. (1989). *Mathematical and theoretical Systems*. North Holland: Elsevier-Science.

Audi, R. (2003). *A contemporary introduction to the theory of knowledge*. (2nd Ed). NY: Routledge.

de Vaus, D.A. (2001). *Research design in social research*. London: SAGE

Fisher, W.P. (2000). Objectivity in psychosocial measurement: What, why and how. *Journal of Outcome Measurement, 4*(2), pp/527-563.

Keeves, J.P. & Alagumalai, S. (1998). Advances in measurement in science education. In. Fraser, B.J. & Tobin, K.G. (1998). International handbook of science education. Dordrecht: Kluwer Academics

Keeves, J.P. & Alagumalai, S. (1999). New approaches to measurement. (Ed). In Masters, G.N. and Keeves, J.P. Advances in measurement in educational research and assessment. Amsterdam: Pergamon

Wright, B.D. & Stone, M.H. (1979). *Best test design*. Chicago: MESA Press.

Appendix

IRT SOFTWARE

1. COMPUTERS AND COMPUTATION

The use of computers in mathematics and statistics has been monumental in providing insights into complex manipulations that were reserved only to university researchers and specialists. The current speed of personal computers' microprocessors and optimised codes in programming languages, provide researchers, graduate students and teachers user-friendly graphical interfaces to input data and to make meaning of the outputs. The conceptually difficulty and practical challenges of the Rasch theory has been alleviated by advent of personal computers.

Traditional comparison of output data against 'standard tables' have been replaced with colour coded outputs with supporting popup help windows. Routines that took a couple of days to run, converge in seconds. As the demand for faster and more meaningful computational programs grow, item analysis program developers are pushed to cater to the demands of current users and prospective clients. Developers of computer programs for fitting item responses have utilised a number of techniques and processes to provide meaningful information. This section reviews some of the available programs, and directs the reader to demonstration/student versions and manuals (if available).

2. BIGSTEPS/WINSTEPS

BIGSTEPS, developed by John M. Linacre, analyses item response data using the Rasch measurement model. "A number of upgrades have been implemented in the recent years, and the focus has shifted to providing comprehensive, flexible, easy-to-use software. An early version of a Windows-95-native version of BIGSTEPS, called WINSTEPS, is available. For those requiring capacity beyond the BIGSTEPS limit of 32,000 persons, WINSTEPS now analyzes 1,000,000 persons. User access to control and output files is also improved. The program allows for a user input and control interface, and there is provision for immediate on-screen graphical displays. A free student/evaluation Rasch program, MINISTEP, is available for prospective users" (Featherman & Linacre, 1998).

Details of BIGSTEPS/WINSTEPS/MINISTEP are available at http://www.winsteps.com/index.htm

3. CONQUEST

ConQuest is a computer program for fitting item response and latent regression models. It provides a comprehensive range of item response models to users and the program produces marginal maximum likelihood estimates for the parameters of the models indicated below. A number of psychometric techniques, namely multifaceted item response models, multidimensional item response models, latent regression models, and drawing plausible values can be examined. It provides an integration of item response and regression analysis (Adam, Wu & Wilson, 1998).

ConQuest can provide fit for a multiple of models including the following:

- Rasch's Simple Logistic Model;
- Rating Scale Model;
- Partial Credit Model;
- Ordered Partition Model;
- Linear Logistic Test Model;
- Multifaceted Models;
- Generalised Unidimensional Models;
- Multidimensional Item Response Models; and
- Latent Regression Models.

A number of research studies, in this volume, have utilised ConQuest in performing item analysis, examining differential item functioning and exploring rater effects. Conquest can also be used for estimating latent correlations, testing of dimensionality and in drawing plausible values.

Details of CONQUEST and student/demonstration version are available at http://www.scienceplus.nl/scienceplus/main/show_pakketten_categorie.jsp?id=38

4. RASCAL

Rascal estimates the item difficulty and person ability parameters based on the Rasch model (one-parameter logistic IRT model) for dichotomous data. The Rascal output for each item include estimate of item parameter, a Pearson chi-square statistic, and the standard error associated with the difficulty estimate. The maximum-likelihood (IRT) scores for each person can also be easily produced. A table to convert raw scores to IRT ability-scores can be generated.

Details of RASCAL and student/demo version are available at http://www.scienceplus.nl/scienceplus/main/show_pakketten_categorie.jsp?id=38

5. RUMM ITEM ANALYSIS PACKAGE

The Rasch Unidimensional Measurement Model (RUMM) software package is a popular item analysis package used by a number of authors in this volume. It is a powerful to facilitate the teaching and learning of Rasch measurement theory. RUMM is Windows-based and has a highly interactive graphical user interface. The multicoloured buttons, and the coloured outputs enable ease of use and interpretation respectively. The graphical display output options include:

- Category characteristic curves;
- Item characteristic curves;
- Threshold probability curves;
- Item/person distributions for targeting;
- Item map;
- Threshold map; and

- Distractor analyses for multiple-choice items.

A number of analyses, by means of the pair-wise conditional algorithm, can be undertaken with RUMM software (RUMM Laboratory, 2003), and allows for the following:

- Analysis of multiple-choice items and distractor analyses;
- Analysis of polytomous items with equal and unequal categories;
- Anchoring some items to fix a scale;
- Deleting items and persons interactively;
- Rescoring items;
- Linking subset of items and combining items;
- Item factor, or facet analysis; and
- Differential item function analysis

Details of RUMM, its manuals and student/demo version are available at http://www.rummlab.com.au/ or http://www.faroc.com.au/~rummlab/

6. RUMMFOLD/RATEFOLD

RUMMFOLD & RATEFOLD computer programs developed by Andrich and Luo (1998a, 1998b, 2002) apply the principles of Rasch model for three ordered categories to advance the hyperbolic cosine function for unfolding models.

RUMMFOLD*ss*/RUMMFOLD*pp* (RUMMFOLD) and RATEFOLD are Windows programs for scaling attitude and preference data. These programs estimate the item location parameters and person trait levels Rasch unfolding measurement model. Unfolding models can arise from two data collection designs — the direct-response single-stimulus (SS) design and the pair-comparison or pair-wise preference (PP) design. RATEFOLD provides a highly interactive environment to view outputs and graphs.

Details of RUMM, its manuals and student/demo version are available at http://www.assess.com/Software/rummfold.htm or from the authors.

7. QUEST

Quest, developed and implemented by Ray Adam and Khoo Siek Toon (Adams & Khoo, 1993), is a comprehensive test and questionnaire analysis program that incorporates both Rasch measurement and traditional analysis procedures. It scores and analyses multiple choice items, Likert-type rating scales, short answer and partial credit items through the joint maximum likelihood procedure. The Rasch analysis provides item estimates, case estimates, fit statistics, counts, percentage and point-biserial estimates for each item.

Item statistics are obtain through an easy-to-use control language. Multiple tasks and routine analyses can be undertaken through batch processing. Quest support the following analyses and examination of data:

- Subgroup and subscale analyses;
- User-defined variables and grouping;
- Anchoring parameter estimates;
- Handling of missing data;
- Scoring and recoding data; and
- Differential item functioning through Mantel-Haenszel approaches.

Item-maps and kid-maps can be produced easily through the control language interface.

Details of QUEST are available at http://www.scienceplus.nl/scienceplus/main/show_pakketten_categorie.jsp?id=38

8. WINMIRA

WINMIRA computer program estimates and tests a number of discrete mixture models for categorical variables. Models with both nominal and continuous latent variables can be estimated with the software. Latent Class Analysis (LCA), the Rasch model (RM), Mixed Rasch model (MRM) and Hybrid models (HYBRID) can be used with WINMIRA for dichotomous and polytomous data.

WINMIRA is Windows-based and provides a highly interactive graphical user interface. It offers full SPSS support, and all help functions are documented. It is capable of estimating the following models:

- partial credit model;
- rating scale model;
- equidistance model; and
- dispersion model

Details of WINMIRA and student/demonstration version are available at http://www.scienceplus.nl/scienceplus/main/show_pakketten_categorie.jsp?id=38

9. REFERENCES

Adams, R.J., & Khoo, S.T. (1993). *Quest: The interactive test analysis system* [computer software]. Camberwell, Victoria: Australian Council for Educational Research.

Adams, R.J., Wu, M.L. & Wilson, M.R. (1998) *ConQuest: Generalised item response modelling software* [Computer software]. Camberwell: Australian Council for Australian Research.

Andrich, D., & Luo, G. (1998a). *RUMMFOLDpp for Windows. A program for unfolding pairwise preference responses.* Social Measurement Laboratory: School of Education, Murdoch University. Western Australia.

Andrich, D., & Luo, G. (1998b). *RUMMFOLDss for Windows. A program for unfolding single stimulus responses.* Social Measurement Laboratory: School of Education, Murdoch University. Western Australia.

Andrich, D., & Luo, G. (2002). *RATEFOLD Computer Program.* Social Measurement Laboratory: School of Education, Murdoch University. Western Australia.

Featherman C.M., Linacre J.M. (1998) Review of BIGSTEPS. *Rasch Measurement Transactions* 11:4 p. 588.

RUMM Laboratory (2003). *Getting Started: RUMM2020.* RUMM Laboratory: Western Australia

Subject Index

GPSR Compliance

The European Union's (EU) General Product Safety Regulation (GPSR) is a set of rules that requires consumer products to be safe and our obligations to ensure this.

If you have any concerns about our products, you can contact us on ProductSafety@springernature.com

In case Publisher is established outside the EU, the EU authorized representative is:

Springer Nature Customer Service Center GmbH
Europaplatz 3
69115 Heidelberg, Germany

Batch number: 10755284

Printed by Printforce, the Netherlands